CHARTING AN EMPIRE

CHARTING AN EMPIRE

Geography at the English Universities, 1580–1620

LESLEY B. CORMACK

The University of Chicago Press ❧ *Chicago & London*

Lesley B. Cormack is associate professor of history at the University of Alberta.

The University of Chicago Press, Chicago 60637
The University of Chicago Press, Ltd., London

Printed in the United States of America

06 05 04 03 02 01 00 99 98 97 1 2 3 4 5

ISBN: 0-226-11606-9 (cloth)
ISBN: 0-226-11607-7 (paper)

Library of Congress Cataloging-in-Publication Data

Cormack, Lesley B., 1957–
Charting an empire : geography at the English universities, 1580–1620 / Lesley B. Cormack.
p. cm.
Includes bibliographical references and index.
ISBN 0-226-11606-9 (cloth : alk. paper).—ISBN 0-226-11607-7 (pbk. : alk. paper)
1. Geography—Study and teaching (Higher)—England—History. I. Title.
G76.5G7C67 1997
910′.71′173—dc21 97-14831
CIP

∞ The paper used in this publication meets the minimum requirements of the American National Standard for Information Sciences—Permanence of Paper for Printed Library Materials, ANSI Z39.48-1984.

For Andrew, Graham, and Quinlan, without whom no empire would be worth the conquest.

Here then at home, by no more storms distrest,
Folding laborious hands we sit, wings furled;
Here in close perfume lies the rose-leaf curled,
Here the sun stands and knows not east nor west,
Here no tide runs; we have come, last and best,
From the wide zone in dizzying circles hurled
To that still centre where the spinning world
Sleeps on its axis, to the heart of rest.

Lay on thy whips, O Love, that me upright,
Poised on the perilous point, in no lax bed
May sleep, as tension at the verberant core
Of music sleeps; for, if thou spare to smite,
Staggering, we stoop, stooping, fall dumb and dead,
And, dying so, sleep our sweet sleep no more.

Dorothy L. Sayers, *Gaudy Night*

CONTENTS

ILLUSTRATIONS

Tables

Figures

ACKNOWLEDGMENTS

This study would have been impossible without access to many college archives and libraries. For their kind assistance, I would like to thank the following: Andrew Cook, India Office, British Library; A. J. Farrington, India Office, British Library; June Wells, archivist, Christ Church, Oxford; Norma Aubertin-Potter, sub-librarian, Merton College, Oxford; Catherine B. Richard, archivist, Oriel College, Oxford; D. A. Rees, archivist, Jesus College, Oxford; Chris Butler, assistant archivist, Corpus Christi College, Oxford; Robin Peedell, assistant librarian, Brasenose College, Oxford; Brian Ward-Perkins, college archivist, Trinity College, Oxford; Malcolm Vale, keeper of archives, St. John's College, Oxford; Penelope Bulloch, librarian, Balliol College, Oxford; John Jones, archivist, Balliol College, Oxford; Helen Powell, assistant librarian, Queen's College, Oxford; John Elliott, professor, Syracuse University; C. M. P. Johnson, senior bursar, St. John's College, Cambridge; Frank H. Stubbings, keeper of rare books, Emmanuel College, Cambridge; Timothy Hobbs, sub-librarian, Trinity College, Cambridge; Pam Cromona, librarian, Peterhouse, Cambridge; C. P. Courtney, librarian, Christ's College, Cambridge; Alison Sproston, assistant librarian, Gonville and Caius College, Cambridge; Michael Hall, archivist, King's College, Cambridge; P. A. Judd, assistant librarian, Pembroke College, Cambridge; Nicholas Rogers, keeper of muniments, Sidney Sussex College, Cambridge; and Ray Plant, librarian, and the library staff at Corpus Christi College, Cambridge.

I have received financial help from the University of Toronto, the Webster Post-doctoral Foundation at Queen's University, and the University of Alberta.

Several people have read all or part of this book in many forms. I thank James McConica and Bert Hall for their support in writing the original thesis, on which this is based. The anonymous readers for the University of Chicago Press provided much invaluable criticism, and I thank them for their careful reading and detailed suggestions. Andrew Ede, my partner in this academic travail, has read many drafts and offered intellectual, emotional, and practical support. He may not have compiled the index (a task grossly undervalued when performed by the proverbial spouse), but without him, this book might have been a lesser (and slower) proceeding. I need hardly add that the faults that remain are all my own.

I would like to thank my parents, for the exciting historical dinner conversations with which I grew up and for their unfailing support. My two sons, while slowing down the completion of this text, have provided me with joy, laughter, and hope, and I thank them for that. I especially thank Quinlan, who delayed his arrival into the world for five days so that I could finish my revisions before rushing to the delivery room.

INTRODUCTION
Charting an Empire

The study of geography was essential to the creation of an ideology of imperialism in early modern England. Large numbers of young men destined to be part of the governing elite began to converge on the English universities just as the English were searching for an identity independent of the Roman Church and focused on the autonomy and superiority of England. These young scholars found in the study of geography a set of attitudes and assumptions that encouraged them to view the English as separate from and superior to the rest of the world. Geography supplied these men with belief in their own inherent superiority and their ability to control the world they now understood. The men who studied geography proceeded from the universities to positions of importance in government, law, mercantile activities, and court positions, and the worldview they had gained at university aided them in their climb through patronage connections. This development of an imperial ideology for a group of men so closely concerned with England's internal and external relations would help create the future history of the English and British empires.

This new imperial ideology, claiming both the supremacy of the English nation and its right to seize control of new trade routes, riches, and foreign lands, was articulated in many of the books of geography written by Englishmen in the last twenty years of Elizabeth's reign and the first twenty of James's. The attitudes in these books reflected the successful lessons of geography, often stemming from a university education, and supplied the raw material for further ideological development by other scholars reading them and assimilating their message.

The clearest articulation of imperial thinking came from the Cambridge-trained mathematician and geographer John Dee.[1] In 1577, in a proposal written for Elizabeth's Privy Council,[2] Dee stressed the potential power and supremacy of England, as well as its ability to achieve a great and lasting empire: "No King, nor Kingdome, hath, by Nature and Hu-

1. Dee called himself "a perfect Cosmographer." *General and Rare Memorials,* 54.

2. Sherman establishes the audience for this book in *John Dee: The Politics of Reading,* 149.

FIGURE 1. Title page from John Dee, *General and Rare Memorials pertayning to the Perfect Art of Navigation* (London, 1577; facsimile, Amsterdam: Da Capo Press, 1968).

mayn Industry (to be used) any, more LAWFULL, and more Peaceable Means (made evident) wherby, to become In wealth, far passing all other: In Strength, and Force, INVINCIBLE: and in Honorable estimation, Triumphantly Famous, over all, and above all other."[3]

Responding to a series of serious threats to maritime security, Dee proposed that Elizabeth establish a Royal Navy to protect England from pirates and the English fishery from incursions and to aid in the establishment of a British maritime empire. Dee directed to the politically powerful men surrounding Elizabeth an overt message of imperialism and of the necessity of using scholarly knowledge for the good of the common weal. These were precisely the lessons to be gained from a study of geography. The illustrated title page to Dee's book, *General and Rare Memorials pertayning to the Perfect Arte of Navigation,* provides a visual representation of this message of power and hegemonic potential, while at the same time warning the English of the danger of ignoring this opportunity (figure 1). In this engraving, Elizabeth commands the ship of state, labeled Europa. The royal coat of arms on the rudder indicates Dee's prophecy (and present claim) of England's supremacy and leadership of Europe. Elizabeth is receiving her advisers, but she looks toward naked lady Occasion (or Fortuna), standing on the fortress to the left. Elizabeth holds out her right hand to grasp Fortune's forelock—undoubtedly by founding her great Royal Navy. Britannia, kneeling on the shore, desires Elizabeth to seize her opportunity with a "fully-equipped expeditionary force," as her scroll states.[4] This navy is to be much more than a coast guard patrolling for pirates; rather, it will begin the divinely sanctioned creation of an English empire. God, Elizabeth, and St. Michael on the right fight back the darkness on the left, and the naval force will soon capture the foreign ships at sea. There is also a more ominous warning here, since the skull on the right acts as a *momento mori* and may be related to the ear of wheat, a Hermetic symbol for man, here somewhat inauspiciously reversed. In other words, if the readers ignore Dee's perspicacious proposal, England's end may be less than felicitous. If there was any doubt of this message, Dee lays it out in the text:

> Why should not we HOPE, that, RES-PVBL. BRYTANICA, on her knees, very Humbly, and ernestly Soliciting the most Excellent Royall Maiesty, of our ELIZABETH (Sitting at the HELM of this Imperiall Monarchy: or, rather, at the Helm of the IMPERIALL SHIP, of the most parte of Christendome: if so, it be her Graces Pleasure) shall obteyn, (or Perfect Policie,

3. Dee, *General and Rare Memorials,* 63.

4. For standard interpretations of this illustrated title page, see Ames, *Typographical Antiquities,* 1:660–62; Corbett and Lightbown, *Comely Frontispiece,* 49–58. French provides a Hermetic twist to this image in *John Dee,* 183–85.

> may perswade her Highnes,) that, which is the Pyth, or Intent of RES-PVBL. BRYTANICA, Her Supplication? Which is, That, ΣΤΟΛΟΣ ΕΞΩΠΛΙΣΜΕΝΟΣ [a fully equipped expeditionary force], may helpe vs, not onley, to ΦΡΟΥΡΙΟΝ ΤΗΣ ΑΣΦΑΛΕΙΑΣ [a citadel of safety, such as Fortune stands on]: But make vs, also, Partakers of Publik Commodities Innumerable, and (as yet) Incredible.

Thus, Elizabeth, already commanding the imperial ship as a leader of Christendom, can employ this new *Navy Royall* both to defend the autonomy of England and, more importantly, to achieve wealth, power, and hegemony.

> Vnto which, the HEAVENLY KING, for these many yeres last past, hath, by MANIFEST OCCASION, most Graciously, not only inuited vs: but also, hath made, EVEN NOW, the Way and Means, most euident, easie, and Compendious: In-asmuch as, (besides all our own sufficient Furniture, Hability, Industry, Skill, and Courage) our Freends are become strong: and our Enemies, sufficiently weake, and nothing Royally furnished, or of Hability, for Open Violence Vsing: Though their accustomed Confidence, in Treason, Trechery, and Disloyall Dealings, be very great. Wherein, we beseche our HEAVENLY PROTECTOR, with his GOOD ANGELL to Garde vs, with SHIELD AND SWORD, now, and euer. Amen.[5]

According to Dee, God takes up the cause of the expansionist English, through the intervention of his sword-wielding angel, Michael.

Dee's book provided a strong pronouncement of the imperial ideology present in geographical discourse. Although the navy was never established, several of Elizabeth's key advisers appreciated and shared this imperial vision for England. The study of geography at university, by Dee and some of Elizabeth's privy councilors as well as by the larger group of men destined for lesser political, judicial, or church careers, provided them with a sense of English superiority and potential hegemony, as well as examples of the heroic feats of those champions of English expansion who had gone before. The vision of geography informed these men that England stood poised on the brink of a great imperial adventure, in which they could and should participate.

Many of the geography books popular with university scholars in the period after Dee's book was published contained this message of English power and potential. Richard Hakluyt, for example, in a widely read compilation of travel narratives and geographical descriptions published in 1598–1600, declared the English most suited to explore and control the eastern trade: "The rude Indian Canoa halleth those seas, the Portingals, the Saracenes, and Moores travaile continually up and downe that reach

5. Dee, *General and Rare Memorials,* 53.

from Japan to China, from China to Malacca, from Malacca to the Moluccaes: and shall an Englishman, better appointed than any of them all (that I say no more of our Navie) feare to saile in that Ocean: What seas at all doe want piracie: What Navigation is there voyde of perill?"[6]

Likewise, in 1598, John Wolfe, translating a work by Jan Huygen van Linschoten describing foreign voyages and places, desired the English to take their rightful place as an ocean-going, imperial nation, both for the riches such action would bring to England and for the civility they would return to inferior parts of the world:

> I doo not doubt, but yet I doo most hartely pray and wish, that this poore Translation may worke in our *English Nation* a further desire and increase of Honour over all *Countreys* of the *World,* and as it hath hitherto mightily advanced the Credite of the Realme by defending the same with our *Wodden Walles.* . . . So it would employ the same in forraine partes, aswell for the dispersing and planting true Religion and Civill Conversation therein: as also for the further benefite and commodity of this Land by exportation of such thinges wherein we doe abound, and importation of those *Necessities* whereof we stand in Neede: as *Hercules* did, when hee fetched away the *Golden Apples* out of the *Garden* of the *Hesperides;* and *Jason,* when with his lustie troupe of couragious *Argonautes* hee atchieved the *Golden Fleece* in *Colchos.*[7]

Proven by its successful defeat of the Armada, employing its famous "wooden walls," the English nation has here become like Hercules, the semidivine, or Jason, with his great seafaring abilities and divinely ordained success.

Indeed, the Englished Linschoten provides another important illustration of English imperialism, while expropriating the valorous deeds of another nation. Linschoten's book, in Wolfe's edition, contains the story of Linschoten's Dutch voyage to the East Indies, a description of Africa by a Portuguese explorer, and a number of Portuguese expeditions to China and the West Indies. Yet, on examining the illustrated title page, a tale of English supremacy is laid before the reader (figure 2). At the top, the royal coat of arms with the garter motto and Elizabeth's motto, "Semper Eadem," anchors the book firmly in an English, courtly context. The translation was dedicated to Sir Julius Caesar, Judge of the High Court of Admiralty and Master of Requests to the queen, thus directed as Dee's work had been to the powerful patronage circles.[8] Below the title, a large and very impressive carrack rides at anchor. The prominent flag jutting

6. Hakluyt, *Principal Navigations* (1598–1600), 3:28.
7. John Wolfe's introduction to Linschoten, *Discourse of voyages,* sig. A4a.
8. Ibid., sig. A1a.

FIGURE 2. Title page from Jan Huygen van Linschoten, *His Discours of Voyages into ye Easte and West Indies* (London, 1598). (BL c.55.g.7. By permission of The British Library.)

into the title space indicates English ownership.[9] The coat of arms on the left belongs to the City of London, announcing that this book was directed to merchants as well as courtiers. Thus, although the book itself described foreign exploits and enterprises, the illustrated title page and introduction show that even these were to be used to the glory of England and to aid in the creation of an English empire.

Reading and studying geography thus encouraged the English to place England in a different context, to begin to envisage an England central to world politics and trade, rather than at its periphery. This movement was similar to the change of perspective made possible by cartographic realignments on sixteenth-century world maps. While the development of these maps was complex, involving many intermediate steps and different perspectives (in all senses of the word), the end points provide a suggestive illustration of the kind of mental shift that allowed the development of an imperial *mentalité* by the last two decades of the sixteenth century.

If we examine two world maps from the beginning and end of the sixteenth century, the radical transplantation of England from a marginal to a central location cartographically illustrates the conceptual shift concurrently taking place. There is a marked contrast between the Ptolemaic world, shown in Sebastian Münster's popular 1540 edition of Ptolemy's *Geographia* (figure 3), and the world of Edward Wright's rendition of Mercator's projection, used to illustrate Richard Hakluyt's collection of 1598–1600 (figure 4). Münster's map, though geometric, was a closed and complete world, with sections based more on myth and tradition than on surveying or exploration. The geometrical grid of Wright's map, with its complex web of rhumb lines, more clearly emphasized the rigid ectoskeleton that enclosed and constrained this newly man-made world. The Indian Ocean in Wright's depiction had been reopened, and a completely new continental group, the Americas, had been added to the picture. Even more important than these changes is one so simple it is often taken for granted: the central focus of Wright's map was no longer Jerusalem and the Holy Land but rather the North Atlantic and northwestern Europe, especially the British Isles.[10] Not only had England become an important part of Europe itself, but the focus of New World trade and exploration now pointed to the western reaches of Europe, particularly to England. Blundeville remarked on this transformation in Britain's status vis-à-vis Eu-

9. Corbett and Lightbown describe this title page in *Comely Frontispiece,* 81–89. They claim the ship is Portuguese but have clearly missed the significance of the flag.

10. As Woodward points out in "Medieval *Mappaemundi,*" 295, the Jerusalem-centered maps were a clear minority of medieval *mappaemundi.* Ptolemy's map is so centered less because of an interest in Palestine and more for reasons of symmetry. That said, the result of these maps focused on the Middle East is to draw the eye and the attention toward that point and thus do function as an implicit sign of centrality.

FIGURE 3. Map of the world from Claudius Ptolemy, *Geographia,* ed. Sebastian Münster (Basle, 1540). (Thanks to the Bruce Peel Special Collections, University of Alberta.)

FIGURE 4. Edward Wright's map of the world, from Richard Hakluyt, *Principal Navigations, Voiages, Traffiques and Discoveries of the English Nation* (London, 1598–1600). (BL 212.d.2. By permission of The British Library.)

rope in his commentary on Plantius's map when he declared that "England together with Scotland making both but one Iland is the greatest and mightiest of all Europe."[11] The reasons behind this alteration of focus were many, including a desire to show properly the new potential colonies in the Americas and an attempt to disguise the lack of knowledge about the Pacific Ocean. The effect of this "re-Orienting," however, is not to privilege America. Rather, it is to focus on northwestern Europe, delivering a message of the power and centrality of these northern nations. For people living in these countries, and most particularly for the English, this image both explained and helped to determine their central role in a nascent imperial history. As this cartographic illustration suggests, the study of geography provided a transfer of focus for English people attempting to control their environment and perhaps to dominate their world. They lived at the center of the world, between a Europe turning increasingly inward as religious and civil wars threatened and a New World that the English could plunder and colonize if they were courageous and clever. With a growing sense of their own importance in the wider world which they gained from a study of geography, the English would be well equipped for the imperial adventure awaiting them. As large sections of the map were colored pink in the eighteenth and nineteenth centuries, indicating the extent of the "empire on which the sun never set," the prophecy of the Wright map and of early modern geographical ideology was to be fulfilled.

The Growth of the English Empire

Many of the episodes that we have viewed as iconic transformations to modernity took place within a relatively short period of time in early modern Europe. However, these developments can no longer be represented by appeals to single master narratives of "Renaissance," "Reformation," or "Scientific Revolution" to explain the transformation taking place.[12] For

11. Blundeville, *His Exercises* (1594), f. 253a.

12. All these terms are now seen as problematic. For example, the English Protestant Reformation, established by Dickens in *English Reformation,* has been extensively challenged, first by Scarisbrick in *Reformation and the English People* and then by Haigh in *English Reformations* and Duffy in *Stripping of the Altars.* Likewise, the existence and stability of the Scientific Revolution has been challenged by Lindberg and Westman, eds., *Reappraisals of the Scientific Revolution.* Cohen, in *Scientific Revolution,* provides a very good summary of the state of the question. For a provocative suggestion that we scrap the very idea of the Scientific Revolution—or the existence of science before the eighteenth century—see Cunningham and Williams, "De-centring the 'Big Picture.'" The "discovery" of the New World also was the subject of many monographs, especially around the 500th anniversary of Columbus, by, for example, Greenblatt in *Marvelous Possessions,* Fernandez-Armesto in *Columbus,* and Stannard in *American Holocaust.* These works are far from being exhaustive, but they suggest the kind of tension under which interpretations of these once-standard events are now placed.

indeed, the world was changing. The European world by 1700 was a different place from that of 1400. There is, however, no easy synthesis to explain the alteration.[13] Many stories and many voices tell different facets of the development; they cannot be blended into a single explanatory tale. One of the illustrative strands that must be emphasized is that of ideological assumptions, those underlying and systemic views of the world that were a powerful source for social, economic, and political change, both as coercive controls on people and behavior and as important ingredients in challenging that authority.[14]

The study of geography provided an important catalyst and ingredient in this changing worldview. Studies of the physical and political globe provided information that challenged ancient and religious authorities and created an ideology that combined national pride with growing distrust of foreigners.[15] This can be seen in the attitudes expressed by Dee, Hakluyt, and Wolfe toward the incursions of the Spanish and Portuguese into new parts of the world and their assertion that the English had a superior claim to these trade routes and to positions of world authority. Geography encouraged a mathematical control of the world and a mentality that sanctioned its exploitation. Geography provided a key to an imperialism that stressed the superiority of English people and customs and the knowability, controllability, and inferiority of the wider world.

Geographical discourse did not provide a single unified explanatory model. Rather, it presented those who studied and lived it with a series of competing interpretations of the world around them. As John Gillies has shown in examining Shakespeare's geography, early modern English thinkers combined older "poetic" geographical approaches, emphasizing connection and hidden meaning, with newer, more "objective" accounts.[16] The constant negotiations between competing geographical explanations, such as those discussed by Stephen Greenblatt and Richard Helgerson, produced a complex pageant of geographical ideologies and allowed individual students of geography to choose explanations and interpretations best suited to their circumstances.[17] Helgerson's analysis of the tension in Richard Hakluyt's work between mercantile practicality and royal nation-

13. Hale, in *Civilization of Europe,* explains this transformation as the discovery of Europe, in part through its divisions.

14. Of course, I am not alone in seeing the importance of these ideological underpinnings. See, for example, Greenblatt, who argues for the struggle between competing ideologies in *Renaissance Self-Fashioning;* Helgerson, *Forms of Nationhood;* and Shapin and Schaffer, *Leviathan and the Air Pump.*

15. Hale claims this distinction between native and foreigner was an integral part of the creation of Europe. *Civilization of Europe,* ch. 2.

16. Gillies, *Shakespeare and the Geography,* ch. 1.

17. Helgerson, *Forms of Nationhood,* esp. chs. 3, 4; Greenblatt, *Renaissance Self-Fashioning.*

alism provides an example of the complexity of this geographical ideology of imperialism.[18] While both the needs and desires of the merchants and the imperial posturing of politicians aided in the development of the English empire, the ideological strands developed from different realities and concerns.

While the contribution of geography to this multifaceted imperialist ideology was important throughout Europe, it can be best investigated in England.[19] Until the sixteenth century, England had been marginal, both on the Ptolemaic map as an insignificant island on the edge of the known world and in the political and intellectual life of Europe. By the mid-seventeenth century, England had begun to develop as a true imperial power, portraying itself as the preeminent Protestant nation and preparing to do battle with other European powers for rights to mercantile, colonial, and imperial ascendancy. In this transformation, geography played an essential role.

The study of geography helped the English identify their country as separate from Continental powers, thereby beginning the movement toward nationalism and imperialism.[20] Since English exploration was increasingly portrayed in geographical treatises such as Hakluyt's as innovative and daring, the English readers of these exploration accounts began to assume the superiority of their country and its ability and inherent right to control the seas. The study of geography provided a means of self-definition for the English people, encouraging them to categorize people from foreign locales as the "other," as well as providing "normal" English standards against which the "other" could be judged. Foreign peoples were almost invariably seen as inferior to the English, and thus geography helped to create an exploitative mentality, leading many English to believe that they had the right to extract from the various sectors of the globe, above and below the surface, what they needed and to expect homage from

18. Helgerson, *Forms of Nationhood,* 171–81.

19. The story I am telling is an English one, although parallels and interrelationships could and should be drawn to similar developments in Continental geography. Especially in the Netherlands, an area similar to England in many ways, geography and cartography developed in tandem with imperial desires and helped fashion the Dutch identity. Still, it would be perilous to make general comparisons without detailed study of each particular location, and the necessary work of comparison must wait until more has been done individually on the relationship between geography and the early modern nation. The first step in this task has perhaps been taken by Buisseret, ed., *Monarchs, Ministers, and Maps;* and Akerman, ed., *Cartography, Statescraft, and Political Culture.*

20. P. Barber, "Evesham World Map," shows that an interest in local and patriotic geographical features was part of at least one fourteenth-century world map, suggesting that geography could have included these messages much earlier. He points out, however, that the Evesham map appears to be quite different in this regard from other so-called Higden maps.

its exotic, foreign, and, by definition, lesser peoples.[21] Geography also helped its scholars and practitioners develop a belief in the inherent superiority of Protestants over Catholics, in terms of both their treatment of other peoples and the climatic determinism that made temperate Protestant northerners more just, industrious, and courageous.[22] This encouraged the English to see themselves as the saviors of war-torn Europe and pagan America alike. Finally, geography provided a mathematical structure within which to contain the natural world.[23] Those who understood this mathematical language were thereby granted mastery over those who did not, as the native Americans were to discover to their cost. (In later centuries, imperial geographers and surveyors were essential in determining which land would belong to the Amerindians and which would be signed away, all based on an ideology of owning and controlling by right of mathematical measurement.) Thus, in order to understand the important changes slowly evolving in England in the sixteenth and early seventeenth centuries, changes that came to fruition in the eighteenth century with the Industrial Revolution and the formation of the British Empire, we must understand the fundamental role played by geography.

The University and a Shared Ideology

Geography became so important in developing an English ideology because many Englishmen destined for public careers were exposed to it in an educational setting. Early modern England had no standing army or police force; the peace, prosperity, and good order of the country depended on the legal superstructure and on the goodwill and interdependence of its people. This goodwill was ensured by a complex if informal institutional network of patronage and reciprocal cooperation, which at its best produced a society in which each individual was implicated in ensuring wider prosperity and order in his or her enlightened self-interest.[24] Thus, early modern English governance depended on the shared ideology of the classes who were acting as members of Parliament, justices of the peace, court

21. Merchant, *Death of Nature,* discusses the development of this exploitative (and masculine) mentality. D. Wood, *Power of Maps,* makes the point with more explicit reference to maps. Gillies, *Shakespeare and the Geography,* discusses the transgressive nature of crossing boundaries to the exotic "other."

22. Carpenter, *Geography delineated forth in Two Books,* 2:40–43. These claims, of course, had ancient precedents.

23. Harley discusses the deconstruction of maps and their underlying ideological power within this mathematical structure in "Deconstructing the Map," "Cartography, Ethics and Social Theory," and "Maps, Knowledge, and Power."

24. This relationship between politics and patronage has recently been discussed by Martin in *Francis Bacon, the State;* and by Peck in *Court Patronage.* Elton, in *Policy and Police,* discusses the constant surveillance necessary to ensure the cooperation of those living at a distance from the central government.

officials, patrons, burghers, and clerics. Increasingly, this shared ideology came from a common education, often at one of the two universities of Oxford or Cambridge. Geography, studied at the universities, became part of that common background of ideas and beliefs about the world.

Education was a rapidly changing institution in early modern England.[25] Entrée into governing and public careers was more and more frequently provided by formal education rather than household apprenticeship. Literacy and knowledge of a number of disciplines were viewed as increasingly important attributes for the ambitious man on his way to the top. Therefore, more and more gentle and mercantile families sent their sons to Oxford and Cambridge, where they would meet the right people and through their studies gain access to the common understanding of the world they would need for governance. As Grafton and Jardine have shown in a different context, these boys were learning to promote and maintain their position in the social hierarchy rather than to think independently or critically.[26] Geography, taught to these aspiring clients, patrons, and governors, provided them with an ideology that justified their place, especially vis-à-vis foreigners and Catholics, and gave them avenues to career advancement and profit-motivated trading expeditions.

Geography as a Discipline

Geography and the study of the natural world in general were an important part of the university curriculum long before the changes introduced into Oxford and Cambridge during Puritan control (1642–60).[27] The scientific curriculum of the universities was deeply affected by and in turn affected the sociopolitical realities of patronage, mercantilism, and the ideology of empire and governance in England before Parliament gained control. Geography was studied as part of the Arts curriculum and therefore reached a large number of future patronage, mercantile, and gover-

25. This is most dramatically portrayed in Stone, "Educational Revolution." For a more nuanced approach, see McConica, ed., *Collegiate University;* Curtis, *Oxford and Cambridge;* Kearney, *Scholars and Gentlemen;* and Frank, "Science, Medicine and the Universities."

26. Grafton and Jardine, *From Humanism to the Humanities.*

27. Historians such as Christopher Hill (*Intellectual Origins*) have argued that Puritanism was an essential ingredient in transforming the English universities from backward bastions of scholasticism into modern centers of scientific interest after the Puritan takeovers of the Interregnum. While I agree that changes in "science" in this period were fundamentally related to the relationship between theory and practice, this was neither religiously exclusive nor outside the university. Thus, while I agree with Hill's view that scientific change was contingent upon social and religious (though not necessarily "Puritan") developments, I would argue that the people responsible for these changes were not outsiders and radicals but part of the larger and relatively conservative ranks of the gentry and merchant groups. For a refutation of Hill's claim, see Feingold, *Mathematicians' Apprenticeship.*

nance players. Indeed, interest in geography expanded and flourished in the late sixteenth and early seventeenth centuries, a period of great social, economic, and political change, when the lessons geography could teach were especially timely and useful.

In the years between the beginning of the war with Spain and the founding of the Savilian chairs in mathematics and astronomy at Oxford (1580–1620), geography grew from a relatively undifferentiated field into three separate subdisciplines, each with its own literature, practitioners, and particular contribution to the ideology of empire. Mathematical geography developed as a branch of mathematics. It allowed scholars to reconceptualize the globe as an abstract, geometrical grid on which they could situate England and the rest of the known world. Its practitioners reduced the world to a carefully bounded and controllable system, one they could understand and begin to subdue. Descriptive geography, the natural and political description of other lands both exotic and well known, provided the English with narratives of exotic and distant peoples and places. It supplied practical economic, political, and social information, while helping to create the "other" against which a typical English person could be assessed. Chorography, or local history, gave these English students a picture of themselves. Through its study, chorographers placed themselves in time and space, named themselves, and began the process of defining who they were.[28]

The presence of geography as part of the curriculum at Oxford and Cambridge demonstrates that interest in the natural world was an important component of university training. The message of geography to those who studied it was one of control and superiority. It provided students with tools and assumptions with which to tackle the ever more complex world they saw before them. Whether they went to court, to Europe, to America, or back to their country estates, these university students took with them an image of England placed prominently in the center of a geometrical world.

Geography and the Scientific Revolution

In addition to supplying a new vision of the world to aspiring English gentry and merchants, geography provided an impetus to the new mathematized vision of the world, part of the transformation of "science" in this period. The changes that are often characterized as the scientific revolution included a new attitude toward methodology based on the behavior of a recently empowered gentry, a strong value placed on mathematics, a

28. For a description of these geographical subdisciplines, see Cormack, "'Good Fences Make Good Neighbors,'" and the following chapters herein.

new, more "objective" way of seeing the world, and an attitude that claimed the possibility of controlling that world.[29] All of these characteristics were present in the study of geography. Geography, as the science of seeing and developing narratives to explain the world, provided a new key for natural philosophers and mechanics alike. Implicit in geographical narratives was the belief that the world was, ultimately, a knowable, describable place, a place that could be predicted, mastered, and governed. This belief was to be central to the impact and evolution of scientific thought in the seventeenth century. Geographical practice developed three ideological components that would become standard for the study of the natural world by the late seventeenth century: an immense value placed on mathematics, an emphasis on the importance of gathering information in an incremental and inductive way, and a desire to make the knowledge so obtained into a science useful for the common weal. Geography thus provides an example of the type of investigation that encouraged its seventeenth-century practitioners to develop a new, engaged approach to natural inquiry; such an approach was essential to the development of the "new science."

Thus, an examination of the role played by this earth science in developing an imperial ideology through education also provides a compelling alternative explanation of how science developed in the years leading up to the scientific revolution. Change in science in the early modern period was largely brought about by new social organizations of natural investigation, which relied on methods adopted from the ideology of new social groups. Geography was studied by these new social groups and helped create a new ideology that was to influence both the social and political sphere and the formation of a new organization, methodology, and value for scientific knowledge.

29. See Shapin, "House of Experiment" and *Social History of Truth,* for a detailed discussion of the socially constructed nature of these hallmarks of scientific modernity. See also Alpers, *Art of Describing.*

CHAPTER ONE

Geography and the Changing Face of the English University

In 1643, estranged from his royalist sons and suffering from a long and ultimately fatal illness, Edward Herbert began to write his autobiography. He recalled fondly his days at University College, Oxford, beginning in 1596, and advised posterity on the type of education most suitable for a gentleman who must "think how you are to behave yourselfe as a publique person or member of the Comon wealth and Kingdome wherein you live."[1] Lord Herbert advised learning just enough logic "to distinguish betwixt truth and Falsehood" and then proceeding to more practical studies: "It will bee fitt to study Geography with exactnes soe much as may teach a Man the scituation of all Countreys in the wholle world together with which It will bee requisite to learne something concerning the Goverments manners and Religions either Anncient or new as also the Interests of state and Relations in Amity or strength in which they stand to their Neighbours. It will bee necessary also at the same time to learne the use of the Celestiall globe, The studyes of both Globes being complicated and ioyned together."[2]

Herbert was part of an illustrious circle of poets, statesmen, philosophers, and courtiers. He considered geography, both as a description of other countries and as the study of the globe itself, to be an essential tool for those about to embark on a similar career. For Herbert, the place to receive this instruction was at university.

As more gentle and mercantile families were sending their sons to university as a prerequisite to higher government as well as ecclesiastical office, the universities and their curriculum changed to meet this influx. This resulted in a new type of student, such as Edward Herbert, more concerned with practical affairs than scholarship for its own sake, and a new relationship between those who studied, the scholars, and those who practiced, the craftsmen. New disciplines were introduced and developed to meet the needs of this new student body. Geography was an important component of this innovation. The growth of interest in geography can be attributed

1. Herbert, *Life of Edward, First Lord Herbert of Cherbury*, 29.
2. Ibid., 19–20.

in part to its new university audience, while at the same time this study of the earth changed these young men's attitude toward England, government, and the wider world.

In sixteenth-century England, geography was developing into a discipline distinct from the older study of cosmography. Although both terms continued to be used, sometimes interchangeably, a distinction was increasingly being made.[3] Cosmography, as John Dee proclaimed, "matcheth Heaven, and the Earth, in one frame," requiring "*Astronomie, Geographie, Hydrographie,* and *Musike*" to be complete.[4] Geography, on the other hand, "teacheth wayes, by which, in sundry formes, (as *Sphaerike, Plaine,* or other), the Situation of Cities, Townes, Villages, Fortes, Castells, Mountaines, Woods, Havens, Rivers, Crekes, and such other things, upon the outface of the earthly Globe . . . may be described and designed."[5] In other words, while the subject of cosmography was the globe and its relationship with the heavens as a whole, picturing the earth as an integral part of the cosmos, geography had a narrower focus, concentrating specifically on the earth itself. Geographers abstracted the globe from its surrounding cosmos and began to classify its parts by separation rather than by union. This newer study of geography was thus more appropriate for people like Edward Herbert, interested first and foremost in the commonwealth of England and needing guidance in their attempts to understand and control their world.[6]

As Herbert indicated, the sixteenth-century English university supplied a venue for learning geography and many other practical arts. The universities provided a far more flexible educational environment than historians once believed. Teaching at Oxford and Cambridge still relied on a scholastic organization and structure established by medieval precedent, but within this basic framework original and innovative teaching and scholarship flourished. The statutory requirements of both universities and colleges did not, as has previously been believed, prevent such flexibility and innovation but allowed and occasionally encouraged it. This, combined with the newly developed tutorial system, permitted the reading and studying of geography within the collegiate system and encouraged a new

3. Lestringant, *Mapping the Renaissance World,* argues that André Thevet worked within this older tradition of cosmography, attempting to describe the entire world and using all available sources to do so. At the end of his life in the 1580s, Thevet was clearly operating in an outmoded genre, as more focused geographies began to take over the field.

4. Dee, *Mathematicall Praeface,* sig. b3a. This distinction is repeated by Blundeville, *His Exercises* (1594), part II; and later by Carpenter, *Geography delineated forth in Two Books,* 1.

5. Dee, *Mathematicall Praeface,* sig. a4a.

6. See R. Porter, "Terraqueous Globe," for a discussion of this shift into the eighteenth century.

interest in the fashioning of an English nation through such study. Judging by the number of geography books owned by students at Oxford and Cambridge, the scores of college men with an interest in geography, and the personal testimony of such men as Herbert, this discipline was an important component of the education received by these young men on their path to political, social, or economic prominence. Geography's study in the changing early modern universities was welcomed by students who sought knowledge useful for engagement with their world.

The "Educational Revolution"

The sixteenth and early seventeenth centuries witnessed a huge influx of students attending the two English universities.[7] Between the 1550s and the 1580s, the number of students beginning their university education at Oxford increased threefold, and, after a slump in enrollment between 1590 and 1603, student admissions at both universities increased to a height that would not be achieved again until the 1870s.[8] By the 1620s, 1,240 young men were entering higher education each year in England, an estimated 2.5 percent of the seventeen-year-old male age group.[9] There are many explanations for this huge demographic surge in the student population, but they are all difficult to prove. There was a general growth in the English population, partially accounting for the rising numbers, but this did not increase as dramatically as student demographics. Perhaps more to the point, careers open to university-trained men became rapidly more numerous under the Tudor monarchs, as men with a university education were sought for positions in government, diplomacy, and even the army.[10] The most important factor, as both Stone and James McConica point out, was the increasing value that the newly empowered classes placed on a university-oriented education, corresponding in part to new career expectations but stemming as well from a desire to bring up their children as cultured gentlemen. The proportion of boys of gentle birth at the universities, as well as those from mercantile origins, seems to have increased greatly in the second half of the sixteenth century.[11] This was especially

7. Stone, "Size and Composition"; Looney, "Undergraduate Education." E. Russell, in "Influx of commoners," argues against Stone's educational revolution. While her caveats are important—record-keeping changed at just the time Stone sees the increasing numbers of nonscholarship students—there still remains a significant change in the composition of the student body.

8. Stone, "Size and Composition," 17; "Educational Revolution," 50–51.

9. Stone, "Educational Revolution," 56, 57. Stone uses Gregory King's figures, drawn up in 1695, to reach this percentage. King's figures have been reprinted in Laslett, *The World We Have Lost,* 32–33.

10. Stone, "Size and Composition," 6–10.

11. With "corrections for misrepresentation of status and for omissions in the matriculation register [which] involve increasing the percentage of plebeian origins," Stone devel-

true at the new Tudor foundations, such as Corpus Christi College, Christ Church, and Trinity College, Oxford, although the medieval foundations also followed this trend.[12] As a liberal education replaced a household one as the sine qua non of the cultured man of early modern England, university education became sought more and more by these men and their sons. As the affairs of state seemed to require more knowledge than could be acquired merely through application, a university education seemed the route to cultural, economic, and political success.[13]

The Changing Collegiate Structure

As a result of this influx of new men at Oxford and Cambridge, the universities began to serve a dual role, catering both to the degree-oriented student with a clerical career in mind and to the student less interested in degree requirements and more anxious to acquire polish and useful knowledge. The university had to cope with a new, more affluent, more independent, and less controllable group of young men and turned increasingly to the colleges to take on this responsibility. As the intellectual and social demands made on the universities changed, a new, lay, college-oriented university emerged, especially at Oxford.[14]

The pre-sixteenth-century English university had been a society of fellows, focused on the university itself rather than the individual colleges, with theology and the training of clerics at its very heart. Undergraduate teaching had little place in the universities; this was handled by the halls that grew up around the town. A scholar's association with the university began only after he had matriculated and been granted the degree of Bachelor of Arts. Men at the colleges were supported by their fellowships or their religious orders. They attended university lectures (at least in theory) and worked to complete their Arts training by obtaining the degree of

---

ops figures of 50 percent gentry, 9 percent clergymen's sons, and 41 percent of plebeian origins at Oxford from 1575 to 1639. "Educational Revolution," 61; McConica, "Scholars and Commoners," 159–81.

12. McConica, "Scholars and Commoners," 171.

13. Elton, in *Tudor Revolution,* first made the claim for "new men," often with humanist training, entering powerful government positions. Elton's thesis has come under attack. For some recent assessments of the Tudor revolution, see Guy, "Tudor Commonwealth: Revising Thomas Cromwell"; Fox, *Reassessing the Henrician Age;* and Coleman and Starkey, eds., *Revolution Reassessed.* The latter two were reviewed by Elton in "Revisionism Reassessed." Despite the attack on Elton's claims for the 1530s, most revisionists agree that men of more education were entering these administrative and governmental positions, especially after the 1550s.

14. McConica, "Rise of the Undergraduate College," 1. I owe much of the description of sixteenth-century university structure to McConica's masterful treatment in *Collegiate University.*

Master of Arts, perhaps later remaining as students of one of the higher faculties of law, theology, or medicine.

As McConica has demonstrated, this began to change in the sixteenth century. At Oxford, from the foundation of Brasenose in 1512 to that of St. John's in 1557, the individual colleges began to cater to the growing company of undergraduates and commoners. In time, this developed into a political structure in which the colleges began to gain control of the university itself.[15] This was partly a result of the changing role of the colleges vis-à-vis undergraduate education and partly a result of the growing interest in liberal education on the part of the newly empowered gentry and mercantile classes. The Faculty of Arts itself was, in fact, gaining autonomy and prestige,[16] as both canon and civil law collapsed in the aftermath of the Protestant Reformation and as the statutory curriculum was revitalized by the introduction of the tutorial system.

Although New College (founded in 1379) was the first officially to introduce scholars (undergraduates) into the colleges,[17] this practice did not become widespread in the other colleges until the early sixteenth century. The invasion of the colleges by these scholars introduced the necessity of teaching them; this task fell to the fellows.[18] The colleges thus began to increase their role in loco parentis, and college regulations began to take into account the fact that there were now young boys present.[19] This, combined with the introduction into collegiate society of commoners willing to pay for personal attention, helped to forge a tutorial system that was highly personal and potentially idiosyncratic.

The first foundation at Oxford explicitly to allow the attendance of commoners, that is, undergraduates who were not supported as scholars on the foundation and therefore paid for their own "commons," was Trinity College in 1555.[20] This increased the number of students with whom the college could deal and increased the amount of money the college would receive, as well as supplying potential income for fellows as tutors for these wealthier students whose parents would often pay for supervision.[21] The

15. Ibid.

16. J. M. Fletcher, "Faculty of Arts."

17. McConica, "Rise of the Undergraduate College," 3.

18. McConica, "Elizabethan Oxford," 693.

19. Stone claims that students in the late sixteenth and early seventeenth centuries ranged in age from eleven to thirty, with the median age of seventeen. "Size and Composition," 32.

20. McConica, "Rise of the Undergraduate College," 43.

21. Of course, this new practice also led to accusations of wealth rather than ability being the final arbiter for admission to a university education, a claim that continued to be made for several centuries (McConica, "Scholars and Commoners," 174). Indeed, this accusation went further, since it was often said that wealthy men took scholarships from

potential for pecuniary remuneration for fellows was increased even more with the introduction of fellow commoners (who paid for the privilege of eating with the fellows). The parents of these boys, who were usually of gentle birth, were anxious to see their sons receive university polish and sometimes even a degree. They were therefore willing to pay a competent fellow to take charge of these young men.[22] Thus, the tutorial system developed around the growing desire for learning on the part of this group of people relatively new to the university system and around a need to look after these young, impressionable, and often wealthy boys.

A number of sixteenth-century educational reformers stressed the importance of a liberal arts education, sometimes explicitly mentioning the need to teach these young gentlemen geography and navigation. Roger Ascham, in *The Scholemaster,* emphasized the importance of education, for both success in government and personal satisfaction, when he supplied advice to Sir Richard Sackville on the education of his son. He was particularly adamant that children be treated gently and encouraged to learn, rather than being beaten.[23] Likewise, Thomas Elyot stressed the need for a light hand and encouragement rather than threat in teaching the potential governors of the state. In a book designed to win favor with Henry VIII, Elyot suggested that "The education or fourme of bringing up of the childe of a gentilman which is to have authoritie in a publike weale" should include an understanding of "the olde tables of Ptolomee where in all the worlde is paynted," as well as "the demonstration of cosmographie," not through travel but through reading: "I can not tell what more pleasure shuld happen to a gentil witte than to beholde in his owne house every thynge that with in all the worlde is contained."[24] Sir Humphrey Gilbert echoed these sentiments in a proposal to educate Elizabeth's wards. Among the instructors to be hired, Gilbert included two mathematicians, one to read cosmography, astronomy, and navigation and the other to teach the art of maps and sea charts.[25] These and several other proposals designed to encourage the education of future gentlemen in search of a government or administrative career stressed the necessity of such an education and its practical applications.[26]

Men with no interest in a clerical career began to see a few years at

---

the poorer applicants. This appears to have been no more the case in the sixteenth century than before, and, in fact, the growing phenomenon of commoners probably relieved some pressure from the scholarship positions (Stone, "Educational Revolution," 67–68).

22. McConica, "Rise of the Undergraduate College," 48.

23. Ascham, *The Scholemaster,* sig. B1a.

24. Elyot, *The Governour,* ff. 15b, 37a, b.

25. Humphrey Gilbert, *Queen Elizabethe's Achademy,* 5.

26. For example, Vives, *Introduction to Wysdome.* See Bantock, *Studies in the History of Educational Theory,* vol. 1; and Simon, *Education and Society,* ch. 3. Henry Peacham also

Oxford or Cambridge as a valuable part of their practical education, aiding them in future careers as merchants, politicians, courtiers, or even country gentlemen. Their needs were different from those of the more clerically oriented scholar or fellow, and it is therefore no surprise that the sixteenth-century universities began to serve two different functions, catering both to those in degree programs and to the potentially less scholarly commoner.[27] Herbert, for example, did not receive a degree while at Oxford. This is not to say that the informal curriculum of the student not aiming for a degree was not rigorous, or even that it did not on occasion follow the prescribed pattern laid out in the statutes. In fact, McConica has demonstrated with his analysis of the diaries of the Carnsew brothers that these two boys, though not sitting for any degree, followed the statutory Arts requirements quite closely, even reading the books specified in the statutes.[28] This more informal branch of university learning, however, did allow a degree of flexibility not available to the scholar, who was more constricted by time and degree requirements.

Curricular Developments

A new group of students with different goals and educational needs, combined with a new, more flexible organizational structure, allowed colleges and tutors to develop an innovative approach to the subjects studied. This change in student demographics and collegiate organization forced a reevaluation of the relationship between theoretical knowledge and practical application, since these new students wished for both. This changing dynamic influenced even the traditional subjects that these men studied. Approaching their studies with an eye to applicability, and with careers ahead of them that would involve engaging in tangible problems of government, law, international affairs, or economic aggrandizement, these young men encouraged the imposition of practical concerns on the theoretical framework of rhetoric, logic, and especially mathematics and natural philosophy. The result of this was a disciplinary evolution that changed the relationship between theory and practice in a new and extremely fertile manner.

Both Oxford and Cambridge had been founded on scholastic models, and this foundation continued to affect the organization of the sixteenth-century institutions. The degree structure and requirements, with lectures, disputations, and declamations, stemmed from the reliance of university

expected his *Compleat Gentlemen* (1622) to learn geography and chorography (Curtis, *Oxford and Cambridge,* 269).

27. McConica, "Elizabethan Oxford," 693.

28. Ibid., 697.

curricula and organization on a scholastic methodology and philosophy.[29] The universities were based on a system that encouraged a logical and detailed exposition of texts, usually, though not necessarily, Aristotelian texts, with the goal of training clerics and the perpetuation of the university structure itself.[30] This emphasis changed somewhat with the advent of the printed book, since it was no longer necessary for masters to lecture strictly from the written text or for students to expend time and energy glossing important passages. The introduction of a new method of exegesis based on humanist innovations also changed the curricular stress from scholastic logic to humanist rhetoric, and to a certain extent this challenged the structure of the universities themselves. In the first half of the sixteenth century, humanists at the universities began to emphasize the *trivium* (logic and grammar, but especially Ciceronian rhetoric) at the expense of the more traditional *quadrivium*.[31]

Historians have long seen this movement away from the *quadrivium* as detrimental to the growth of the "new science" in England and as partly responsible for the putative scientific backwater in which England found itself in the early seventeenth century.[32] This stagnation of scientific investigation, claimed by Continental contemporaries such as Giordano Bruno[33] and modern historians such as Hill, was more imagined than real, however. The movement away from the sciences toward the *trivium* at the universities was neither so extreme nor so devastating as has been supposed.[34]

By 1575, there was a return to the rigor of the *quadrivium* and to the high standards of the best in scholastic scholarship. As Charles Schmitt has effectively demonstrated, this late-sixteenth-century scholasticism was not

29. Costello, *Scholastic Curriculum,* 11.

30. Mallet, *History of the University of Oxford;* Mullinger, *University of Cambridge;* Costello, *Scholastic Curriculum.*

31. McConica, "Humanism and Aristotle," 294. Oxford especially had been famous throughout Europe in the fourteenth century for its school of mathematical natural philosophy at Merton College. It continued to rest on these laurels for the next two centuries, to the detriment of new and interesting scientific investigation. See, for example, Wallace, "Philosophical Setting of Medieval Science," 111–13.

32. See especially C. Hill, *Intellectual Origins;* also Merton, "Science, Technology, and Society"; and Jones, *Ancients and Moderns.*

33. See Yates, *Giordano Bruno;* Weiner, "Expelling the Beast"; and Gatti, *Renaissance Drama of Knowledge.* See also McMullin, "Bruno and Copernicus." Bruno, on the occasion of his visit to England in the 1580s, condemned English science as outmoded and unsophisticated; his criticisms seem to have influenced much subsequent scholarship. It should be remembered, however, that Bruno was not especially well received by his English hosts, even by Fulke Greville, in whose house he stayed (Weiner, "Expelling the Beast," 1). As an outsider, Bruno saw little of the inner workings of either the universities or the scientific community (McConica, "Elizabethan Oxford," c721).

34. This is the main thesis of Feingold's *Mathematicians' Apprenticeship.*

simply a resurrection of its medieval progenitor but a newly established discipline based on humanist textual analysis and concern with original sources while retaining the structure and rigor of scholasticism.[35] This complementarity of medieval logic and humanist rhetoric helped to invigorate the university curriculum while maintaining hard-won standards of scholarship.[36] In fact, to distinguish this new methodology and philosophy from medieval scholasticism, Schmitt labels it "Aristotelianism."[37] "Aristotelianism" benefited from new studies of the Greek (as opposed to medieval Latin) Aristotle and of other Hellenists, while at the same time it promoted a renewed interest in the natural world, which was to prove extremely fruitful, even after Aristotle and his logic had outlived their usefulness by the mid-1650s.[38]

Thus, the mathematical *quadrivium,* and with it the study of the natural world, was an extremely important part of the curriculum of the early modern universities. The new students, eager for practical knowledge, could find it not only in the beautiful phrases of a Ciceronian oration but more particularly in the understanding of the structure of the natural world. Geography, as a scientific investigation with real practical applications, wedded utility and natural philosophy, provided an important avenue for such students, and fit well into the changing curriculum.

The Scholar and the Craftsman

This new concern with utility was encouraged by the necessary connections developing between scholars and the more practical men of affairs. This relationship between scholars and craftsmen was a fertile ground for changing standards of knowledge, methods of analysis, and reasons for investigating the natural world. For geography was not studied in isolation at the universities. Even the most theoretical geographer required the information and insight of navigators, instrument makers, cartographers, and surveyors in order to understand the terraqueous globe. This can be seen in the work of Richard Hakluyt, who used sailors' tales to construct a description of the world and England's role in its discovery, or that of Edward Wright, a serious mathematical geographer whose firsthand experience on voyages of discovery deeply affected his research program. In addition, the collaboration between John Dee, a university-trained mathematician and geographer, and Henry Billingsley, a London merchant, in the 1570 translation of Euclid indicates the fruitful exchange between the life of the mind

35. Schmitt, "Renaissance Aristotelianism" and *John Case.*
36. McConica, "Humanism and Aristotle," 294.
37. Schmitt, "Renaissance Aristotelianism," 159, 168.
38. Ibid.; Bennett, "Mechanic's Philosophy"; Watson, *Beginnings of the Teaching.*

and that of the marketplace.[39] This interplay of theoretical and practical issues ensured that the geographical community included the universities but was in no way limited to those provincial towns. This relationship between scholar and craftsman was important for other scientific endeavors as well, and an analysis of geography can help historians of science begin to develop a new picture of the contribution of practical men and issues to the changing scientific enterprise.[40]

The English geographical community, in fact, was an extremely complex one, in large part because of this peculiar characteristic of geography. Geographers had to develop a theoretical superstructure for their knowledge, which was initially supplied and legitimated by the classical underpinnings of the subject. At the same time, their knowledge could never be complete without reliance on the skill and information of navigators and travelers, or of personal experience. This geographical investigation was encouraged by and in turn encouraged an exploration of the world more rapid, nationalistic, and potentially economically advantageous than any earlier English generation had seen. Thus, geographers, who were interested in increasing and developing their knowledge of the world, had to be theoretical, practical, and political. Men interested in geography developed multiple and overlapping roles as scholars, craftsmen, and statesmen, or various combinations of the three.

The investigation of geography in the late sixteenth century embodied that dynamic tension between the world of the scholar, since geography was clearly an academic subject legitimated by its classical, theoretical, and mathematical roots, and the world of the artisan, since it was inexorably linked with economic, nationalistic, and practical endeavors. It provided a synthesis that enabled its practitioners to move beyond the confines of natural philosophy to embrace a new ideal of science as a powerful tool for understanding and controlling nature. The usefulness of geographical study was of paramount importance to the new men attending the universities in ever greater numbers, and it was this concept of utility to the state and to the individual that drove these new university men to investigate and appreciate geography. The geographical community, then, was a wide-ranging group, with many different concerns and goals, but with a desire to be useful to the nation and to their own self-interest and a vision of England as an increasingly illustrious player on the world stage.

39. See chapter 2 below for a fuller treatment of Dee and Billingsley.

40. This whole relationship between scholar and craftsman was first articulated by Zilsel, especially in "Sociological Roots of Science." A. R. Hall refuted this theory in "Scholar and the Craftsman." Zilsel's position is beginning to receive renewed attention. See, for example, P. H. Smith, *Business of Alchemy.* For a full treatment of this issue and its implications for interpretations of the Scientific Revolution, see Cohen, *Scientific Revolution.*

Geography in the Universities and Colleges: Statutes

This new cohort of young men, eager for useful information and personal and national aggrandizement, found in the developing discipline of geography a field of study particularly suited to their needs. It belonged in the traditional *quadrivium* as an offshoot of astronomy and cosmology, yet it engaged with the practical world, was potentially useful, and helped to create a shared image for these students of England's preeminence on the world stage.

Geography was studied at the English universities, both in the new informal tutorial setting and in the more formal Arts curriculum. This can be demonstrated to a certain extent by the flexibility and at least limited encouragement of geography and the mathematical sciences within the new sixteenth-century statutes; it is much more apparent in less formal evidence, such as the reading material of serious undergraduates and fellows, commonplace books, and the networks of geographers attending the two universities.

The curricular statutes at Oxford and Cambridge, of both the universities and their colleges, were periodically revised in the sixteenth and seventeenth centuries to deal with changing demographic, religious, and philosophical developments. They should not, however, be taken as the absolute guide to what was taught in lectures or tutorials, any more than modern university statutes disclose modern course contents. While these statutes have often been cited by historians to demonstrate the lack of scientific education available in the universities before the Civil War,[41] this is an inadequate interpretation of these documents. A growing number of men attending Oxford and Cambridge, of course, were not bound by the statutes, and even those who were experienced a degree of latitude in their application. Yet even allowing the statutory curriculum to give some indication of the relative importance of certain subjects, the statutes did allow the study of science in general and of geography specifically. In fact, the Oxford and Cambridge statutes mentioned the study of geography several times and indicate a university superstructure in no way obstructive to the study of science, mathematics, or, more particularly, geography.[42]

41. Mullinger, *University of Cambridge,* 2:402–3; Watson, *Beginnings of the Teaching;* Morison, *Founding of Harvard College,* 76–77; Costello, *Scholastic Curriculum,* 102–6; C. Hill, *Intellectual Origins;* Mallet, *History of the University of Oxford;* Frank, "Science, Medicine, and the Universities," 201; Westfall, *Construction of Modern Science,* 107–8.

42. Feingold, *Mathematicians' Apprenticeship,* 23. Much of the foundation for this investigation of the scientific content of the statutory curriculum has been laid by Feingold. In *The Mathematicians' Apprenticeship,* he has discovered many references within various statutes, at both the college and the university levels, to the study of the mathematical and physical sciences. I add to this specific detail concerning the place of geography within the statutory requirements.

The *Nova Statuta* of 1564/5 for Oxford and the Elizabethan statutes of 1558 and 1570 for Cambridge are often cited to demonstrate a decrease in interest in mathematics and the *quadrivium* in the second half of the century.[43] The Edwardian statutes at Cambridge, for example, had specified that arithmetic, geometry, and cosmography (and therefore, to an extent, geography) be studied in the first year,[44] while the Elizabethan statutes were much less expansive, only stipulating that students study rhetoric in the first year, logic in the second and third, and philosophy (presumably including natural philosophy) in the fourth.[45] The *Nova Statuta* more explicitly includes mathematical studies, requiring three terms of arithmetic and two of music, although the *trivium* seems to have gained in importance over previous statutes, with four terms of rhetoric, two of grammar, and five of dialectics, as compared with one year of dialectics in the Edwardian statutes.[46] On the surface, there does seem to have been a diminution of emphasis on scientific teaching in the statutes of the Elizabethan era.

This, however, is a limited reading of these statutes. The *Nova Statuta* reflects a large measure of continuity with previous statutes in its mention of mathematics, and the Cambridge statutes, although no longer stipulating mathematical topics in its section on "method of study," continues to indicate in other clauses that scientific topics were still integral to the B.A. curriculum.[47] Thus, mathematics and, by implication, applied mathematics including geography maintained a presence in the universities, demonstrable by their inclusion in the Elizabethan university statutes. This presence was confirmed in Oxford in the seventeenth century by Henry Savile's 1619 foundations of the Savilian chairs of Geometry and Astronomy. Savile specified that land measurement be taught as part of the duties

43. Gibson, ed., *Statuta Antiqua; Collection of Statutes for the Universities and Colleges of Cambridge,* trans. J. Heywood (hereafter referred to as Heywood); C. Hill, *Intellectual Origins,* uses these statutes, among other evidence, to claim the poverty of sixteenth-century university science.

44. Heywood, *Collection of Statutes,* 1; for the Latin text, see J. Lamb, ed., *Collection of Letters,* 125.

45. Heywood, *Collection of Statutes,* 290, app. (The 1570 Elizabethan Statutes), 6; Lamb, *Collection of Letters,* 319.

46. Gibson, *Statuta Antiqua,* 390.

47. Feingold, *Mathematicians' Apprenticeship,* 26. For example, mathematics and natural or moral philosophy were both subjects to be disputed by undergraduates in the course of their studies. In the 1570 statutes, students were required to pay the mathematical lecturer eight pence in order to receive their degrees (Heywood, *Collection of Statutes,* 34; Lamb, *Collection of Letters,* 346). Further, the provision for college admission stated that "none shall be admitted into places of this kind who have not acquired a proficiency therein [in Latin grammar] *sufficient to learn mathematics* and logic" (Heywood, *Collection of Statutes,* app., 40; emphasis added). Feingold discusses this provision, which remains unchanged from the 1549 and 1558 statutes (*Mathematician's Apprenticeship,* 27).

of the Professor of Geometry and that the Professor of Astronomy teach geography and navigation as part of the application of astronomical theory.[48]

Turning to college instruction, provisions for the teaching of mathematics and applied mathematics appear even more clearly. There were university lectureships established in mathematics at Cambridge, while mathematical instruction at Oxford existed on a more informal level, generally being supplied by regent masters. Although many of the medieval foundations did not revise their statutes to reflect the inclusion of undergraduates and therefore did not mention the B.A. curriculum, a significant number of the sixteenth-century foundations and those older foundations that revised their statutes in the sixteenth century did indeed make mention of mathematics lectures and even more specifically of geography lectures.

The statutes of several Cambridge colleges specified engaging mathematical lecturers, especially Jesus (in its 1559 statutes), Queens' (1559), Trinity (1560), and St. John's (1545).[49] St. John's College followed this statute most closely, engaging no fewer than four mathematical lecturers to teach arithmetic, geometry, perspective, and, most significant for this study, cosmography.[50] In addition, one of the four examiners of the college was designated as a mathematical examiner. The mathematical sciences were clearly an important part of the curriculum at these Cambridge colleges and were promoted in an explicit manner.

Oxford colleges tended to be more informal in their encouragement of mathematical studies. There were fewer endowed lectureships in *quadrivium* subjects, though instruction in these topics was soon required. The earliest college to include mathematics was Magdalen, which in 1487 specified that mathematical lectures were to be delivered during each long vacation.[51] Corpus Christi and St. John's followed suit, specifying that their students should attend the Magdalen lectures, as did Wadham College, which copied the Statutes of Corpus at its foundation in 1612.[52] Perhaps the clearest indication of interest in these subjects came from the Trinity College statutes of 1556, which stated that each morning during term, the lecturer in logic was to read to the members of the college arithmetic, logic, philosophy, and geometry, "while three times a week during the long vaca-

48. Curtis, *Oxford and Cambridge,* 116–17.

49. *Documents relating to the Universities and Colleges of Cambridge,* 3:109; see also Feingold, *Mathematicians' Apprenticeship,* 35.

50. Mayor, ed., *Early Statutes of St. John,* 105–17, 245–47.

51. Ward, trans., *Statutes of Magdalen College,* 61–62; Feingold, *Mathematicians' Apprenticeship,* 36; McConica, "Rise of the Undergraduate College," 5.

52. Feingold, *Mathematicians' Apprenticeship,* 37.

tion he was to deliver lectures in astronomy and *geography*."[53] Exeter modeled its 1560 statutes on Trinity's and stated that during the long vacation, the college lecturer who had been responsible for reading the classical authors during term time was to read arithmetic, geometry, or *elementary cosmography*.[54]

Other evidence points to the existence of lectures in mathematics and geography, even when not specified in the statutes. For example, the Magdalen College, Oxford, Archives lists charges for a lecturer in geography in 1540–41, although there is no mention of such a lecturer in the statutes.[55] The 1591 Magdalen Register states that President Bond issued a decree that "all bachelors deliver lectures in *geography* and *cosmography* in a system of rotation."[56] We also have Richard Hakluyt's own testimony regarding the geographical lectures that he delivered at Christ Church in the period from 1577 to 1582:

> According to which my resolution, when, not long after, I was removed to Christ Church in Oxford, my exercises of duty first performed, I fell to my intended course, and by degrees read over whatsoever printed or written discoveries and voyages I found . . . and in my public lectures was the first, that produced and showed both the old imperfectly composed, and the new lately reformed maps, globes, spheres, and other instruments of this art for demonstration in the common schools, to the singular pleasure, and general contentment of my auditory.[57]

Likewise, Sir Thomas Smith, whose globes still survive in the Queens' College, Cambridge, library, set up two lectureships in mathematics at the college in 1573, and there is evidence that there was a mathematical lecturer at King's College, Cambridge, in the 1550s, during the provostship of Sir John Cheke.[58] It is thus clear that much mathematical and geographical instruction was carried out at Oxford and Cambridge outside the statutory curriculum for the Bachelor of Arts.

The same picture emerges from an examination of the requirements for

53. *Statutes of the Colleges of Oxford,* 4:44–45. In Feingold, *Mathematicians' Apprenticeship,* 37; emphasis added. See also McConica, "Rise of the Undergraduate College," 44.

54. Exeter College Archive A.1 (2), f. 37. In Feingold, *Mathematicians' Apprenticeship,* 37.

55. Macray, *Register of the Members,* 2:69. See Feingold, *Mathematicians' Apprenticeship,* 39.

56. Feingold, *Mathematicians' Apprenticeship,* 39. From H. A. Wilson, *Magdalen College,* 137; emphasis added.

57. Hakluyt, *Principall Navigations* (1589), sig. *2a.

58. Feingold, *Mathematicians' Apprenticeship,* 40; Rose, "Erasmians and Mathematicians at Cambridge."

the Master of Arts degree. For example, the *Nova Statuta* specified that for this degree, the student must complete two terms of geometry and astronomy (which usually included geography as applied astronomy), three terms of natural and moral philosophy, and two of metaphysics.[59] Thus, the requirements for the Master's degree, as was the case with those for the Bachelor's, both allowed and encouraged the study of mathematical and applied mathematical topics, which could include geography.

Geography was a possible area of study for students at Oxford and Cambridge, and, as we can see from college and more informal evidence, interested a significant number of students.[60] Although it is not often noted specifically in the statutes of the two universities or of their colleges, these statutes did not oppose and occasionally encouraged geographical investigation. Indeed, geography often took its place within the requirements for the teaching of mathematics and of applied mathematics, as well as providing subject matter for the practice of rhetoric within the statutory curriculum. Also, geography and its cousin cosmography were specifically mentioned in several of the college statutes, including St. John's at Cambridge and Trinity and Exeter Colleges at Oxford. Still, any attempt to see the statutory requirements as representative of the complete reality of university instruction must be strenuously resisted. The statutes supply only one small piece of the puzzle, the rest of which is made up of the more personal contacts made by the student in his college with tutors, friends, and books. Geography could easily be taught in the more informal setting of the collegiate system, which also encouraged the development of informal reading programs and networks of like-minded scholars, innovations that were to prove especially important for mathematical geography and chorography. Since geography as a discipline encouraged the interaction of theoretical and practical concerns, it was a study best suited for, though in no way limited to, this more informal method of instruction.

We can gain some insight into this informal pedagogy by examining student commonplace books. Most students at the universities were encouraged to keep notebooks in which they could record interesting facts or pertinent quotations. These commonplace books were often written with an eye to future sermon or speech material. Vives suggested that the virtuous young man would want to keep such a book: "Thou shalt have alwayes at hand a paper boke, wherin thou shalte write such notable thynges, as thou redest thy self, or herest or other men worthy to be noted."[61] This was part of the pedagogical method, especially of rhetoric,

59. Gibson, *Statuta Antiqua*, 390.
60. McConica suggests this of Oxford in "Elizabethan Oxford," 716–20.
61. Vives, *Introduction to Wysdome*, sig. C7b.

at Oxford and Cambridge,[62] but these books could also be used as a lifelong aide-mémoire for important facts or theories. These commonplace books, as well as less formulaic notebooks of Oxford and Cambridge students, provide information on the sources that undergraduates consulted and reveal their general geographical interests.

The commonplace book of Sir Julius Caesar (alias Dalmarius or Adelmare) is a wonderful example of this form. Caesar, an important politician and courtier in Elizabeth's and James's court, began compiling his commonplace book at Oxford in the 1570s and continued to add to it throughout his life.[63] This book is doubly interesting because it is a *printed* commonplace book: the *Pandecte Locorum Communium.* This book, published in 1572 with an introduction by John Foxe, contains a title page with edifying verse, running heads throughout the book, and an index at the end, while the majority of the book is left blank for the use of the owner.[64] Given John Foxe's hand in the production of this volume, it is no surprise to see the preponderance of religious and moral topics implied by the various headings. What is more interesting for our purpose is the illustrated title page, which suggests that the *quadrivium* generally and geography particularly were of prime importance to the student compiling his commonplace book (figure 5). This title page was first used by publisher John Day to illustrate William Cunningham's *Cosmographical Glass* (1559) and Dee and Billingsley's 1570 *Euclid,* so had wider currency than simply this commonplace book. It is therefore striking that John Day chose to use such a mathematically and geographically oriented title page when he published this commonplace book, emphasizing the importance of these areas for all students of the commonplace. In the bottom half of the page sit the female personifications of the four mathematical arts: Geometria and Arithmetica on the left and Astronomia and Musica on the right. Each

62. Costello, *Scholastic Curriculum,* 56–62. The standard source of such pedagogy is Lechner, *Renaissance Concepts of the Commonplaces.* Also useful is Blair, "Humanist Methods in Natural Philosophy." See also Kearney, *Scholars and Gentlemen,* for an analysis of several students' commonplace books.

63. Sir Julius Caesar (M.A. 1578, Magdalen Hall, Oxford) was an English-born son of the Italian doctor Caesare Adelmare, who was physician to Elizabeth and Mary. After his stay at Oxford, he became a student at the Inner Temple and received his L.L.D. from Paris in 1581. He became Judge of the Admiralty and Master of Chancery under Elizabeth. With James's accession, Caesar became Chancellor of the Exchequer and, in 1614, Master of the Rolls, both positions being held until his death in 1636. See *Dictionary of National Biography* (hereafter *DNB),* 8:204–7.

64. Sir Julius Caesar's Commonplace Book. British Library (hereafter BL) Add. MS 6038. This is described for some political and religious detail in L. M. Hill, *Bench and Bureaucracy.* Although Moss, in "Printed Commonplace Books in the Renaissance," addresses the issue of commonplace books printed in their entirety (with no blank space for personal additions), she does not mention this form, with printed running heads and most of the book left blank.

Post Tempestate Tranquillitas.
9. Decembris, 1577.

PANDECTAE
Locorum communium,
præcipua rerum capita & titu-
los, iuxta ordinem elemẽ-
torum complectentes.

Cum adiunctis locorum ac paginarum numeris, tum charta insuper ad manũ parata, quanta cuiq; loco sufficere visa est. In quã velut in proprias sedes ac nidos, quicquid usquã ex omnis autorum lectione, memoria dignum occurrerit, studioso lectori, pro suo cuiusq; delectu, reponere, rursusq; indidem tanquam ex memoriæ penuario expromere, quæ libeat, licebit.

Per *Ioan. Foxum.*

LONDINI,
Excudebat Iohannes Dayus.
1572.
Cum gratia & Privilegio Regiæ Maiestatis.

FIGURE 5. Title page from John Foxe, *Pandecte Locorum Communium* (London, 1572). (BL ms. ADD 6038. By permission of The British Library.)

holds the instrument traditional to her art. In the absence of their sisters of the *trivium,* this certainly suggests a strong emphasis on those studies more closely identified with mathematical studies of the natural world. Even more striking is the top of the page. We might expect the great classical authors, whose works the attending students would surely be recording. We are not disappointed—except that these are all *geographers,* a fact made more evident by the instruments they hold. At the top, on either side of a terrestrial globe, stand Ptolemy and Marianus. Below them are Aratus and Hipparchus on the left and Strabo and Polibius on the right. Strabo is even engaged in drawing a map, an interesting endeavor for the father of descriptive geography. Even more interesting, it is a map of England. Look at the stars, Ptolemy seems to say, but with Strabo you know where you are. This message of geographical emplacement, at the very start of an important published commonplace book, demonstrates the importance of geographical thought and study to serious students.

Geography in the Universities and Colleges: Book Ownership

With the exception of a limited number of surviving commonplace books, informal instruction is extremely difficult to uncover, since by its very nature it left few traces. Little traditional source material exists to describe exactly what was taught by lecturers and tutors at the universities and thus to discover just what geographical ideas were passed on to interested Arts students. Tools of teaching and university administration survive in an extremely fragmentary form, and the possibilities of the few guides to students and letters of praise and censure have been exhausted long since.[65] Two new, virtually untapped sources of information succeed in telling us far more about the intellectual lives of students and fellows: the lists of books owned by students, masters, and colleges, which provide an insight into the typical geographical ideas with which these students were in contact; and the collective biographies of students of geography and geographical practitioners, which allow the less important men to play their parts alongside their more famous and flamboyant contemporaries. In examining the discipline of geography at Oxford and Cambridge using these two collective sources, the mundane experiences of the "normal" student reveal a much more detailed portrait of geography as a university discipline.

Most bibliographers of the Renaissance have examined individual libraries or collections in their entirety. This is an important approach and

65. See, for example Curtis, *Oxford and Cambridge,* especially 131, where he discusses Richard Holdsworth's "Directions for a Student at the Universitie." The use of this and other documents is questioned in C. Hill, *Intellectual Origins,* 301–14. See also Webster, ed., *Intellectual Revolution.*

has yielded much new information. It has tended, however, to favor the exceptional bibliophile rather than the more average student. And the examination of famous libraries, such as those of Sir Walter Raleigh, William Camden, or John Dee,[66] traces the development of a complete career, leaving the subject's university experience rather hazy. In order to examine the actual experience of early modern Oxford and Cambridge, and in order to appreciate the mental world of a larger group of more average scholars, it is necessary to take a different approach from that of the single library or collection. Instead, a composite picture must be drawn, using a sampling of smaller individual collections. There is much to be learned from an accumulated bibliography selected topically, since it allows us to develop a picture of the usual reading or collecting habits of a whole group of people. There is, of course, the danger in this approach of assuming too much homogeneity among the group members owning these books, but this can be overcome by creating a group biography of geographically inclined students. The construction of a collective bibliography is, in fact, necessary if we are to get beyond the "great men" of English intellectual history and see instead the ranks of average scholars who peopled the intellectual landscape. Therefore, I have created such a bibliography of geography and related books owned by students and colleges in the period from 1580 to 1620 in order to focus more clearly on the larger group of scholars. This book list has been drawn from a variety of sources: benefaction books from Oxford and Cambridge colleges, inventories of Oxford and Cambridge students who died intestate, and university and college library catalogs.[67] Clearly, not all students who attended Oxford and Cambridge during this period are included; the selection has been made on the basis of beneficence or death. The result of an examination of all extant documents is a list representing a wide range of interests and ownership patterns. This book list, in its final form, contains 2,825 books owned in the five decades from 1576 to 1625 and reveals the changing patterns and foci of concentration on geographical topics.

All the sources used in this analysis of book ownership have certain limitations. The wills and inventories of the Vice-Chancellor's Court are an excellent source of books owned by students who died while at Oxford and Cambridge, especially for the years before 1590. From the 1590s on, however, as students personally owned more and more books, the inventories stopped itemizing complete libraries, often listing only the number or

66. Oakshott, "Sir Walter Ralegh's Library"; DeMolen, "Library of William Camden"; Roberts and Watson, eds., *Dee's Library Catalogue.* The recent production of *Private Libraries in Renaissance England,* edited by Fehrenbach and Leedham-Green, will greatly increase our understanding of such libraries.

67. See appendix A for a list of sources consulted.

value of books owned.[68] Library benefaction books, which recorded gifts of books and money to the college libraries, were often begun in the middle to late seventeenth century (Jesus College, Oxford, for example, did not initiate its benefaction book until 1649),[69] and although many recorded early gifts post factum, some simply provide no information concerning the period in question.[70] College library catalogs, too, while relatively frequent in the late seventeenth century, did not become common before 1620 (Emmanuel College and Peterhouse, Cambridge, and the Bodleian Library, Oxford, are exceptions to this).[71] College libraries often were open only to fellows, not to undergraduates, and thus reveal the books to which these more seasoned members of the community had access, rather than those available to the seventeen-year-olds invading the universities. Thus, the vagaries of survival have tended to slant this study in favor of those colleges with early catalogs and benefaction books and away from others that might also have had geographical interests. The books listed in the Vice-Chancellor's Court probate inventories help correct this bias.[72] The men who died while at Oxford and Cambridge constitute a small group, likely characterized more by frailty or misfortune than by their mental prowess, yet this collection probably portrays better than the benefaction books the average student or the man who chose to make the scholarly life his career. The book ownership of these unfortunate students or perennial scholars counterbalances the selective, lavish gifts of the more well-placed students who donated books to their colleges.

While lists of books owned by students and libraries cannot tell us whether those books were read or the ideas digested, they do provide a new and fruitful method for learning something about changing disciplinary interests of those at university. These books often represented a serious financial outlay, since many of the volumes were illustrated and often in folio format. By examining these books, owned by a large heterogeneous

68. Ker, "Provision of Books," 473.

69. Jesus College Benefactors' Books, 1626–1712; 1624–84. Although the dates indicate that these books begin in the 1620s, the first entries date from 1649.

70. This was especially true for late foundations, such as Christ Church, which recorded the founding library donation of its founder, Richard Fox. Christ Church Donor's Register 1614–1841.

71. Emmanuel College, Cambridge, Library Catalogue, 1597; "Nomina librorum qui erant in Bibliotheca Collii ante Doctorum Perne mortuum [1588]"; Peterhouse MS 400; and Peterhouse MS 405. This library list, badly damaged, must date from before 1634, and probably dates from before 1610. An unnumbered Peterhouse manuscript, which Jayne, in *Library Catalogues,* claims is a copy of MS 400, is actually a close, though not identical, copy of MS 405; Bodl. MS Rawl. Qe 31. "Catalogus in Librorum in Bibl. Bodl. Thos. James 1602–3," now printed as *First Printed Catalogue of the Bodleian Library 1605. A Facsimile.*

72. For a complete discussion of the Oxford Vice-Chancellor's Court inventories, see Ker, "Provision of Books," 471. For Cambridge lists, see Leedham-Green, *Books in Cambridge Inventories.*

group of students at Oxford and Cambridge, we can see that an interest in geography was widespread at both universities in this period. Geography constituted a significant portion of the book acquisitions of these university men and was thus an important part of the curriculum as they experienced it. Also, their interest in specific geographical topics, as evidenced by books purchased, shifted during the period under investigation from an emphasis on mathematical geography in the early years to a greater concern with modern descriptions of the world both familiar and novel as the seventeenth century began.

Three Branches of Geography

Changing book ownership suggests a real development in geographical understanding and sophistication. During the late sixteenth century at Oxford and Cambridge, three main divisions emerged in the study of geography. Before 1580, geography had been a relatively unexamined discipline, investigating the earth in all its facets in a somewhat haphazard way, but by 1620, the categories of mathematical, descriptive, and chorographical geography had developed almost into separate subdisciplines. Although there continued to be a large amount of overlap, both in subject matter and in the people interested in these geographical topics, the three emerging subdisciplines can be identified by a different emphasis and a different social structure of study.

Mathematical geography, influenced by the ancient descriptive and mathematical geography of Ptolemy, emerged as mathematics developed at the universities. This branch of geography was most closely akin to the modern study of geodesy, that branch of applied mathematics that determines the exact positions of points and the figures and areas of large portions of the earth's surface, the shape and size of the earth, and the variations of terrestrial gravity and magnetism. Closely related to mathematical geography was the practical art of cartography, the study of maps and mapmaking, although cartographers depended far more on guild methods of transfer of knowledge and less on any systematic development of theories or models.[73] Academic mathematical geography aided in the popularization of mapmaking for social and administrative purposes already under way by familiarizing the new men of the universities with the concept of maps and map projections.[74] The development of mapping techniques in this period focused on translating coordinates and measurements of actual

73. T. Smith, "Manuscript and Printed Sea Charts"; T. Campbell, "Drapers' Company and Its School."

74. P. Barber argues that by 1580, "map-consciousness—the ability to think cartographically and to prepare sketch maps as a means of illuminating problems—was becoming ever more widespread" ("England II," 58). See also Harvey, *Maps in Tudor England.*

coastlines and country estates onto charts suitable for use by navigators and government officials, while mathematical geography at the universities imagined the globe as a theoretical construct, consisting of an exact grid of coordinates and properties, which necessitated the use of exact mathematical formulae. The two are interrelated, but relatively few men who read or wrote mathematical geography treatises also drew maps and charts for a living. Mathematical geography was the most rigorously theoretical form of geography and was studied by a small group of men who were also interested in other mathematical topics.

The second subdiscipline, descriptive geography, developed from the classical tradition of Strabo. This branch of geography portrayed the physical and political structures of other lands, usually in an inductive and relatively unsophisticated manner. Maps might form part of the material used by students of descriptive geography to understand foreign locales, not from any interest in map construction but for the illustrations of foreign lands these maps contained. Because of its relative lack of rigorous analysis, and because its primary goal, as defined by Strabo, was to promote useful knowledge,[75] descriptive geography was the most easily accessible of the three geographical subdisciplines. It encompassed everything from practical descriptions of European road conditions to outlandish yarns of exotic locales, providing intriguing reading and practical information alike. It was also the area most useful for aspiring statesmen, secretaries, or intelligencers, since they would be expected to produce physical, political, and cultural descriptions of the foreign lands they were to visit.

The final geographical subdiscipline, chorography, developed in the course of the late sixteenth century, combining a medieval chronicle tradition with the Italian Renaissance study of local description. Chorography (from the Greek *khora,* meaning "region") was the most wide-ranging of the geographical subdisciplines, since it included an interest in genealogy, chronology, and antiquities, as well as local history and topography. Chorography united an anecdotal interest in local families and wonders with the mathematically arduous task of genealogical and chronological research.[76]

These three geographical subdisciplines attracted different groups of students and scholars, as we will see in the next chapter. These groups often had close ties with one another, and, indeed, occasionally scholars can be identified with more than one area. Obvious examples include Thomas Harriot, who was both a mathematical and descriptive geographer, and Thomas Lydiat, a mathematical geographer and chorographer. In such an

75. Strabo, *Geography,* Book 1.1.21: 45.

76. See Mendyk, "Painting the Landscape," now published as *"Speculum Britanniae."* See also Levy, "Making of Camden's Britannia."

interrelated discipline, interest in one branch of geography could lead to an interest in geography generally and to investigating different problems in other subdisciplines. This was the exception rather than the rule, however, and the majority of students of geography identified themselves most strongly with one of the three subdisciplines.

Changing Geographical Trends

Men at university were interested in all three subdisciplines of geography, as their ownership of geography books demonstrates. This interest encouraged these men to absorb a whole set of underlying attitudes and assumptions concerning England and its place in the world, attitudes articulated by books such as Dee's or Hakluyt's. The purchasing trends of this geographically minded group of scholars changed over time, demonstrating that these men found different areas of geography more compelling and relevant at different periods. While an interest in mathematical geography remained constant among a small group of serious, mathematically inclined scholars, an interest in descriptive geography increased dramatically from 1580 to 1620. This was partly a result of the growing availability of such books, but, more significantly, it reflects the changing demographics of the university world and the growing emphasis on practicality and the glory of England on the part of these new students.[77] What had begun in 1580 as a relatively homogeneous study of the world had clearly emerged, by 1620, as three separate subdisciplines, each with its own literature, its own aspirations, and especially its own distinct group of practitioners.[78]

The wide variety of books of geography owned by students, fellows, and colleges suggests that neither university had a single accepted text for

77. There is no study of geography book production with which to compare these statistics. The closest comes from Parker, in *Books to Build an Empire,* who lists all "English books" published between 1481 and 1620 concerning parts of the world later of interest to England's empire. He omits all books on Europe and on navigation and clearly does not include foreign books, all of which make up a large proportion of the books on my book list. He finds a relatively steady book production from 1580 to 1605 of an average of 3.6 books per year. This increases dramatically for the next fifteen years, with the publication of an average of 6.5 books per year. This indicates that the increase in geography book ownership generally (see figure 6) is in part accounted for by increased book production, although it does not tell us anything about the changing distribution among the three subdisciplines.

78. Given the indeterminate survival pattern of book lists and inventories, the conclusions outlined here are admittedly tentative. Not only had more books been published by 1620, but more lists survived, and a somewhat skewed picture results from any attempt to rely only on the number of texts that libraries or students owned. Quantification, however, whatever its flaws, does provide a terminus a quo from which to grasp the extent of geographical knowledge and concerns. Thus, an analysis of changing ownership patterns, based on a significant number of geography books (2,825 books, 1,286 of which were strictly geographical), reveals suggestive patterns of changing interest from mathematical to descriptive geography over the forty-year period.

the study of geography. Unlike the case of logic, a few books did not dominate the field, and there were too few multiple copies of any particular book to make an assigned text probable.[79] Thus, geography book ownership probably indicates a genuine and extended interest in the subject, rather than the existence of prescribed lectures or assigned readings.

Most of the volumes owned in the 1580s had been published before 1560. Sixty-eight, or 17 percent, were classical; 12 percent were written before 1480; and 40 percent were written between 1480 and 1560. These figures indicate that early Elizabethan scholars relied on pre-sixteenth-century ideas, a supposition borne out by an examination of the most popular geographical, historical, and mathematical titles. In 1580, the most commonly owned book on this list was Aristotle's *Physica* (forty-two copies owned) and this "best-seller" list contained relatively few actual geographical texts. The exceptions included five copies each of Ptolemy's *Geographia,* Pomponius Mela's *De Situ Orbis,* Strabo's *De Situ Orbis,* Peter Apian's *Cosmographia,* and Sebastian Münster's *Cosmographia,* as well as four copies of Ortelius's *Theatrum Orbis Terrarum.* With the exception of Ortelius's book of maps, these books supplied well-established background rather than new ideas or theories.

The picture does not change substantially in the 1590s. Books written after 1560 account for rather more of the overall list (23 percent), but the fare remains relatively old-fashioned. Aristotle's *Physica* retains its most-owned status (twenty-six copies), closely followed by Euclid's *Elementa* (twelve copies).[80] The increase in the number of this standard mathematical text demonstrates the rising importance of mathematics and of Euclid's sophisticated and reliable treatment of arithmetic and geometry as the century progressed. Although this text had little direct relation to geography, it was important in establishing a familiarity with spherical geometry and with a mathematical view of the world. Geography was again represented by Ptolemy (seven copies), Strabo (six), Münster (five), and Apian (five copies), while William Camden's *Britannia* made its first multiple appearance with four copies. Ortelius's *Theatrum* (six copies) and Krantz's *Chronica Regnorum Aquilonarium* (1562) (three copies) show more modern treatments were available, but Camden's work is remarkable as the only multiply owned geography book published within five years of 1590 and

79. Schmitt names, for example, Richard Stanyhurst's *Harmonia* (1570) as a logic text that "set a standard to be aimed at in later productions" (*John Case,* 33–35). See Costello, *Scholastic Curriculum,* 46.

80. If these books included Dee's 1570 edition of Euclid, with its important new introduction, this ownership would indicate a much more modern contribution. Unfortunately, the book lists do not supply that kind of exact information except in rare cases.

suggests that owners of these books were rapidly gaining an interest in local history.

By the turn of the century, Oxford and Cambridge students began to alter their choices of geographical literature dramatically. This was true in all three subdisciplines and demonstrates a genuine shift in interest of geographically inclined book owners. Classical and medieval authors were losing popularity, perhaps because most libraries had already obtained copies and fellows therefore had access without purchase. These authors were still owned in multiple copies, with Apian's *Cosmographia* topping the 1600 list with eight copies, followed by Ptolemy's *Geographia* (seven copies) and Strabo's *De Situ Orbis* (six copies). At the same time, while these common works continued to appear as the "best-sellers" of the decade, a full 40 percent of books owned were published after 1560. Innovative Continental works appeared for the first time. For example, Copernicus's *De Revolutionibus,* which had been a point of contention on the Continent for almost sixty years,[81] appeared on this list for the first time (four copies), as did J. J. Scaliger's *De Emendatione Temporum* (1583) (three copies), while Ortelius's *Theatrum Orbis Terrarum* (1570) (five copies) and Camden's *Britannia* (1586) (four copies) were becoming better known. Although Camden was English, his interest in antiquities and chorography had in part a Continental basis, and his decision to write in Latin, unique among British chorographers, placed his work in the Continental theater. Thus, by 1600, students and fellows were beginning to experiment with new and often Continental ideas, while reducing their reliance on classical sources.

By 1620, this trend had gained momentum. Newly rediscovered classics such as Theodosius's *De Sphaera*[82] appeared (two copies in 1610, five in 1620); students of geography showed a new interest in medieval chorography, with works by such authors as Matthew of Westminster and Johan Funck. At the same time and at a rapidly accelerating rate, these students were becoming aware of new geographical information, with multiple copies in 1610 of such works as de Bry's *America* (1590), Maffei's *Historia Indica* (1588), Possevino's *Moscovia* (1587), Savile's *Rerum Anglicarum Scriptorum* (1596), and Camden's *Britannia* (1586); and, in 1620, of *Purchas His Pilgrimage* (1613), Serres's *General Inventorie of the History of France* (1607),

81. For discussion of the reception of Copernicus in Europe, see Westman, "Melanchthon Circle"; Westman, "Astronomer's Role in the Sixteenth Century"; and Gingerich and Westman, *Wittich Connection.* For an analysis of the reception of Copernicus in England, see J. L. Russell, "Copernican System in Great Britain."

82. Theodosius of Tripoli, *De Sphaera,* was first printed in Latin (Vienna, 1529), followed by editions in Messana, 1558, and Rome, 1586. It was usually included with other classical geometrical astronomical or geographical texts. It first appeared with Greek and Latin versions in Paris, 1558.

Knolles's *Generall Historie of the Turks* (1603), Acosta's *Natural History of the East and West Indies* (1596), and Hakluyt's *Principal Navigations* (1589). By 1610, more than 11 percent of titles owned had been published after 1600; by 1620, that number had risen to 26 percent.

Thus, the geographical interest of university students shifted from 1580 to 1620, beginning with a reliance on fifteenth- and early-sixteenth-century mathematical authors, long surpassed on the Continent by 1580,[83] and developing into a desire for up-to-the-minute descriptions of new lands and explorations. This transformation indicates an English intellectual community that was becoming aware of the geographical knowledge of its Continental rivals and beginning to challenge their control over both ideas and trade routes.[84] It also argues a movement from mathematical to descriptive geography, which can be better understood by examining the changing proportions of geographical books in the three categories of mathematical, descriptive, and chorographical over the five decades of the book list.

If the list of geography books is divided into the three main categories of mathematical (including maps and navigational treatises), descriptive, and chorographical, and traced on a percentage basis over each decade, a distinct pattern emerges of increased ownership in descriptive works over the fifty-year period, corresponding to a steadily decreasing ownership of mathematical geography over the same period (figures 6 and 7).[85] In 1580, mathematical geography accounts for more than 46 percent of all geography books, with descriptive geography accounting for 35 percent. This distribution began to change in 1590, and, by 1600, mathematical texts represent less than one-third of the total, as compared to descriptive geography's share of 42 percent. By 1620, descriptive works account for nearly 60 percent of the geography books owned, more than three times as many as mathematical geography books.

83. Westman, "Melanchthon Circle."

84. D. W. Waters contends that this same transformation was occurring in the realm of navigational understanding (*The Art of Navigation in England in Elizabethan and Early Stuart Times*, 196).

85. Figure 6 represents a raw graphing of the books owned in each of the three categories of mathematical, descriptive, and chorographical geography, over each ten-year period, while figure 7 shows these same books as a percentage of each decade's total. The normalized curve of the graph in figure 7, created by taking each decade's books as 100 and comparing different categories as a percentage of this total, is necessary given the differing numbers of books mentioned for each decade. Since raw numbers of books per decade tell more about the survival of book lists than about changing tastes, a comparison within decades is more fruitful for the purposes of this analysis. In each graph, the category of mathematical geography includes maps and navigational treatises. These graphs do not include the related disciplines of astronomy, mathematics, and history.

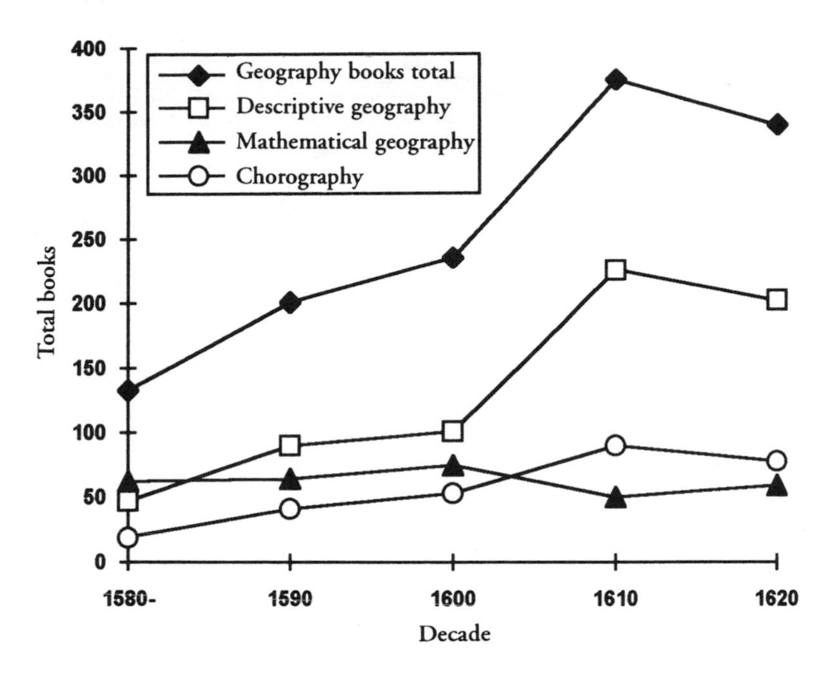

Figure 6. Geography book ownership, 1580–1620, raw numbers.

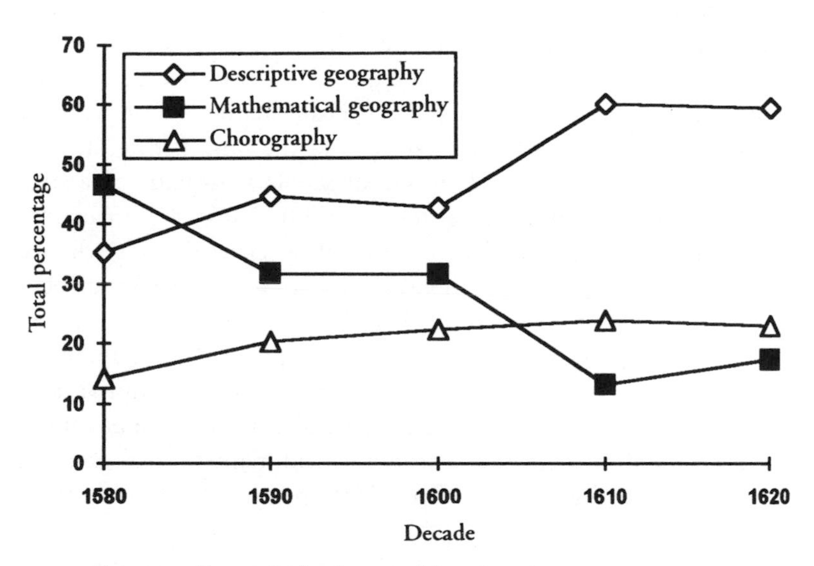

Figure 7. Geography book ownership, 1580–1620, percentages.

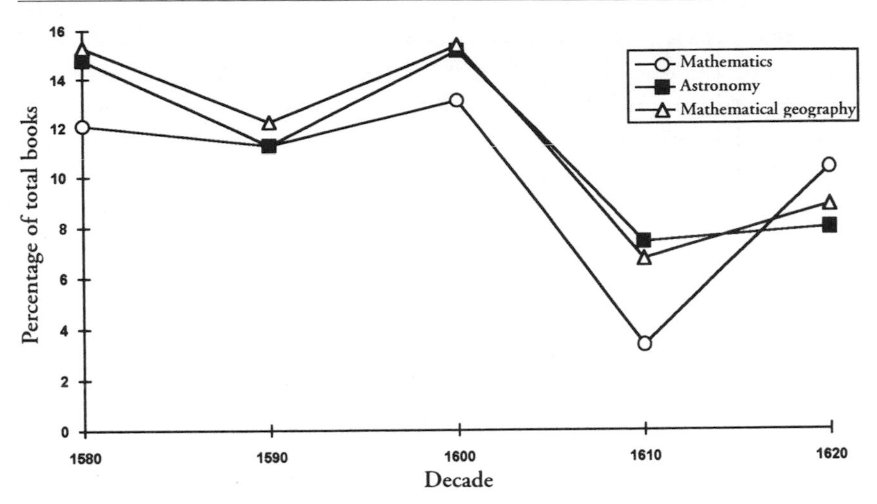

FIGURE 8. Book ownership: mathematical geography and related subjects.

This change in emphasis is deceptive. If the ownership of mathematical geography books is compared with that of works of mathematics and astronomy, its sister disciplines, a different picture emerges (see figure 8).[86]

The ownership in all three areas (mathematics, astronomy, and mathematical geography) follows approximately the same curve in this fifty-year period, with roughly the same percentage of books owned in each area. This indicates that mathematics and especially mathematical geography were not becoming less frequently studied at the universities but rather that a relatively small and stable group of scholars perpetuated a strong interest in books on mathematical topics. In other words, scholars interested in mathematical geography were a small, specialized group who were also interested in and owned books on other mathematical subjects.[87] These men often knew one another and corresponded by manuscript rather than by printed book, which helps to explain why the number of mathematical texts is relatively low in comparison to descriptive literature.[88] They remained a relatively stable group of university men, and the actual numbers of mathematics, astronomy, and mathematical geography books demonstrate this stability. Overall, the percentage of mathematical

86. Figure 8 pictures mathematical geography and its related fields of astronomy and mathematics, again using a normalized curve by assuming each decade to represent 100.

87. See chapters 2 and 3 below.

88. Several prominent mathematicians, in fact, in the sixteenth and seventeenth centuries showed a great reluctance to publish, including Thomas Harriot, whose work was not published in coherent form until this century (J. W. Shirley, *Thomas Harriot. Renaissance Scientist*), and Sir Isaac Newton, who was only persuaded to publish to prevent intellectual theft.

geography compared to geography books appears to have declined because the number of descriptive geography books increased rapidly. The reasons for this changing trend in the distribution of geographical interest, then, must be sought in an increasing intellectual "market" for descriptive geography rather than in any decreasing interest in mathematical geography.

Part of the burgeoning numbers of descriptive geographical books in the hands of Oxford and Cambridge students was because of the fact that new discoveries were constantly being reported and descriptive books therefore were published much more frequently (and lucratively) than mathematical ones. But, more significantly, this increasing interest in descriptive geography occurred at the same time as the new men were entering the universities. While enthusiasm for mathematics and applied mathematics was common for those contemplating a serious scholarly life, with aspirations to teach mathematical subjects at university, Gresham College, or privately, men with more diverse interests and goals were drawn to descriptive geography. The influx of gentry and of merchants' sons as commoners at Oxford and Cambridge occurred concurrently with the increase in descriptive geography book ownership, suggesting that these new entrants into university life were those most concerned with knowledge to be gleaned from descriptive geography sources. This contention is confirmed by the fact that a Tudor foundation such as Christ Church, with a student population drawn in large numbers from urban merchant backgrounds,[89] showed a special interest in descriptive geography. The study of descriptive geography equipped men for careers that matched the aspirations of gentry and merchants; it helped teach them to invest in mercantile enterprises, to make wiser trade decisions, to appear better educated at court, to make diplomatic forays to Europe, and to develop a shared ideology of superiority and imperialism, all in a manner that mathematical geography could not. Mathematical geography led the practitioner to fundamental truths about the globe and its relationship to the cosmos; descriptive geography led to temporal knowledge of cultures and byways. The first discipline equipped the academic, the second the man engaged in more practical and political endeavors. As more members of those eminently practical classes, the gentry and the merchant class, began to attend Oxford and Cambridge in greater numbers, descriptive overtook mathematical geography as the subdiscipline best represented in the book lists.

At the same time, books on chorography displayed a much less exciting pattern, showing a slight increase in ownership in 1590 and a largely stable level through the rest of the period (see figure 7). The audience for chorographical ideas might have been theology students, men who eventually

89. McConica, "Elizabethan Oxford," 682.

became parish priests patiently working on local history, or the more serious antiquarians, such as William Camden, discovering an absorbing vocation for the rest of their lives. Although the popularity of local history topics did not increase dramatically as did that of descriptive geography, it was maintained at a significant level throughout the period.

Both the increase in descriptive geography book ownership and the stability and slight increase in the local history category suggest a growing identification of Englishmen with their own country.[90] These students of geography read books that described European countries as distinct from England, while coupling this curiosity about other places with a growing awareness of the richness of British history and countryside. As Edward Grimstone said in his introduction to John de Serres's *General Inventorie,* "Where you may see the sundry Battailes woon by our Kings of England against the French, and the worthie exploits of the English, during their warres with France, whereby you may bee incited to the like resolutions upon the like occasions." This was a call to English arms, combined with disdain for another nation, since the author "you must consider . . . was a Frenchman."[91] This all encouraged the English populace to identify and define its uniqueness vis-à-vis the Continental experience. Equally, the increase in the number of domestically produced books in this Oxford and Cambridge book list, from 15 percent in 1580 to 23 percent in 1620, and the growth in the number of books written in English, from 7 percent in 1580 to 17 percent in 1620, suggest not only the vibrancy of English book production but also a growing interest in an English point of view.

❦

THE SHARED WORLDVIEW established by a common university education included a geographical framing of that world. The new classes of students invading the English universities in the late sixteenth and early seventeenth centuries discovered in the three branches of geography a series of lessons that equipped them well for their future lives and helped alter the way they thought about their world. This interest in geography was more focused at some colleges than at others, perhaps because of the idiosyncrasies of personal contacts or the momentum of a critical mass of like-minded scholars. These college connections will be investigated in the following chapter. The result was a new generation of young men, using

90. Helgerson traces the growth of this sense of self-contained loyalty to the land in "The Land Speaks," claiming the growing number of chorographical studies and maps of England shows a displacement of loyalty from the monarch to the land. This argument is extended in Helgerson, *Forms of Nationhood.* See chapter 5 below.

91. Grimstone, introduction to Serres, *General Inventorie of History of France,* sig. ¶ 4a.

the lessons of geography to help them invest wisely, commission maps, and envisage a world in which England was both morally and actually superior and where other parts of the globe were inferior and available to be controlled. This was an essential psychological support for the ideology of empire.

CHAPTER TWO

The Social Context of Geography

In 1599, Sir Anthony Sherley, his brother Robert, and twenty-five gentlemen set off on a hazardous mission to Persia (see figure 9). Sir Anthony, a Bachelor of Arts from All Soul's College, Oxford, and a client of the Earl of Essex, was determined to contact the Persian sultan, establish trade, and persuade him to join Christian Europe in its war against the Turk. In the process, this band of English adventurers encountered the "merry Greeke[s] . . . [who] alwayes sit drinking and playing the good fellowes before there dowres," the perfidious Turks, and the more honorable Persians.[1] The Turks, according to William Parry, one of Sherley's entourage, "are beyond all measure a most insolent superbous and insulting people. . . . They sit at their meat (which is served to them upon the ground) as Tailers sit upon their stalls, crosse-legd." They pass the day "banqueting and carrowsing, until they surfet, drinking a certaine liquor which they do call *coffe,* which is made of a seede much like mustard seede, which wil soon intoxicate the braine, like our *Petheglin.*"[2] The Persians, by contrast, were seen to be less odious, shown, for example, by the fact "that though the king have a large increase of Issue, the first borne only ruleth; & to avoyd all kind of cause of civill dissention, the rest are not inhumanly murthered, according to the use of the Turkish government, but made blind with burning basons: & have otherwise all sort of contentment and regard fit for Princes children."[3]

Although clearly for the Sherleys and their company, these foreign peoples behaved in a fashion far below the standards to be expected of the English or Europeans, they provided excitement for Sir Anthony's entourage and fascinating reading to those interested in geography who remained at home. Indeed, at least six separate accounts of Sherley's expedi-

1. W. Parry, *New and large discourse,* 6.
2. Ibid., 10.
3. A. Sherley, *His Relation of his Travels,* 30.

FIGURE 9. Sir Anthony Sherley, from Richard Hakluyt, *Principal Navigations, Voyages, Traffiques and Discoveries of the English Nation* (Glasgow, 1904), vol. 10, facing p. 272.

tion were published in England in the early seventeenth century, indicating the growing audience for such fare.[4]

For the study of geography attracted a large number of individuals, both within and outside the universities, in the period from 1580 to 1620. Part of the cause of this interest is obvious: gold and silver were flooding the European market from the Spanish Main; intriguing tales of cannibals, noble savages, and Amazons returned with mariners of all nationalities;[5] and lucrative trading routes were being established to both the East and the West.[6] Some people read geographical texts describing these adventures or other expositions of the terraqueous globe as a subsidiary interest, while others either at home or abroad devoted their lives to the investigation of the geographical world. Many of the men interested in geography attended university, where they may well have gained their knowledge and enthusiasm for the topic; of these, the vast majority were serious scholars, devoting at least four and often seven or more years of their lives to academic training. A significant portion went on to professional careers, as teachers, clerics, physicians, or common lawyers, while many were drawn to service for the state. Thus, geography was attractive to men with aspirations to serving the commonweal and achieving status within the networks of power. Reading geography helped these future governors develop a view of themselves as English and the rest of the world as foreign and yet accessible.

The English Geographical Community

More than seven hundred Englishmen in the period under discussion can be identified as having had some interest in geography, either in a practical or theoretical aspect of the subject. Much of this group was first

4. These included those by Sherley himself and W. Parry; also *A True Report of Sir A. Shierlie's Jouney . . . by two Gentlemen* (London, 1600); Anthony Nixon, *Three English Brothers . . . Sir Anthony Sherley his Embassage to the Christian Princes* (London, 1607), from which issued a play by John Day, William Rowley, and George Wilkins, *Travailes of the Three English Brothers, Sir Thomas, Sir Anthony, Mr. Robert Shirley* (1607); George Manwaring, in John Cartwright, *Preacher's Travels* (London, 1611); and Robert Cottington, *A true Historicall discourse of Muley Hamets rising and the three kingdoms of Morocco, Fes and Sus . . . the adventures of Sir Anthony Sherley, and divers other English Gentlemen . . .* (London, 1609).

5. For discussions of the impact of New World contact on Old World explorers and reporters, see Brandon, *New Worlds for Old.* For treatments emphasizing the intellectual influences, see Pagden, *Fall of Natural Man;* H. C. Porter, *Inconstant Savage;* Pagden, *European Encounters;* Grafton, *New Worlds, Ancient Texts;* and Greenblatt, *Marvelous Possessions.* For an interesting discussion of the image of the Amazon, see Lestringant, *Mapping the Renaissance World,* ch. 4.

6. K. R. Andrews, in *Trade, Plunder and Settlement,* demonstrates that the primary focus of the "age of discovery" was trade rather than colonization or missionary zeal. Indeed, he argues that the failure of the English to realize imperial expansion before the late seventeenth century was because they chose to concentrate on trade and plunder rather than on settlement.

brought to light by E. G. R. Taylor, who identified the vast majority of individuals writing or publishing geographical books in England from 1485 to 1650.[7] Added to this group are all the students, fellows, and masters at Oxford and Cambridge who were known to have owned geography books or to have donated them to their college libraries between 1580 and 1620.[8] Finally, correspondence of scientific men on geographical topics has added several names to the list,[9] as has recent scholarship on mapmaking and estate surveying.[10] While such a compilation can never be exhaustive, it provides a large sample of Englishmen interested in geographical topics from which to assess the depth and breadth of the geographical community.

This geographical community included both scholars and craftsmen. There was often a close connection between the navigators, surveyors, and instrument makers on the one hand and the university-trained geographers on the other. Anthony Sherley, for example, was a man with university training who then ventured out to explore the world, with explicit help from navigators, mapmakers, and guides. Given this link between the practical and the theoretical, it is not surprising to find that 48 percent of the geographically inclined men identified did not attend university, either in England or abroad (see figure 10).

This does not mean that the study of geography was somehow incompatible with university regimes or that practical, middle-class concerns alone inspired a revolution in scientific thinking.[11] Rather, it shows the mixed nature of geographical investigation and reminds us of the integral role played by both scholars and craftsmen in the development of the "new science." Of course, many of the non-university-affiliated men did not actively participate in constructing an English vision of the globe and its inhabitants. Some were East India Company employees who came to the attention of late-sixteenth-century England and E. G. R. Taylor alike by writing reports from various points East that were published under the auspices of the company.[12] Others were sailors whose accounts of sea battles and shipwrecks found their way into Hakluyt's and Purchas's an-

7. E. G. R. Taylor, *Tudor Geography; Late Tudor Geography;* and *Mathematical Practitioners.*

8. These have been drawn from wills and inventories (Oxford Vice-Chancellor's Court [VCC] inventories and Cambridge VCC probate inventories) and from library benefaction books. See appendix A for a full list of sources.

9. See Rigaud and Rigaud, *Correspondence of Scientific Men;* and Halliwell, *Collection of Letters.*

10. Thrower, ed., *Compleat Plattmaker;* Eden, "Three Elizabethan Estate Surveyors." See also Eden, ed., *Dictionary of Land Surveyors.*

11. See this argument in Drake, "Early Science and the Printed Book."

12. Danvers, ed., *Letters Received by the East India Company.*

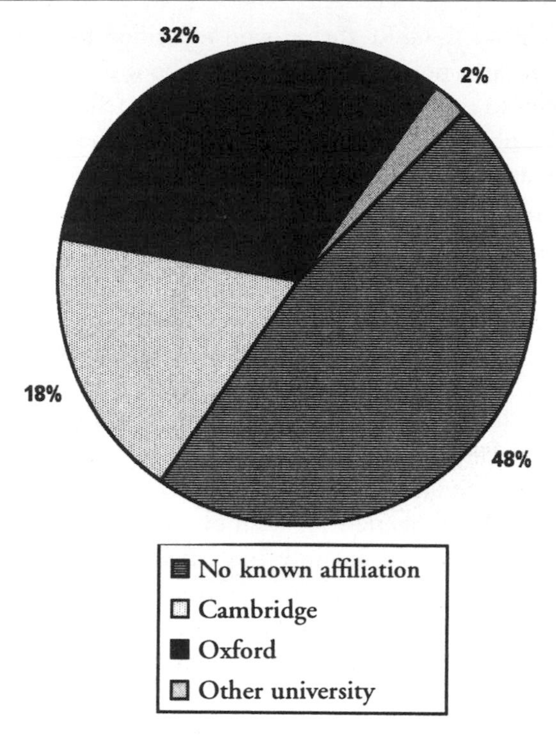

FIGURE 10. University affiliations of geographically inclined Englishmen.

thologies.[13] For these men, an interest in geography was part of their employment, as well as an exigency of survival. Although they contributed to the improved picture of the world as perceived by the geographical community that remained at home, they were not part of that interpretive

13. For example, "The first [second, and third] voyage of the right worshipfull and valiant knight sir John Hawkins, somtimes treasurer of her Majesties navy royall, made to the West Indies in the yere 1562" (Hakluyt, *Prinicipal Navigations,* 1598–1600, 3:500–524); "The voyage and valiant fight of The Content, a ship of the right honourable sir George Carey knight, . . . 1591" (3:555–566); and "The voyage of Henry May one of M. James Lancaster his company, in his navigation of the East Indies, 1591, and 1592; who in his returne with M. Lancaster by the yles of Trinidad, Mona, and Hispaniola, was about Cape Tiburon taken into a French ship under the conduct of Capitan de la barbotiere, which ship was cast away upon the yles of Bermuda: where all the company that escaped drowning remained for certain moneths, built themselves a barke, sailed to Newfoundland, and so home 1593" (3:573–74). Likewise, "The voyage of Master Beniamin Wood into the East Indies, and the miserable disastrous successe thereof" (Purchas, *Hakluytus Posthumous or Purchas his Pilgrimes,* 1625, hereafter cited as Purchas 1625; Part I:110–13); "The travailes of John Mildenhall into the Indies, and in the countries of Persia and the Greatt Mogul (where he is reported afterwards to have died of poyson) written by himself in two Letters following" (Part I:114–15); and "A true reportorie of the wracke and redemption of Sir *Thomas Gates,* Knight, upon and from the Ilands of the Bermudas: his comming to Virginia, and the estate of that Colonie then, and after, under the Government of the Lord *La Warre,* July 15, 1610. written by Wil. Strachy, Esq." (Part II:1734).

community. While shipwrecked sailors contributed colorful and often important stories and descriptions, the members of the interpretive community attempted to construct a complete cosmography of the globe, with multiple and often conflicting goals of personal, patronage, national, and intellectual profit. Eliminating those peripheral men, a much smaller proportion of the engaged English geographical community (112, or 15.5 percent) was self-taught or lacked a university affiliation. The English universities supplied an important ingredient in the development of the geographical community and the discipline.

Geographers and the Universities

Several Oxford and Cambridge colleges provided a nurturing environment for the study of geography. Oxford, especially, seems to have been home to a large number of students of geography, since nearly twice as many geographically inclined students had ties to Oxford as to Cambridge. The books owned by their affiliates tell us, for example, that Corpus Christi College, Oxford, and Peterhouse, Cambridge, were important centers of geographical study. Even more significant, Christ Church and New College, Oxford, and Trinity and St. John's colleges, Cambridge, stand out as centers of geographical interest, taking into account the careers of the large numbers of geographers and students of geography present within their walls and the books owned by their students, fellows, and libraries.[14]

At Oxford, geographical studies were pursued in at least twenty-two halls and colleges. Four foundations housed a significant number of students with an interest in geography: Christ Church, New College, Corpus Christi College, and St. John's College (see table 1).

There were also geographically inclined men at Magdalen College, Balliol, and Brasenose College. Christ Church provided the clearest opportunity for those with geographical inclinations to pursue the study; thirty-three people affiliated with Christ Church had some known interest in geography, fourteen through their book ownership and nineteen through their future careers or publishing. Twenty-five people with a known geographical proclivity had connections with New College, and eighteen students or members of Corpus Christi College owned or published geography texts.

The collection of books owned by students, fellows, and libraries of

14. See appendix B for a list of geography books owned by students, fellows, and libraries of Christ Church and Corpus Christi College, Oxford, and Peterhouse and St. John's College, Cambridge. St. John's, Cambridge, was also an important breeding ground for Elizabethan statesmen, especially the "Athenians," including John Cheke, William Cecil, Thomas Smith, and Roger Ascham, so important to early Elizabethan government. See Hudson, *Cambridge Connection,* 54. This suggests a connection between geographical education and service to the state.

TABLE 1 Geographically Inclined Men at Oxford Colleges

College	*Geographers*	*Book Owners*	*Total*
Christ Church	19	14	33
New	12	13	25
Corpus Christi	9	9	18
St. John's	9	9	18
Magdalen	12	4	16

these same four foundations was also significant, suggesting that these four provided centers for detailed geographical investigation.[15]

As table 2 shows, the four colleges of Christ Church, New College, Corpus Christi, and St. John's do not by any means exhaust the locations of geographical study at Oxford, although they were the most significant. Evidence of substantial geographical reading can be found in the book lists for All Souls, Oriel, University College, Merton, Queen's, and Balliol, while Magdalen, Trinity, Brasenose, Lincoln, Exeter, Broadgates Hall, Gloucester Hall, and Hart Hall all have some geography books listed. The opportunity for geographical investigation was thus widespread at Oxford, more so than could be accounted for by the presence of single tutors or even coteries, though the existence of these groups might help explain the relatively greater emphasis on geography in some colleges than in others.

At Cambridge, sixteen colleges and halls housed at least one person with a geographical bent. Four colleges contained a significant number of geographical enthusiasts: Trinity College, St. John's, Peterhouse, and Christ's College (see table 3).

Secondarily, Queens' and King's College students showed an interest in geographical topics. Trinity College was clearly the most geographical of the Cambridge colleges: seventeen people can be identified as having pursued active geographical careers from Trinity, and a further fourteen owned geographical books. St. John's was scarcely less prolific, with fifteen men actively involved in geographical study and a further six owning books on the subject. Peterhouse housed twelve men known to have been interested in geography, four of whom demonstrated this by their book ownership.

Three of the colleges housing geographers stand out as centers that en-

15. McConica, in "Elizabethan Oxford," 717–18, singles out Christ Church, Exeter, Broadgates Hall, and Gloucester Hall as centers of geographical coteries, all of which were represented in this geographical book list. Indeed, owners in only three halls at Oxford had geographical texts: Broadgates, Gloucester, and Hart halls. Since these halls did not have benefaction books and were losing status and students in the sixteenth century (McConica, "Rise of the Undergraduate College," 52), the book list represents only the tip of the iceberg of interest in geography present in these halls. The existence of even a few geography books at these three halls thus indicates that there was significant interest in the study of the globe among the members of Broadgates, Gloucester, and Hart halls.

TABLE 2 Geography Books Owned by University Affiliates, Oxford Colleges

College	Mathematical Geography	Descriptive	Chorography	Miscellaneous	Total
Corpus Christi	24	101	35	3	163
Christ Church	15	62	20	4	101
St. John's	21	37	18	3	79
New	12	40	13	1	66
All Souls	12	16	9	1	38
Oriel	3	19	6	1	29
University	16	5	1	—	22
Merton	9	11	1	—	21
Queen's	3	9	6	1	19
Balliol	6	8	1	—	15
Magdalen	6	6	1	—	13
Trinity	3	3	1	—	7
Brasenose	3	2	1	1	7
Lincoln	1	1	3	1	6
Exeter	—	2	2	—	4
Broadgates Hall	—	2	2	—	4

TABLE 3 Geographically Inclined Men at Cambridge Colleges

College	Geographers	Book Owners	Total
Trinity	17	14	31
St. John's	15	6	21
Peterhouse	8	4	12
Christ's	7	5	12

couraged ownership of geography books as well: Peterhouse, St. John's, and Trinity College.[16] Along with Corpus Christi College, these colleges provided support for the study of geography within that East Anglian university. In contrast to these relatively high church foundations,[17] Emmanuel College, that self-consciously Puritan foundation, had little interest in geography or mathematics. Emmanuel College's preference for theology over geography is significant, considering claims by Hill and others for a putatively close connection between science and Puritan ideology.[18]

As with Oxford, many Cambridge colleges showed an interest in geog-

16. Feingold relates that Trinity engaged a mathematics lecturer in 1546, and this college, of course, would become famous scientifically in the 1660s as the college of Isaac Newton. In a similar vein, the Master of Peterhouse from 1554 to 1580, Andrew Perne, was a "keen mathematician and astronomer" (*Mathematician's Apprenticeship,* 35, 59). As seen in chapter 1 above, this college and St. John's both evidenced mathematical leanings, apparent in their statutes and regulations.

17. H. C. Porter, *Reformation and Reaction.*

18. The Emmanuel College Library did contain a number of geography books, including one copy of Ptolemy's *Geographia* and two of Strabo's *De Situ Orbis* (Emmanuel College

TABLE 4 Geography Books Owned by University Affiliates, Cambridge Colleges

College	*Mathematical Geography*	*Descriptive*	*Chorography*	*Miscellaneous*	*Total*
Peterhouse	28	75	29	3	135
St. John's	14	57	31	2	104
Corpus Christi	15	34	13	2	64
Trinity	12	26	17	—	55
Gonville and Caius	13	24	10	1	48
Sidney Sussex	10	19	13	—	42
King's	3	8	10	—	21
Emmanuel	8	6	1	—	15
Christ's	6	3	2	1	12
Pembroke Hall	3	3	1	—	7
Queens'	4	1	1	—	6
Trinity Hall	1	4	—	—	5
Jesus	2	2	—	—	4
Clare Hall	3	—	1	—	4

raphy, seen through the book ownership of their students and members. Geography appeared in ownership and donation lists of several Cambridge colleges aside from those of Trinity, St. John's, and Peterhouse, the three most significant. Most notably, Gonville and Caius, Sidney Sussex, King's, Christ's, and Queens' were locations of geography book ownership (see table 4). Thus, geographical interest was widespread, if slightly less prevalent at Cambridge than at Oxford. Again, the number of Cambridge colleges at which geographical interest was found suggests that, just as at Oxford, the study of geography was not dependent on a single college or tutor.

The book list of Corpus Christi College, Oxford, demonstrates the wide range of geographic interest shared by college members (table 2 above). Corpus Christi had a huge list of geography books owned by fourteen separate owners or donated to the college library.[19] This is particularly remarkable given its small foundation of twenty fellows and is in part accounted for by the presence of John Rainolds, a strong book collector.[20]

Library Catalogs 1597 and 1621). It also contained one description of Turkey, one of Greece, and one of New World explorations. Its only truly innovative geographical text was Mercator's *Atlas,* which, when included in the Library Catalog of 1620, had been well established on the Continent for twenty years. Thus, while Emmanuel and its students did have some interest in geography, this interest seems to have been antiquarian in nature and certainly did not rival the explosion of geography books on the Oxford college book lists. Puritanism does not seem to have encouraged the mathematical or investigative spirit to develop more freely, at least in this important locale.

19. See appendix B for a full list.

20. See McConica, "Scholars and Commoners," 153; and McConica, "Elizabethan Oxford," 669–71, for the size of the Corpus Christi foundation.

Corpus students demonstrated a tremendous interest in all three types of geography but especially in descriptive geography, which accounted for nearly two-thirds of all geography books owned at Corpus. The mathematical geography books owned by Corpus fellows and library were relatively conservative, including a classical exposition by Ptolemy and several late-medieval cosmographies. Pierre d'Ailly's book, *Imago Mundi,* for example, had been used by Columbus in his argument about the size of the world, but by the late sixteenth century, this book was more a collectors' item than representative of new knowledge. Corpus men did own William Gilbert's path-breaking work on magnetism, which was both current and controversial. Their collections also included some modern maps, both of the wider world and of England. The latter could indicate an interest in both the mathematical emplacement of the globe and a descriptive image of the world. The descriptive geography section, on the other hand, was full of current and exotic descriptions. Many books reported on Europe, especially Muscovy, Hungary, and the Low Countries. There were descriptions of the New World as well, especially Brazil and the West Indies. André Thevet's *France Antarctique* (1558), for example, provided much fascinating detail, both about the religious strife of the abortive French colony in Brazil and about the anthropophageous customs of the Tupinamba tribe, later immortalized by Montaigne. The Far East was also represented, as were collections of explorations. The exploration accounts of Linschoten supplied a wealth of information about Dutch and Portuguese forays into the West and East Indies and, as we have seen, in its English version supplied a more palatable message of English ascendancy. In all, this is a truly spectacular collection of descriptive geography, containing some of the best contemporary wisdom about nearly every corner of the globe. The chorography section is nearly as impressive. It is most remarkable for the wide representation of contemporary English chorographers. These included John Stow's *Survey of London,* which was soon to become a classic of urban description. Both the descriptive geography and the chorography sections of this list indicate that Corpus fostered a genuine and deep interest in understanding the people and places of many parts of the globe, both exotic and mundane.

The situation at Peterhouse, Cambridge, was similarly exciting (table 4 above). Students there took an active interest in geographical information. The college library and its student holdings contained twenty-eight mathematical geography books, seventy-five descriptive, and twenty-nine chorographical, showing an inclination toward mathematical geography not seen in other colleges.[21] The mathematical section included three co-

21. See appendix B for a list of Peterhouse geography books.

pies of Ptolemy's *Geographia,* as well as contemporary works by important sixteenth-century Continental mathematicians and geographers. The book list also contained several modern atlases, indicating a real interest in up-to-date geographical knowledge of the wider world. The descriptive geography texts included many contemporary and challenging accounts, most notably descriptions of the Turks and Saracens, a copy of the Koran, and several accounts of exploration and new worlds. The descriptive discourse appears to have been slightly more innovative than the mathematical. Monardes' *Ioyfull Newes out of the Newe Founde Worlde* (1577), for example, contained important natural historical and medical information about New World plants. Most commercially important for posterity (if somewhat ruinous to health) was "the Tabaco, and . . . his greate vertues."[22] Monardes was not especially keen on the smoking of this weed, which caused those who partook of it to fall "downe uppon the grounde, as a dedde manne."[23] Rather, he promoted the manifold healing properties of the herb as a hot poultice. Local history was also well represented; most notably, medieval English chorographers were included, as were more modern local descriptions by English and Continental chorographers. Thus, all three types of geography, with an unusually large collection of mathematical geography texts and an emphasis on innovative descriptive geography, were important topics of study at Peterhouse.

A Network of Geographers

The same Oxford and Cambridge colleges that housed active geographical investigators (both theoretical and practical) were home to men who collected large numbers of geographical works. The two sources of evidence for the pursuit of geographical studies thus reinforce each other and, combined, point to the colleges most receptive to geography. Christ Church and New College, Oxford, and Trinity College and St. John's, Cambridge, emerge as colleges that actively encouraged the study of geographical topics. Housing as they did significant numbers of geographical book owners and practitioners or students of geography, these four colleges represented a hub of geographical interest at the two universities, as well as a significant source of theoretical geographical inspiration for the country as a whole. While it would be a mistake to limit the scope of geographical investigation at the universities to these four colleges, given the incomplete nature of sources available, they demonstrate the significant presence of geography in the universities as a whole. Although these four colleges should not be seen as an exhaustive list of the place of geographical concen-

22. Monardes, *Ioyfull Newes,* f. 34a.
23. Ibid., f. 39a.

tration, they do offer concrete examples of the opportunity that existed at Oxford and Cambridge to study geography. It is thus fruitful to examine these colleges in some detail, specifically looking at the geographical scholars connected with each one.

Christ Church, Oxford, was the most obvious nucleus of geographical activity at either university. Richard Hakluyt, one of its fellows, delivered early geography lectures there, possibly as early as 1574 and certainly from 1577 to 1579,[24] and his influence as well as William Camden's are seen in the list of geographically inclined students and the books owned on this topic. Of the thirty-three names associated with Christ Church, several stand out as preeminent geographers of their day: Hakluyt, of course, whose immense *Principal Navigations* (1589, revised and enlarged 1598–1600), though owing much to the work of such Europeans as Peter Martyr and Giovanni Ramusio, set the standard for descriptive literature of English exploration and discovery; William Camden, whose *Britannia* (first edition 1586) sparked an interest in British local history and antiquities that continued long after his death; Robert Burton, with his path-breaking exposition of medicine and the natural world, *The Anatomy of Melancholy* (1621); Richard Carew, the local historian, who wrote a *Survey of Cornwall* (1602); Richard Rowlands, alias Verstegen, the noted antiquarian and publisher; and Edmund Gunter and Nathaniel Torporley, two distinguished mathematicians of their generation.[25] With such material, it is no surprise that Christ Church, a large if atypical foundation, provided an arena for those students with a predilection for geographical knowledge.[26]

The presence of an active program of geographical investigation is confirmed by the Christ Church book list, containing works in all three branches of geography (see table 2 above).[27] These books, some owned by individual students and the majority donated to the Christ Church Library, include Ptolemy's and Strabo's geographical works, typical for the classical foundations of mathematical and descriptive geography. Apart from this small classical representation (8 of 101 geography books listed), most of the geography books in Christ Church students' hands were contemporary descriptions of other parts of the world. Mathematical geogra-

24. Hakluyt, *Principal Navigations* (1589), sig. 2a. He was not the first geography lecturer at Oxford, since Baldwin Norton is known to have taught geography at Magdalen College in the 1540s (Baker, *History of Geography*, 120).

25. The basic biographical information for this section comes from four main sources: *The Dictionary of National Biography;* Venn, *Alumni Cantabrigienses;* Clark and Boase, *Register of the University of Oxford;* and J. Foster, *Alumni Oxonienses.* All additional sources are cited separately.

26. McConica describes the convoluted path of the foundation of Cardinal College before Wolsey's disgrace and thereafter that of Christ Church, which did not follow any normal foundation as a college ("Rise of the Undergraduate College," 29–42).

27. See appendix B for a list of Christ Church geography books.

phy accounted for only fifteen books, as compared with sixty-two descriptive books, making it clear that the forte of this college, like that of its famous alumni, was descriptive geography. Many of these books told of New World explorations. Others provided news of eastern Europe, important as the English launched successful trading ventures to Russia through the Muscovy Company.[28] New maps and atlases were prominent on the list, reflecting concern for up-to-date visual depiction of new lands. There was also a collection of chorography, including both Continental and English authors.

The majority of these Christ Church men are known to have completed at least one degree while at Oxford. More than two-thirds of them achieved the degree of B.A.; 64 percent received an M.A. One M.D. was awarded, to Robert Fludd, the well-known alchemist and putative Rosicrucian,[29] who began his studies at St. John's in the 1590s, received his M.D. from Christ Church in 1605, and thereafter went abroad to study chemistry. Ten students attended the faculty of theology, receiving either a Bachelor or a Doctor of Theology degree. The five Doctors of Theology all owned books of geography but showed no indication of having gone farther in pursuing the study.

Most of these Christ Church men entered a profession suited to their many years of academic training. Six became theologians: Tobias Matthew (D.Th. 1574) became Archbishop of York in 1606, after having attracted the notice of Queen Elizabeth and served as her chaplain in ordinary in the late 1570s; Herbert Westphaling (D.Th. 1566) and Robert Chalioner (D.Th. 1584) became Professors of Theology at Oxford (Westphaling was the Lady Margaret Professor of Divinity at Oxford before becoming Vice-Chancellor of the University in 1576); William Godwin (D.Th. 1602) became dean of Christ Church and in his turn Vice-Chancellor from 1614 to 1618; Simon Juckes (D.Th. 1618) and William Watkynson (B.Th. 1587) preached and wrote theological treatises.

Four more men entered the ranks of the lesser clergy, in various parochial livings. They may well have maintained close ties (and possibly residency) with Christ Church after their appointments, though there is no reason to assume that all were absentee holders.[30] Thus, these men were

28. For studies of the Muscovy Company, see, for example, *Studies in the History of English Commerce;* Willan, *Muscovy Merchants.*

29. See Huffman, who argues in *Robert Fludd* against the earlier Rosicrucian label applied by Yates in *Rosicrucian Enlightenment.*

30. Many held multiple livings and so must have paid curates to perform their duties in at least some of their parishes. This, of course, was one of the abuses of the church that reformers were always eager to eradicate, but the practice continued well into the nineteenth century, providing convenient support for scholars. Haigh, in *English Reformations,*

perhaps both clerical and scholarly, providing links between the study of geography and both the colleges and the church.

Eight men on the Christ Church roster earned their livelihood or their fame as geographers, often publishing their results. Richard Hakluyt was the earliest and most famous of these. He arrived at Christ Church in 1570 as a scholar from Westminster and remained to earn both a B.A. and an M.A.; at Christ Church, he lectured in geography as regent master from 1577 to 1579. He held a number of parish livings, as far afield as Bristol, Westminster, Suffolk, and the Savoy[31] (implying a nonresidency status with some of them), but spent his life scouring libraries for historical accounts of famous and obscure explorations; gathering sailors', merchants', and other travelers' tales of English voyages; and persuading friends and associates to translate descriptions of foreign travels and discoveries. His enthusiasm for geographical description clearly began at Christ Church; for example, while in college, he persuaded a fellow Oxonian, John Florio, to translate Jacques Cartier's *Voyages* into English (1580).[32]

Two other translators of geographical material were alumni of Christ Church as well. Philip Jones of Bristol translated *Certaine briefe, and speciall Instructions for Gentlemen, Merchants, Students, Soldiers, Mariners, . . . employed in services abrode,* by A. Meierus (1589), an instruction book for travelers on what to observe and record, almost certainly at Hakluyt's suggestion. As he remarks in his dedicatory preface to Sir Francis Drake, "I was motioned to remember your self in this impression of this Method, by my very good and learned friend, M. Richard Hackluit, a man of incredible devotion to your selfe, and of speciall carefulnesse for the good of our Nation."[33] In a similar manner, John Bingham translated the *Historie of Xenophon containing the ascent of Cyrus into the higher countries* (1623). Jones matriculated at Christ Church in 1581, two years after Hakluyt had officially finished his university studies, and Bingham did not attend Christ Church until 1604, twelve years before Hakluyt's death. The geographical nature of these translations suggests that Hakluyt maintained close connections with his college and always kept an eye open for bright young men willing and able to translate the huge Continental literature on the New and Old Worlds.

---

argues that absentee priests were not the problem earlier historians of the English Reformation, such as Dickens, in *English Reformation,* held them to be.

31. J. Foster, *Alumni Oxonienses.*

32. Jacques Cartier, *A shorte and briefe narration of the 2 navigations to Newe Fraunce: first tr. out of the French into Italian, by G. B. Ramutius, and now turned into English by J. Florio* (London, 1580). McConica, "Elizabethan Oxford," 717.

33. In Meierus, *Certaine Instructions for Gentlemen,* sig. A3b. Emphasis in original.

Hakluyt came from a merchant background; his father had been a member of the Skinner's Company.[34] Hakluyt himself was a chief adventurer in the South Virginia Company (Raleigh's second attempt to find a southern route to Cathay) and acted as the historiographer for the East India Company on its formation in 1601.[35] He was thus very interested in the practical trading potential of geographical knowledge; many of his unpublished manuscripts were explicitly promotional of the plantation of North America and of any attempt to discover a northwest passage, while the implicit message of the *Principal Navigations* was one of imperial expansion for the glory of England.

His Christ Church protégé, Philip Jones, also reflected this combination of interests. In a curious manuscript, "Concerninge a passage to be made from our Northe Sea into the southe sea as shall appeare," written in 1586, Jones argued that the northwest passage, open to the English, was a much more certain route than the northeast, citing the glosses from the Geneva Bible and the map of Gemma Frisius showing the Strait of the Three Brothers to prove his point.[36] The reasons for such a venture were primarily mercantile, with national glory thrown in for good measure.

William Camden was another alumnus of Christ Church who can clearly be identified as a professional student of geography. Like Hakluyt the son of a merchant (his father was a painter/stainer), Camden entered Magdalen College in 1566 as a servitor but soon migrated, first to Broadgates Hall and finally to Christ Church, where he received his Bachelor of Arts degree in 1574. While at Christ Church, he received much encouragement in his geographical enthusiasms from a friend and fellow student, Philip Sidney.[37] Before being granted his degree, however, Camden left Oxford in 1571, possibly since, as a result of religious controversy, he was unsuccessful in obtaining an All Souls fellowship.[38] Camden spent the next four years traveling throughout England, gathering material for his antiquarian research. He may have been created M.A. by the Oxford Convocation in 1614.[39] Camden earned his living as a teacher and as headmaster of Westminster School from 1593 to 1610, but his first love was always chorog-

34. E. W. Gilbert, *British Pioneers,* 36.

35. India Office Marine Records, Introduction, iv.

36. BL MS Harley 167, ff. 106–7.

37. Trevor-Roper, *Queen Elizabeth's First Historian,* 6; Piggott, *Ruins in a Landscape,* 35. For a more popular biography of Camden, see McGurk, "William Camden: Civil Historian."

38. Piggott, *Ruins in a Landscape,* 34. Piggott is incorrect, however, in stating that Camden was denied a B.A.

39. "Convocation granted that he might be created M.A., but no record has been discovered of his having been actually created" (J. Foster, *Alumni Oxonienses,* 232). Clark and Boase, *Register of the University of Oxford,* 2.1:237.

raphy and the investigation of British antiquities. He spent each vacation continuing his perambulations of Britain (traveling to Wales as well as England, and learning Welsh and Anglo-Saxon), sometimes in the company of younger protégés, such as Sir Robert Cotton. In 1597, he was created Clarencieux King of Arms in the College of Heralds, a mark of some distinction in antiquarian and heraldic circles.[40] His network of correspondents was large, and he single-handedly raised the prestige and practice of British local history to unprecedented heights.[41]

Camden may have encountered Richard Carew while both were at Christ Church, and it is likely that they discussed their mutual interest in local history. Carew, born on his father's estate at Antony in Cornwall, attended Christ Church, probably in about 1569 or 1570.[42] While at Oxford, probably in 1569, he was called to dispute "with the matchless Sir Ph. Sidney" before Leicester and several other visitors.[43] He performed the typical duties of a member of the landed gentry, serving as M.P. for various ridings in Cornwall in 1584 and 1597 and acting as the local Justice of the Peace on several occasions. It is for his *Survey of Cornwall* (1602), however, that he is best remembered, and this work establishes him as an important local historian. Although he did not receive a degree at Oxford, it is quite likely that his interest in chorography was sparked by his studies and contacts in his student days at Christ Church.

Two other antiquarians and chronologers attended Christ Church, although they could not have come under Camden's direct influence. William Harrison, a London native, received his B.A. in 1556 and his M.A. in 1560, before moving to Cambridge for his B.Th. in 1569. He thus predates Camden's arrival by six years. He was, however, extremely interested in both chronology and topography, two aspects of local history that place it in the realm of applied mathematics.[44] Harrison wrote *An historicall description of the Islands of Bretayne,* which was printed with Holinshed's *Chronicles* in 1577. He also wrote "The Great English Chronology," an attempt to place English history in biblical chronological perspective, which he never published.[45] Likewise, Richard Rowlands, alias Verstegen,[46] at-

40. He was appointed to this position as chief herald of England under the patronage of Fulke Greville, Lord Brooke (Piggott, *Ruins in a Landscape,* 36).

41. See chapter 5 below.

42. J. Foster, *Alumni Oxonienses.* Carew does not appear in Clark and Boase, *Register of the University of Oxford.*

43. Carew, *Survey of Cornwall,* f. 102b.

44. G. J. R. Parry, *A Protestant Vision.* For discussion of chronology, see chapter 5 below. Wilcox, *Measure of Times Past;* Grafton, "Joseph Scaliger and Historical Chronology."

45. Trinity College Dublin MS 165.

46. He assumed his grandfather's name of Verstegen when he set up his printing business in Antwerp (*DNB*).

tended Christ Church around 1565 as a servitor,[47] before his recusant sympathies made England too hot and he moved to Antwerp, becoming an internationally known antiquary and printer. He corresponded with Sir Robert Cotton about British antiquities and manuscripts[48] and wrote chorographical tracts such as *Antiquities concerning the English Nation* (1605), as well as more generally descriptive works such as *The Post for divers parts of the World: to travelle from one notable citie unto another* (1572). The presence of these two antiquaries predating Camden by several years strongly suggests that a tradition of local geographical study existed at Christ Church before Camden attended this college. Although Camden was arguably Christ Church's most famous and important product, he was educated and shaped by a geographical framework already in place on his arrival.

Finally, Edward Terry might be called a geographical practitioner, in that he wrote travel literature and traveled to Turkey in his capacity as chaplain to the East India Fleet. He attended Christ Church in the early seventeenth century; he matriculated in 1608, receiving his B.A. in 1611 and his M.A. in 1614. Terry became a priest but did not have a normal clerical career, since his travels took him to exotic locales, such as India and Mandoa, which he recorded in *Voyage to East India, 1616–19,* printed by Samuel Purchas in 1625.[49] His association with Christ Church may have been a contributing factor in his gaining such an appointment, or at least the tales of faraway places that he might have consumed at college could have made him curious to seek them out. This is, of course, speculative, since no connecting link has yet been discovered between his university and East Indian careers. Still, it is significant that Christ Church was the incubator for this world traveler and writer of descriptive geography.

The other geographically inclined students of Christ Church turned to service for the state, a career that benefited from the implicit ideology geography provided. Two achieved military careers, and one became a politician. Thomas Aylesbury, a Londoner of gentle birth, served his country in a variety of political roles after graduating from Christ Church with his M.A. in 1605. He was Master of Requests and Master of the Mint, before being cashiered as a royalist in 1642 and retiring to Antwerp on the death of the king. Before this sad end, however, he manifested a sincere enthusiasm for problems of mathematics and mathematical geography, evidenced

47. Listed as a servant of Mr. Barnarde at Christ Church in 1565 (Clark and Boase, *Register of the University of Oxford*). J. Foster, *Alumni Oxonienses,* has not identified his college.

48. For example, Oxf. Bodl. MS Smith 71.

49. "A Relation of a Voyage to the Easterne India: Observed by Edward Terry, Master of Arts and Student of Christ-Church in Oxford" (Purchas 1625, Part I:1464–82).

by his correspondence with that famous mathematical geographer from St. Mary's Hall, Oxford, Thomas Harriot.[50] The other two politically motivated associates of Christ Church were much more flamboyant. Sir Philip Sidney attended Christ Church, probably in the 1560s, though he never earned a degree, and went on to become one of the foremost soldiers, statesmen, and poets of the age.[51] Indeed, Richard Helgerson claims that Sidney was one of the generation of men, born in the period from 1551 to 1564, who together helped to fashion and to possess a kingdom of their own language, a nation truly English.[52] His claim to geographical fame lay in his own extensive firsthand knowledge of Europe, demonstrated by his own travels, by the geographical features of *Arcadia,* by his *Letter to Robert Sidney [his brother] on how to Travel* (written in 1580 and printed in 1633), and by his interest in the colonization of America, shown in part by Hakluyt's dedication to Sidney of his *Diverse Voyages to America* (1582).[53] Sidney's interest in geography had been influenced by his family, since both his father, Sir Henry Sidney, and his grandfather, John Dudley, Duke of Northumberland, had been very keen on mapping, colonization, and exploration in general.[54] While at Christ Church, he was able to pass on that enthusiasm, especially to his close friend William Camden, whom he probably met while both were in college.[55] Just as Sir Philip is better remembered for his poetry than for his college affiliations and geographical interests, his cousin Sir Robert Dudley, styled the Duke of Northumberland and Earl of Warwick, is better known for his notorious second marriage and defection to Florence than for his Christ Church connections. He was a first-rate inventor and mathematician, however, and many new designs for ships and artillery manufactured in Florence were executed by him.[56] Near the end of his life, he wrote a popular compendium of naval knowledge, *Arcano del Mare* (Florence, 1646), which included several important maps using the Mercator projection. He matriculated from Christ

50. Halliwell, *Collection of Letters,* 43, 44.

51. Greville, *Life of the renowned Sr Philip Sidney;* Buxton, *Sir Philip Sidney;* Dorsten, *Poets, Patrons, and Professors.*

52. Helgerson, *Forms of Nationhood,* ch. 1.

53. "To the right worshipfull and most vertuous Gentleman master Phillip Sydney Esquire . . . I conceive great hope that the time approcheth and nowe is, that we of England may share and part stakes (if wee will our selves) both with the Spaniarde and the Portingale in part of America, and other regions as yet undiscovered" (Hakluyt, *Divers Voyages,* sigs. 1a–4a). See also L. B. Wright, *Middle-Class Culture,* 529. This dedication suggested that colonization of America might relieve overpopulation in England. In his dedication of volume 2 of *Principal Navigations* (1598–1600) to Robert Cecil, Hakluyt is even more explicit that such colonization would result in great financial reward and national glory (sig. 3a).

54. P. Barber, "England II," 67.

55. *DNB,* 3:730. Camden mentions this association in *Britannia.*

56. There is no good modern biography of Warwick. See Leader, *Life of Sir Robert Dudley;* Ritchie, "Sir Robert Dudley."

Church in 1588, under the tutorship of Thomas Chaloner, later Governor of Henry Prince of Wales,[57] and was thus a contemporary of Nathaniel Torporley, Christ Church's most important mathematician.

Torporley is best remembered as Thomas Harriot's intellectual executor,[58] publishing a few of Harriot's papers in rather abbreviated form after his death. Torporley attended Shrewsbury Grammar School and matriculated to Christ Church in 1581, shortly after Hakluyt's official departure. He received his B.A. in 1584 and his M.A. in 1591.[59] Torporley then became involved with the coterie surrounding Henry Percy, ninth Earl of Northumberland (the "Wizard Earl"), receiving a pension from him and associating with Harriot and Walter Warner, all of whom worked on mathematics and mathematical geography problems, especially the question of solving the longitude.

Edmund Gunter, the other mathematician of note hailing from Christ Church, achieved his B.A. in 1603, his M.A. in 1606, and his B.Th. in 1615. During his relatively short life of forty-five years, he held several church livings, although he remained a scholar, holding the position of Professor of Astronomy at Gresham College from 1619 until his death in 1626.[60] There he investigated the calculation of logarithms, devised a sort of prototypical slide rule, and dealt with many geographical problems, under the auspices of applied astronomy. Some of this interest in geography was probably developed while he was a student at Christ Church, given that such a large portion of his adult life was spent at the university.

A considerable number of people with geographical interests spent their formative years at Christ Church. They had a variety of specializations, from descriptive geography and antiquarianism to mathematics and mathematical geography, though the subdiscipline that attracted the most Christ Church scholars was descriptive geography. The majority of these men attended the Faculty of Arts at both the Bachelor's and Master's levels, while a few went on to the theology faculty. Only one of the other faculties is represented, with Fludd's foray into medicine. Thus, the full seven-year Arts degree, up to the M.A., seems to have been the norm, rather than either a few years of noncurricular study or the specialized investigation of the higher faculties. The study of geography, at least at Christ Church, was most compatible with the Arts curriculum, and Christ Church provided a congenial atmosphere for the pursuit of these geographical investigations.

57. For further discussion of Chaloner, see chapter 6 below.

58. J. W. Shirley, *Thomas Harriot: A Biography*, ch. 2.

59. "The M. A. suppl. and lic. are put as from Brase. but this must be in error, as the inc. is from Christ Church and also a dispensation in 1591" (Clark and Boase, *Register of the University of Oxford*, 2.3:118).

60. Adamson, "Foundation and Early History of Gresham College," 142–43.

New College, Oxford, a foundation of seventy scholars,[61] was only slightly less prolific than Christ Church in the number of geography books owned or the number of its geographically minded students, though it lacked Christ Church's truly innovative geographers. There were twenty-five geographical men associated with New College, including such people as William Lambarde, author of that often-reprinted chorographical work *A Perambulation of Kent* (1570); Thomas Lydiat, that impecunious divine and mathematician, whose importuning letter to James I suggesting the establishment of observatories in southern Africa fell on deaf ears; and the divine George Coryate, a descriptive geographer in his own right and father of Thomas, whose *Crudities* became extremely popular.[62] None of the guiding lights of geography, such as were found at Christ Church, appeared here; rather, this was a collection of middle-ranked academics, interested in new and old geographical information on a modest and unexceptional level.

Nearly two-thirds of these geographically inclined men achieved a B.A. Almost as many (fifteen, or 60 percent) received an M.A. Thirteen attended the higher faculties, two receiving M.D.'s, five B.Th.'s or D.Th.'s, and six B.C.L.'s or D.C.L.'s. The high proportion of law degrees is especially interesting, given the nearly moribund state of the Faculty of Civil Law.[63] Of these civilians, only two pursued the study of geography to the point of publication, but even ownership of geography books suggests that these students' interest was piqued by this attractive sideline to their lengthy degrees. Since civil law relied on European rather than English models, and since the careers of civilians often tended to be diplomatic and concerned with foreign relations, it makes sense that the men studying it would have wanted to learn more about the world outside their own small island.

In keeping with this emphasis on civil law, it is not surprising to see professions represented that would reflect this training.[64] Three recipients of D.C.L.'s went on to become advocates and judges. Two studied civil law overseas as well as at Oxford. Thomas Martin, a fellow of New College from 1538 to 1553, received his D.C.L. at Bourges and later had it incorporated at Oxford (1555). He worked as an advocate, especially in connection with the court of Queen Mary, and is important for this analysis because

61. Rashdall and Rait, *New College,* 40.

62. See Strachan, *Life and Adventures of Thomas Coryate;* Severin, *Oriental Adventure.*

63. Barton, "The Faculty of Law," 257–93.

64. "Edward VI's professor of civil law at Cambridge, the learned Sir Thomas Smith, in an inaugural lecture which was intended to give his auditors the rosiest possible view of the prospects of the legal graduate principally stressed the opportunities of diplomatic employment and the value of a legal degree as a qualification for the more desirable benefices in the church" (ibid., 271).

he donated several geography books to the New College Library in 1588, after he had retired from active law practice.[65] Sir Thomas Ryves was also a graduate in civil law at New College. A student from Winchester, he received his B.C.L. in 1605 and his D.C.L. in 1610, studying overseas at some point during this time. He became a judge, a master in chancery, and an Advocate General to Charles I, though he ended his life in relative seclusion, having supported Charles in the 1640s.[66] His interest in geography took a practical turn, evident in his book *Historia Navalis* (1629), which concerned the legal aspect of oceanic travel and was often reprinted. This link between royalist sympathies and geographical interest, seen earlier in the career of Thomas Aylesbury, suggests that an imperial point of view gained through geographical themes would have been attractive to those sympathetic to a strong monarch. Finally, Richard Zouche, a scholar from Winchester, was a fellow of New College from 1609 to 1622. He received his B.C.L. in 1614 and his D.C.L. in 1619 and went on to become the Regius Professor of Civil Law at Oxford from 1620 to 1661. He, too, was a royalist (perhaps this went hand-in-hand with being a civilian) but took a more active interest in geography than his civil law colleagues. This interest was manifested in *The Dove: or passages of Cosmography* (1613), a long descriptive poem styled after the classical work of Dionysius Periegetes,[67] which described Europe, Asia, and Africa, although it ignored the New World. One final New College associate interested in geography had a quasi-legal career: George Turberville became secretary to Thomas Randolph, ambassador to Russia, in 1568. Although Turberville himself is not known to have taken a degree while at New College (he was a fellow in 1561),[68] a diplomatic appointment was an important step in increasing the role of young university-trained gentlemen in government.

Only two of the geographically disposed young men at New College occupied the majority of their time with geographical endeavors. These were William Lambarde and Thomas Lydiat. Lambarde, the "historian of Kent," began his New College career in 1558 as a fellow from Buckingham. In 1561, he was granted his B.A. and in 1565 his M.A. He then became Keeper of Records in the Tower of London and devoted the rest of his life to his *Perambulation of Kent* (first published 1570) and to the vastly ambitious undertaking, never completed, *Dictionarium Angliae: Topographicum, et Historicum: an alphabetical description of the chief places in England*

65. New College Library Benefaction Book 1617–1909, 39–40.

66. *DNB*, 17:562, states that "Ryves was an able civilian, and his works evince considerable learning; but Archbishop Ussher had no high opinion of his honesty."

67. Parker, *Books to Build an Empire*, 233. Dionysius Periegetes was an Alexandrian poet mentioned by Strabo as one of the most famous tragedians of that city. *Nouvelle Biographie Generale* (Paris: Firmin Didot, 1853–66), 114:297.

68. J. Foster, *Alumni Oxonienses*; *DNB*.

and Wales (first printed 1730). He was deeply influenced by the elder chorographer and founder of Old English studies, Lawrence Nowell, and was part of the circle of scholars William Camden turned to when beginning to learn about Anglo-Saxon antiquities.[69] Although both Camden and Lambarde were at Oxford, their residences did not overlap, and they must have become friends after Lambarde's departure for London. Lambarde was well respected by William Camden, who sent him a manuscript version of *Britannia* for his comments.[70]

Thomas Lydiat was a more colorful character than Lambarde, gaining a wide circle of mathematical friends during his lifetime and losing their support by the time of his death. Lydiat attended New College as a scholar from Winchester.[71] He was granted his B.A. in 1595 and his M.A. in 1599. Thereafter, "his defective memory and utterance led him to relinquish both the study of divinity and his fellowship in 1603, in order to devote himself to mathematics and chronology."[72] He did manage to support himself through a number of small livings and through a pension from Henry, Prince of Wales, but he constantly complained of his poverty and tried to devise schemes whereby he could use his mathematical and geographical skills to achieve wealth and fame. The most spectacular of these was a plan to establish an observatory below the equator in Africa in order to gain, in a spirit of true Baconian induction, information concerning southern hemispheric astronomy and navigation, as well as to provide English mariners with a safe harbor before rounding the Horn.[73] This potentially practical suggestion shows the direct link between the growth of the English empire and geographical investigation. Unfortunately, this and other plans came to nothing, and Lydiat died in 1646 in great poverty.

These two men found inspiration at Oxford, even if they were influenced by students of geography outside the walls of New College. Despite a lack of famous or important geographers at New College, the number of people interested in geographical topics in the college demonstrates that a basic grounding in geography was possible and common.

Four of the geographically oriented men at New College pursued teaching careers of one type or another. Sir George Rives and Arthur Lake, for example, shared both college and university careers and an interest in geo-

69. Piggott, *Ruins in a Landscape,* 36.

70. Mendyk, "Early British Chorography," 473.

71. Winchester was the common grammar school for these geographical students, largely reflecting the fact that New College actively recruited students from Winchester, awarding many of its scholarships to Winchester boys (Leach, *History of Winchester College,* 90).

72. *DNB,* 12:316.

73. Letter from Thomas Lydiat to East India Company, Nov. 22, 1628, Oxf. Bodl. MS Bodl. 313, f. 32.

graphical books. Rives received his B.A. at New College in 1582, his M.A. in 1586, his B.Th. in 1594, and his D.Th. in 1599. He was Warden of New College from 1599 to 1613 and Vice-Chancellor of the university from 1601 to 1613. With such a prestigious academic career, it is not surprising that he donated books to the New College Library in 1600, probably on the occasion of his election as Warden.[74] One of these books was an important descriptive geography work, Philip Lonicer's *Chronica Turcica* (Frankfurt, 1578). Arthur Lake was Rives's successor as Warden and a huge benefactor to the New College Library in 1600.[75] Lake was a much more prominent scholar, expert in Hebrew, and a translator of the Authorized Version of the Bible.[76] As was true of many New College fellows, Lake had attended Winchester Grammar School, arriving at New College in 1588. He was granted his B.A. in 1591, his M.A. in 1595, his B.Th. and D.Th. in 1605, and he followed Sir George as Warden from 1613 to 1617 and as Vice-Chancellor from 1616 to 1626. While Warden, he established lectureships in Hebrew and mathematics at his own expense,[77] and in 1617, at the end of his wardenship, he donated a substantial number of books to the New College Library. Included in this donation were many geography books, from all three geographical subject areas.

Two other geographical New College graduates became schoolteachers. Simon Harward attended New College from 1577, when he became college chaplain, until 1578, when he received his M.A. He had previously earned his B.A. at Christ's College, Cambridge, and migrated to Oxford in 1577. Thereafter, he lived out his life as a vicar and schoolteacher, as well as practicing physic on the local inhabitants of Tandridge in Hampshire. He wrote miscellaneous works, including a geographical/meteorological tract entitled *A Discourse of Lightnings* (1607). Another college man turned schoolteacher was John Owen, who donated a number of geographical books of all three types to the New College Library.[78] He attended New College from 1583 to 1591, achieving as his sole degree his B.C.L. in 1590. He was known as "the epigrammatist"[79] and spent his later life in Wales, teaching school and writing pithy sayings. Owen donated several geography books to the library in 1590, demonstrating that his geographical knowledge and interest came from his university years.

Three New College graduates pursued careers in medicine. Walter Bailey (B.A. 1552, M.A. 1556, M.B. 1559, M.D. 1563) was a prominent physi-

74. New College Library Benefaction Book, 43.
75. Ibid., 53–64.
76. Paine, *Men Behind the King James,* 185.
77. McConica, "Social Relations of Tudor Oxford," 120.
78. New College Library Benefaction Book, 41.
79. Clark and Boase, *Register of the University of Oxford.*

cian in the second half of the sixteenth century.[80] He was Regius Professor of Medicine at Oxford from 1561 to 1582 and became personal physician to the queen in the 1580s. In 1587, Bailey wrote an interesting description of a medicinal spring, *Discourse of certain baths or Medicinale Waters . . . nere Newnham Regis* (1587), which belongs with local histories of Britain. In his will of 1592, he bequeathed to New College medical texts and geographical books describing New World flora.[81] Thomas Hopper (B.A. 1595, M.A. 1599) was a less prominent doctor and geographer than Bailey. He was not known to have pursued actively any geographical investigation after his departure from New College but, in 1623, donated several descriptive geography texts to the New College Library.[82]

Most of the other men at New College interested in geography pursued the more usual careers of theology and the priesthood. Henry Ball, for example, became a theologian at New College, receiving his B.Th. in 1588 (when he donated a number of geography books to the New College Library, probably in honor of the granting of his degree)[83] and his D.Th. in 1594. He probably remained with the Faculty of Theology until his death in 1603. John Meyrick (B.A. 1558, M.A. 1562) held the bishopric of Sodor and the Isle of Man from 1577 until his death in 1599. He was an active chorographer, corresponding with William Camden and writing a local description of the Isle of Man, *Epistola de antiquitatibus insulae Manniae* (ca. 1580),[84] which he circulated in manuscript form to Camden.[85] At least five men who donated geography books to the New College Library or who died with geography books in their possession were clerics. Since the Arts curriculum was most geared toward those studying for the priesthood, the presence of these books in the effects of clerical students argues for the inclusion of geography in the official Arts curriculum. Although geography could be used for added worldly polish, it was also an important part of the more formal studies in the Arts faculty.

Although the New College slate of geographically minded scholars is less impressive in breadth and depth than that of Christ Church, it confirms a number of conclusions. As was the case at Christ Church, the majority of these men remained in college for several degrees: only four of the twenty-five, or 16 percent, achieved one degree (either a B.A. or a B.C.L.), and a further two are not known to have taken any degree. New College differed from Christ Church in its emphasis on the Civil Law Fac-

80. For a more complete study, see Horton-Smith, *Dr. Walter Bailey.*
81. Bailey's will is described in Lewis, "Faculty of Medicine," 236.
82. New College Library Benefaction Book, 66–76.
83. Ibid., 39.
84. BL Cotton MS Julius F.10.
85. *DNB,* 37:319.

ulty and correspondingly lighter emphasis on theology. But the higher faculties were all well represented; the study of geography evidently complemented well their protracted and specialized endeavors. Geography was a pursuit that attracted those in the first stages of their Arts training and also those who remained longer at university and passed on to more specialized studies. New College graduates seldom pursued the more exciting geographical careers of several Christ Church alumni, but their interest was sincere and was reflected by the large number of New College men who owned geography books in all three subdisciplines. Likewise, the high proportion of New College men who served the English state in their varied and various careers bears witness to the utility of geographical training for such vocations and thus to a shared collection of ideas and assumptions about the world that would have informed their actions within that world.

The emphasis on descriptive geography book ownership seen in the book list for New College reflects this developing imperialist ideology. Nearly two-thirds of the geography books owned by New College students or donated to its library were descriptive.[86] These books included exploration accounts, which told tales of brave and foolhardy ventures of European explorers to the New World, often with an agenda of national aggrandizement. Hakluyt's *Principal Navigations* was the most obvious example of this genre, while the encroaching expansion of the other European nations was told by José de Acosta concerning the Spaniards and Portuguese in the West and East Indies and Lescarbot concerning the French in America. There were also descriptions of northern and eastern Europe, Asia, Turkey, Italy, and the Netherlands. Both Old and New Worlds thus received substantial treatment. Local history, as well, was given considerable attention, with representatives of both medieval and modern chorography. Although the number of books about mathematical geography was small, at New College it was significant as a branch of applied astronomy. In fact, New College seems to have been an important center for astronomy, as its book list included the latest astronomical work of Copernicus and Tycho Brahe. New College students thus had access to innovative material in all three areas and read geography that encouraged them to see the controllability of the globe and the riches awaiting the English as they set out to conquer it.

The emphasis on geography was less pronounced at Cambridge than at Oxford. Few of the Cambridge colleges rival the geographical stature of Christ Church. Trinity College, however, was an important center for

86. Sources for this book list include Richard Secoll, Inventory 1577, Oxford VCC.; New College Library Benefaction Book 1617–1909.

geographical study at Cambridge, as demonstrated by the analysis of books owned by its associates and by an investigation of geographically inclined men who graced its halls.

Trinity was an important center for the study of natural philosophy and the "new science" before and after the Civil War, and this is borne out by the list of geography books owned or donated by its students during this period (see table 4 above).[87] All three categories of geography were present, with exempla from both classical and contemporary sources. Mathematical geography books included classics and medieval cosmographies which had informed the late-medieval scholar about the place of the earth in a Ptolemaic universe and of its mathematical construction. A Mercator *Atlas* provided modern views of European countries and the New World, while the presence of two navigational treatises suggest that two students at least pursued practical interests. Contemporary astronomical ideas were represented by the works of Copernicus, Kepler, Mercator, and Thomas Digges. The appearance of these controversial works was extremely unusual and argues for a group of men eager for the latest astronomical theories. Perhaps this affected their view of the terrestrial globe as well. Descriptive geography included many tales of exploration, as well as descriptions of Spain, Hungary, Russia, France, and the Orient. Most of these tomes described parts of Europe, but the New World was thoroughly represented as well. Finally, the local history section was quite extensive. It included all the major medieval British chorographers,[88] plus William Camden and most of the major European chorographers. Trinity was thus an important center of geography at Cambridge in all three areas. As well as demonstrating real strengths in mathematical and descriptive geography, Trinity students took a special interest in local history and in history in general. This is made more evident by the presence of all the major contemporary descriptive historians of the period.[89]

The majority of geographically inclined men at Trinity received a B.A. and an M.A.; twenty-three, or 74 percent, were granted a B.A., and

87. Sources for the Trinity College, Cambridge, book list: Nicholas Abythell, Inventory 1586, Cambridge VCC (4); William Anger, alias Angier, Inventory 1589 (5); William Ball, Inventory 1601 (6); Edward Liveley, Inventory 1605 (7); Nicholas Sharpe, Inventory 1576 (3); John Shaxton, Inventory 1600/1 (6); "Memoriale Collegio Stae. et Individuae Trinitatis in Academia Cantabrigiensi Dicatum 1614," Trinity College, Cambridge, R.17.8.

88. Gildas, Geoffrey of Monmouth, Gervase of Tilbury, Gerald of Wales, William of Malmesbury, and Ranulf Higden. Mendyk, in *"Speculum Britanniae,"* sees these as the most important forerunners of sixteenth-century chorography, with Higden's *Polychronicon* especially setting the standard for Lambarde and Camden. See chapter 5 below.

89. These include Paolo Giovio, *Historiarum sui Temporis* (Florence, 1550); Raphael Holinshed, *Chronicles of England* (London, 1577); Francesco Guicciardini, *Historia d'Italia* (Basel, 1566); Speed, *Historie of Great Britaine* (London, 1611); and Antonio Sigonio, *Historia de Regno Italiae* (Venice, 1574).

twenty-four, or 77 percent, were granted an M.A.[90] A large proportion continued to the higher faculties: three received an M.D.; eight received divinity degrees; and one was granted a Bachelor of Civil Law. The proportion of these degrees (twelve out of thirty-one) demonstrates a predisposition of those in the higher faculties at Trinity toward the study of geography.

Trinity was a college heavily patronized by the nobility, and two of its noble students were interested in descriptive geography and navigation. George Clifford, third Earl of Cumberland, began his academic career as a fellow-commoner at Trinity in 1571. He was granted his M.A. directly in 1576 and then migrated to Oxford, where he probably studied geography and navigation.[91] He soon became an important naval commander under Elizabeth, involved, as was every able-bodied English navigator, in the defeat of the Armada. As a naval commander and adventurer, he had a practical interest in mathematical geography and navigation all his life, and he may first have discovered his inclination toward mathematical geography while at Trinity College.

Robert Devereaux, second Earl of Essex, also attended Trinity College in the 1570s and '80s. He was, of course, more famous for his exploits with Elizabeth and his disastrous campaign in Ireland,[92] but he did study at Trinity College for four years, from 1577, when he was admitted as a fellow-commoner, until 1581, when he was granted his M.A. He, too, was a geographical practitioner, especially interested in settling, surveying, and describing Ireland (the first expansion of the English empire),[93] and likely gained some of his curiosity and facility in this while a student at Cambridge. Indeed, Essex had a keen interest in extending the influence of England (and Essex) into new parts of the world. William Barlow's 1597 dedication to Essex of a proposal to seek a northwest passage to Cathay shows this.[94] It was also Essex who encouraged his client Sherley to travel to the East, indicating for both men strong political ambitions and a deep curiosity about the rest of the world.[95]

90. Because Trinity College catered to the nobility, the three noblemen's sons interested in geography were granted their degree of Master of Arts without proceeding B. A., by reason of their nobility. Thus, not all the Bachelors at Trinity went on to M.A.'s, though the numbers of B.A.'s and M.A.'s in our sample appear to be nearly the same.

91. Curtis, *Oxford and Cambridge*, 235. For biographies, see G. C. Williamson, *George, Third Earl of Cumberland;* E Wright, *Certaine errors in navigation* (1599); added to it, *The Voyage of the Right Ho. George Earle of Cumberland to the Azores* (London, 1599).

92. Strachey, *Elizabeth and Essex;* Harrison, *Life and Death of Robert Devereux;* Lacey, *Robert, Earl of Essex.*

93. For example, Robert Devereux, *A Proposal touching the inhabiting of the North of Ireland* [1593]; mentioned in E. G. R. Taylor, *Late Tudor Geography.*

94. Barlow, *Navigator's Supply,* sig. b1a.

95. Letters from Anthony Sherley to Essex, concerning Persia, ca. 1600. BL Add. 38,139, ff. 15, 16.

Two other Trinity men interested in geographical topics achieved careers in government service. One, Francis Bacon, should be categorized first as a natural philosopher with geographical interests, rather than simply as a geographer.[96] He was, after all, most concerned about questions of the reliability of natural knowledge and how to achieve true understanding, rather than the more down-to-earth descriptions of peoples and places. He was admitted as a fellow-commoner at Trinity in 1573 and received his M.A. in 1594. Bacon also enjoyed an extremely long and influential career as a politician, serving as M.P. for many years before he was raised to the peerage with the title of Baron Verulam and Viscount St. Albans, and serving two monarchs in a variety of positions, including Lord Chancellor. His geographical pursuits included an interest in local history, corresponding with Sir Robert Cotton, his contemporary at Jesus,[97] and a concern with descriptive geography. Bacon corresponded with Anthony Sherley concerning Persia, showing an interest in both political relations and trade.[98] He also actively encouraged English plantations in North America and the search for a northwest passage to Cathay.

Sir Edward Stanhope was another Trinity man who graduated from his academic studies to become a political figure, though on a much less exalted level than Bacon. Stanhope arrived at Trinity as early as 1560, achieving his M.A. in 1566 and his D.C.L. in 1575. He was at Cambridge for four years with Cumberland and two years with Bacon. He went on to become an M.P., Chancellor of the Diocese of London, and a judge at Sir Walter Raleigh's trial. Though his role in Raleigh's trial perhaps demonstrated some hostility toward navigational and geographical enterprises,[99] he was a generous donor of geography books to the Trinity College Library, giving books in all three subdisciplines.[100] Stanhope's taste in reading and study was probably influenced by the company of such men as Bacon, Edward Lively, and, for a short period, Thomas Hood during his time at Trinity.

A large proportion of graduates of Trinity College who showed an interest in geography remained at the university throughout their lives. Six of the thirty-one held academic positions, several of very high level, and four of these were closely involved in the translating of the Authorized

96. The literature on Bacon is immense. For a recent exposition of Bacon's natural philosophy and his plans to reform the law, see Martin, *Francis Bacon, the State.* For a deeply provocative discussion of the term *natural philosophy* and its relationship with *science,* see Cunningham and Williams, "De-centring the 'Big Picture.'"

97. For example, Oxf. Bodl. MS Smith 71, 39.

98. BL Add. 38, 139, f. 15a–b.

99. *DNB,* 44:7.

100. "Memoriale Collegio Stae. et Individuae Trinitatis in Academia Cantabrigiensi dictatum 1614," Trinity Coll. R.17.8, 94.

Version of the Bible.[101] The compilation of the six translation committees, each designated a particular section of the Bible, seems to have been based on a combination of linguistic, biblical, and scholarly skill, with proper established church credentials.[102] Thus, these four geographically inclined scholars must have combined these attributes with a keen interest in both biblical and European geography. Indeed, the development of this new and authorized translation of the Bible was an important hallmark in the creation of an English identity and suggests a link between this fundamental self-fashioning and the study of geography.

Edward Lively (Trinity B.A. 1569, M.A. 1572) was one such academic geographer. Lively is best remembered for his translation of part of the King James Bible, from 1 Chronicles through the Song of Solomon, a section requiring much chronological and geographical knowledge.[103] He was Regius Professor of Hebrew at Cambridge from 1575 to 1605, as well as son-in-law to Dr. Thomas Lorkin, Regius Professor of Physic from 1564 to 1591. Perhaps more important for our purposes, he also wrote and owned books of mathematical geography and chorography. His book on Persia, *A true Chronologie of the . . . Persian Monarchie* (1597), dealt with a part of the world fascinating to Elizabethans as a center of intrigue, wealth, and the Infidel. It combined this interest in the Middle East with an attempt to determine the chronology of the world, necessary to salvation and fascinating for a man translating the chronicles of the Old Testament. The inventory of books owned at his death indicates that Lively had a strong interest in astronomy, mathematical geography, chronology, and descriptive geography.[104]

William Bedwell was another translator of the Authorized Version who hailed from Trinity and was interested in geography. As nephew of the Cambridge mathematician and military engineer Thomas Bedwell, it is not surprising that William's thoughts ran to mathematical and geographical topics. He attended Trinity from 1579, when he matriculated as a sizar, achieving his B.A. in 1585 and his M.A. in 1588. He may have remained at Cambridge later than this; it was certainly at Trinity that he met his fellow biblical translators, who all attended the college at approximately the same

101. In all, ten men interested in geography were involved in this biblical translation. Elton wrote: "Above all, one must here mention the Authorized Version of the Bible (1611), whose perfection, achieved by a committee of rather dreary divines, is as good a proof of direct inspiration as one can find" (Elton, *England Under the Tudors,* 441).

102. The "hot Protestant" Hugh Broughton, for example, was excluded, although he was one of the greatest Hebraists in England (Opfell, *King James Bible Translators,* 8). A definitive history of the translation still waits to be written. The two modern accounts, by Paine and Opfell, seem little interested in questions of context, religious controversy, or political machinations.

103. Opfell, *King James Bible Translators,* 44.

104. Cambridge VCC probate inventories (7).

time. William Bedwell was an orientalist, the "father of Arabic studies in England."[105] Indeed, he may have come to his interest in Arabic through his mathematical studies of numbers and symbols. Besides a Persian dictionary and an Arabic lexicon, Bedwell wrote works on mathematics, such as *Mesolabium Architectonicum, that is a most rare instrument for measuring* (1631), and local history, including *A brief description of the town of Tottenham* (1631). He is also reported to have given communion to Henry Hudson and his crew in 1607, before they departed for North America.[106]

John Layfield and John Overall both attended Trinity College at approximately the same time—Layfield from 1578 to 1603, Overall from 1575 to 1607—and must have met both of their fellow translators, Lively and Bedwell, early in their academic careers. Layfield went through the usual early degrees, gaining his B.A. in 1582, his M.A. in 1585, and his B.Th. in 1592. His career then took a curious twist as he became chaplain and attendant to George Clifford and traveled with him to the West Indies. On his return, he wrote a manuscript description of this adventure, *Relation of the Earl of Cumberland's Voyage 1596–98,* which was published in abridged form by Purchas.[107] He returned to Trinity, became the grammar examiner, and received his D.Th. in 1603. Just after this, he was solicited, probably by Lancelot Andrewes, to join the First Westminster Company of the Authorized Version translation, joining both William Bedwell and John Overall. This group was given the task of translating the section from Genesis to 2 Kings. This portion of the Old Testament, of course, dealt with geographical creation, and thus Layfield's grounding in geography would have aided him in sorting out the biblical geography. He divided his time between scholarship and active geographical experience, making him a most instructive example of the interplay between scholarly concerns and practical interests inherent in a serious study of geography.[108]

John Overall, who later became Bishop of Coventry, Lichfield, and Norwich, received his B.A. in 1579, his M.A. in 1582, his B.Th. in 1591, and his D.Th. in 1596. He held the chair of Regius Professor of Divinity from 1596 to 1607, as well as being Master of Catherine Hall at that time. Like

105. Venn, *Alumni Cantabrigienses,* quoted without attribution in Opfell, *King James Bible Translators,* 36.

106. Opfell, *King James Bible Translators,* 36.

107. Purchas 1625, Part II:1141–49.

108. His life could be compared with that of Lawrence Kemys, a brilliant scholar from Balliol College, Oxford, who, after completing his M.A., joined Raleigh's expedition to Orinoco in 1595–96 as captain of one of Raleigh's ships. This was part of Raleigh's attempt, also seen in his consultation with Thomas Harriot, to bring university learning to practical enterprises. Unfortunately, Kemys's story has a tragic ending; he captained Raleigh's flagship on his last ill-fated voyage to Guiana in 1617, and when it was clear that the return to England would bring disgrace to them all, Kemys committed suicide (J. W. Shirley, "Science and Navigation," 80).

his fellow translator Edward Lively, Overall was interested in local history and chronology, both of which were closely connected with Old Testament studies. In fact, just as Bishop Ussher's search for the true chronology of the created world issued from a desire to understand the biblical context in a rational, mathematical way, so, too, Overall's interest in medieval and ancient chronologies had its origin in biblical studies and in an attempt to fit these various chronological systems into absolute time.[109]

Five men associated with Trinity College have some claim to fame as geographers who combined theory and practice in a fruitful way. The most famous of these was Thomas Hood, mathematical lecturer to London merchants and to the East India Company. Hood arrived at Trinity from the Merchant Taylors School in 1573, receiving his B.A. in 1578 and his M.A. in 1581. By 1582, he was in London, teaching mathematics (in fact, mathematical geography and navigation) at the home of Sir Thomas Smith in Gracechurch Street. This lectureship was established by Sir Thomas, merchant and later Governor of the East India Company, and was intended to educate those involved with overseas ventures, possibly employees of the Virginia Company (whose expeditions Hood underwrote). The makeup of the audience is now unknown, although from the tone of his introductory remarks, published under the title *A Copie of the Speache made by the Mathematicall Lecturer, unto the Worshipfull Companye present . . . in Gracious* [sic] *Street: the 4 of November 1588,* Hood seemed to be talking to his mathematical colleagues and mercantile patrons, rather than to the mariners who he was so insistent needed training.[110] The contents of Hood's lecturers are unknown, although the treatises bound with the British Library copy indicate that he stressed navigational techniques, instruments, astronomy, and geometry, all of which he might have learned while at Cambridge.[111]

Both Abraham Hartwell (the younger) and Sir Henry Spelman (the elder) were Trinity men with a lifelong interest in local history and British antiquities. Hartwell attended Trinity from 1568 until at least 1575, when he was granted his M.A. He then took a number of clerical positions, including secretary to Archbishop Whitgift, attended Gray's Inn, and possibly stood for Parliament on two occasions.[112] He was a member of the Society of Antiquaries, demonstrating his interest in antiquities. Hartwell's published work ventured into the realm of descriptive geography, with a number of translations of French and Italian works describing the Middle

109. See Wilcox, *Measure of Times Past,* as well as chapter 5 below.

110. Hood, *Copie of the speache,* sig. A2a.

111. Thomas Hood, *The Use of the two Mathematical Instruments the crosse Staffe . . . And the Iacob's Staffe* (London, 1596); and *The Making and use of the Geometrical Instrument, called a Sector* (London, 1598). BL 529.g.6.

112. "One of these names M.P. for East Looe, 1586; for Hindon, 1593" (Venn, *Alumni Cantabrigienses*).

East and northern Africa: for example, *History of the Wars between the Turks and the Persians, from the Italian of John Thomas Minadoi* (1595) and *A Report of the Kingdom of Congo . . . drawn out of the writings . . . of Odoardo Lopes* (1597). The latter Hartwell translated at the insistence of Richard Hakluyt, showing that Hakluyt's influence was in no way limited to his alma mater.[113] This early description of Africa, setting out to prove that people could live in the torrid zone, that the Nile did not rise in the Mountains of the Moon, and that blacks were not colored by the rays of the sun, provides important early impressions of a part of the world important economically and politically to England in the centuries to come.

Sir Henry Spelman arrived at Trinity just after Hartwell had left, attending from 1581, when he matriculated as a pensioner, until 1583, when he received his B.A. While there, he apparently pursued astrological studies,[114] and shortly after leaving, he became a founding member of the Society of Antiquaries (1593).[115] He probably developed his love of geography and especially of chorography at Trinity. He divided his later life between his estate in Norfolk and London, where he studied with his friend Sir Robert Cotton, compiling several chorographical works, including notes on ancient English genealogies[116] and *Icenia: A Description of Norfolk* (written before 1611 and passed on to John Speed).[117] In 1635, he founded a short-lived Anglo-Saxon readership at Cambridge, demonstrating both his interest in British antiquities and his conviction that the university was the place to teach the subject, rather than through more private study.

Two men from Trinity College demonstrated their interest in geography mainly by translating geographical works. Philemon Holland (B.A. 1571, M.A. 1574, M.D. 1597) was a Master of the Free School at Coventry for twenty years, but the achievement for which he was dubbed the "Translator General" of his age[118] was the translation of both Latin classics and modern English geographical authors into English. He translated Livy, Plutarch, and especially Pliny and Martianus (*A Summarie collected by John Bartholomew Martianus, touching the topography of Rome in ancient time* [1599]), as well as Camden's *Britannia, or a Chorographicall Description* (1610), into English. He also translated John Speed's *Theatre of the Empire of Great Britaine* into Latin (1616). Holland's training in translation came from his time at Trinity, and probably the subject matter he chose was influenced by his college as well. Ralphe Carre was less famous but was also

113. Abraham Hartwell, trans., *Report of the Kingdom of Congo,* dedication, sig. *1a.
114. Feingold, "Occult Tradition," 93.
115. See J. Evans, *History of the Society of Antiquaries;* and chapter 5 below.
116. Oxf. Bodl. MS Dodsw. 141.
117. Speed, *Theatre of the Empire,* 35–36.
118. A generally agreed title, quoted in Simon, *Education and Society,* 384; *DNB;* and L. B. Wright, *Middle-Class Culture,* 350, where it is attributed to Thomas Fuller.

a geographical translator hailing from Trinity. The coincidence of Carre's translation of *Mahumetan or Turkish History* appearing in 1600, the same year he received his B.A. at Trinity, indicates that he gleaned his interest in descriptive geography, as well as the impetus to undertake this translation, while he was attending undergraduate lectures there.

Thus, Trinity College encouraged its students to study geography in all three subdisciplines. Though none of Trinity's alumni (with the exception of Thomas Hood) went on to fame and fortune as a professional geographer, many were interested enough to own a book or translate a foreign work on the subject. In fact, what distinguished Trinity from the colleges previously discussed was the prevalence of translators of geography. Part of this emphasis on translation can be traced to the Authorized Version, the creation of which intimately involved Trinity; in addition, Trinity students' desire to learn more about the world being discovered, especially to the east, must have encouraged the investigation of non-English sources. North America seems to have attracted the attention of fewer men from this college (though two of them did visit America) than did the riches and exotica of the Middle East, demonstrated by the prevalence of Eastern rather than Western adventures in their writing and translating.

St. John's College, Cambridge, housed a slightly smaller though enthusiastic group of geographically minded students. Earlier in the sixteenth century, St. John's supported the scholarship of John Dee and Henry Billingsley. Later, Henry Briggs, William Gilbert, and that most famous successor to Hakluyt, Samuel Purchas, graced its hallowed halls.

This college had a large number of geography books on its list, many of which included the best contemporary knowledge (see table 4 above).[119] The geography books owned by St. John's students also demonstrated a remarkably English character, in terms of language and place of publication (27 percent of these geography books were of English origin). The fourteen mathematical geography books included Leonard Digges's *Tectonicon* (1556; on surveying) and his *Prognostication* (1555), Blundeville's *Exercises* (1594), William Gilbert's *De Magnete* (1600), and Warner's *Albions England* (1586). Six of the descriptive geography books were written in English and published in England. Fourteen of the thirty-one chorographical works were published in England as well, indicating that there was generally an emphasis at St. John's on English sources and even on English topics. At St. John's, then, a relative balance existed among the three geographical subdisciplines, with an interesting shift toward English texts and topics.

Of the St. John's men interested in geography, almost three-quarters

119. See appendix B for a list of St. John's geography books.

(fifteen of twenty-one) obtained the rank of Bachelor of Arts, and two-thirds (fourteen) received an M.A. as well. A smaller proportion of students achieved degrees in the three higher faculties than was the case at the other three colleges examined above. Two students obtained degrees in medicine, and four received the rank of either Bachelor or Doctor of Law. Thus, the study of geography is placed once again in the Arts curriculum, more explicitly at St. John's than at the other colleges.

The most common career pattern for St. John's men interested in geography was to remain at the university as scholars, fellows, or masters. Five of these men obtained important academic positions at St. John's or at Cambridge, while two men interested in geography remained students there until their untimely deaths. Of the academics, only John Hatcher (B.A. 1532, M.A. 1535, M.D. 1544), Regius Professor of Physic and Vice-Chancellor of Cambridge, demonstrated his geographical proclivities solely by the works of geography in his possession at his death in 1587.[120] His books covered all three areas of geography, but especially mathematical geography, which was a specialty of the college. The other academics displayed a much more active participation in geographical pursuits.

John Dee (B.A. 1545, M.A. Trinity 1548) was the most notorious of this group. Even while at Cambridge, his interest in astronomy, astrology, and geography was established,[121] and it was this initial spark that sent him on to Louvain to study with those great mathematicians and geographers Gemma Frisius and Gerard Mercator, among others. He later returned to England, where he set himself up as an astrologer and geographical adviser, becoming astrologer to Elizabeth (in which capacity he advised her on hydrographical and geographical matters) and advising the Muscovy Company, Humphrey Gilbert, and numerous other practical geographers.[122] Though in recent years Dee's name has become synonymous with occultist practices and Hermeticism, in his own time he was much better known as a learned and practical geographer.[123] As was demonstrated in the intro-

120. Cambridge VCC, Box 4, in Plan Press II.

121. Curtis, *Oxford and Cambridge,* 242; French, *John Dee,* 28.

122. Dee, in *Private Diary,* records many instances of these men coming to consult with him before undertaking hazardous voyages.

123. French, *John Dee;* Deacon, *John Dee;* and Yates, especially *Theatre of the World,* have done much to encourage this tendency to see Dee as an Elizabethan "magus." While it is true that Dee was very interested in alchemy, numerology, astrology, and, in the end, crystal ball gazing, most of these activities were legitimate sixteenth-century pursuits, and the focus on this has obscured his important geographical work. Even historians such as Sir Roy Strong, in *Henry, Prince of Wales,* are not immune to this tendency, claiming Prince Henry's interest in Dee was Neoplatonic rather than exploring the more obvious geographical link. See chapter 6 below. In an interesting twist, Livingstone, in *Geographical Tradition,* claims a direct relationship between geography and magic, especially as seen in Dee's work. Clulee provides a much more balanced view of Dee's work, arguing in *John Dee's Natural*

duction above, he was equally an overt promoter of English imperialism, especially although not exclusively in *General and Rare Memorials Pertayning to the Perfect Arte of Navigation* (1577).[124] Although Dee did not stay at St. John's, he remained a scholar all his life, with a private library that rivaled any in Britain.[125] Such an example must have inspired many a St. John's student, as is evident from the large number of students owning geography books or pursuing geographical studies.

One man who may have been influenced by Dee, though the two did not quite overlap at St. John's—he attended as a scholar from 1550 to 1551—was Henry Billingsley. Billingsley could not be categorized as a scholar, since he earned his money and spent much of his time in mercantile pursuits in London, where he was Lord Mayor for a time, but his collaboration with Dee on an English translation of Euclid's *Elements* (published 1570) marks a turning point in English mathematical sophistication. The instance of a university-educated merchant aiding in the translation of a scientific classic makes an informative connection between theory and practice. Clearly, mathematical competency was important to Billingsley as a merchant, but the theoretical possibilities expressed by Dee in his *Praeface* suggest that he at least had a far grander agenda in mind.

Another mathematician from St. John's who spent his life in academic circles was Henry Briggs (B.A. 1582, M.A. 1585).[126] He was Linacre Lecturer at Cambridge for a time (ca. 1592), then the first Professor of Geometry at Gresham College (1596–1620), and finally the first Savilian Professor of Astronomy at Oxford (1619–30). His most lasting achievement was his English exposition of Napier's logarithms, making them accessible to a wider public. The fact that this translation was printed in the second edition of Edward Wright's *Certaine Errors in Navigation* (1610) demonstrates that Briggs, like Napier, saw logarithms as a practical aid to navigation, and thus his explication and translation of them should be seen as part of his

---

Philosophy that as Dee grew more interested in natural magic, he became less concerned with natural philosophy. Sherman, in *John Dee,* has begun to set the record straight on Dee, showing his close interconnection with court and government and establishing the primacy of his geographical work.

124. Sherman does a close reading of *General and Rare Memorials,* as well as *Of Famous and Rich Discoveries* (ms. 1577) and *Brytanici Imperii Limites* (1576–78). In so doing, he makes explicit the link among Dee, imperialism, and the Privy Council (Sherman, *John Dee,* ch. 7).

125. Wormald and Wright, eds., *English Library,* 9; Roberts and Watson, *Dee's Library Catalogue.* See also "Catalogus librorum bibliothecae externae Mortlacensis D. Joh. Dee, Ao 1583, 6 Sept. [Transcribed from the manuscript in the library of Trinity College, Cambridge]," pp. 65–89 of Dee, *Private Diary.* Translated by James in "Lists of Manuscripts Formerly Owned."

126. For more information on Briggs, see T. Smith, "On the Life and Works of Henry Briggs"; and Hallowes, "Henry Briggs, Mathematician."

work with mathematical geography. Edward Wright, in fact, was at Cambridge concurrently with Briggs, attending Gonville and Caius College from 1576 to 1596, and these two keen young men, both interested as they were in mathematics and geography, may well have met while at university. Briggs also investigated the other two branches of geography, corresponding with Bishop Ussher probably concerning chronologies[127] and writing "A Treatise on the North-west Passage to the South Sea, to the continent of Virginia, and by Fretum Hudson."[128] The latter developed from his interest and membership in the Virginia Company. In fact, he left ten pounds in his will to aid in the discovery of the northwest passage.[129]

Both Alexander Neville and Hugh Broughton were St. John's men who pursued academic careers. Neville attended St. John's from 1559, when he matriculated as a pensioner at the age of fifteen, until 1581, when he received his M.A. He then became secretary to Archbishop Parker, edited for him the *Tabulae Heptarchiae Saxonicae,* and wrote a Latin account of Kett's rebellion, *De Furoribus Norfolciensii Ketto Duce,* accompanied by a description of Norwich and its antiquities.[130] This latter work was written while he was ostensibly at Cambridge, indicating that Neville became interested in history, local history, and antiquities while at St. John's. Equally, Hugh Broughton seems to have become interested in that combined branch of mathematics and local history, chronology, while at Cambridge. Broughton had a more varied academic career, migrating from Magdalen, where he received his B.A. in 1570, to St. John's, where he was a fellow from 1570 to 1572, and again moving to Christ's, where he received his M.A. in 1573, remaining a fellow until 1578. Part of his movement may have been a result of his sometimes unpopular religious convictions. Although a recent chronicler of the King James translation argues that he was excluded from the translation committees "because of his violent temper,"[131] it was far more likely to have been because of his "hot Protestant" sympathies. Thus, his focus on biblical chronology had a very immediate salvational purpose.[132] Broughton also seems to have had some contact with Thomas Lydiat, who, as noted above, was equally concerned with establishing a biblical chronology.[133]

127. Adamson, "Foundation and Early History of Gresham College," 132.

128. Published in Purchas 1625, Part II:848–54.

129. Adamson, "Foundation and Early History of Gresham College," 133.

130. Alexander Neville, *De Furoribus Norfolciensii Ketto Duce. Lib. I Ejusdem Norvicus* (London, 1575). Translated into English 1623.

131. Opfell, *King James Bible Translators,* 9.

132. G. J. R. Parry, in *A Protestant Vision,* makes this claim about Harrison's chronology.

133. "Extracts of Hugh Broughton's works," in a book with chronological papers of Thomas Lydiat, Oxf. Bodl. MS Bodl. 666, f. 112 ff.

Four St. John's men had interests in the Virginia Company, either as subscribers or as colonists. Henry Briggs, as mentioned above, was one of the original subscribers to the scheme. So, too, were Sir Thomas Wilson (B.A. 1584, M.A. Trinity Hall 1587) and William Crashaw (B.A. 1592, M.A. 1595). Crashaw, furthermore, was a publicist for the company, writing *A Sermon preached on 21 February 1609/10 before the . . . Lord Delaware, Lord Governor and Captain General of Virginia* (1610),[134] and publishing a sermon by Alexander Whitaker (son of Crashaw's former master at St. John's) in *Good Newes from Virginia* (1613).[135] Finally, Gabriel Archer (matriculating at St. John's as a pensioner around 1591) was a secretary or recorder in Virginia, writing several manuscript descriptions of the colony, including *A Letter touching the Voyage to Virginia* (1609), which was printed by Purchas, a fellow St. John's alumnus.[136]

Samuel Purchas was Cambridge's answer to Richard Hakluyt, taking up Hakluyt's manuscripts and geographical schema after Hakluyt's death.[137] Purchas attended St. John's from 1594, when he matriculated, through his B.A. in 1597 and his M.A. in 1600, until he was ordained priest in 1609. He became chaplain to George Abbot, Archbishop of Canterbury, who had been a translator of the Bible and the author of a book of geography before becoming archbishop.[138] Thereafter, Purchas devoted the rest of his life to compiling a huge compendium of voyages and descriptions of faraway places. *Hakluytus Posthumus: or, Purchas His Pilgrimes* (1625) (see figure 11) followed a much more religious and less imperially promotional path than Hakluyt had done, perhaps in keeping with an age more sensitive to religious fervor than military conquest.

Several other St. John's men actively pursued the study of geography,

134. This sermon "casts both King and Prince [Henry] in an imperial messianic role as 'new Constantines or Charles the Great'" (Strong, *Henry, Prince of Wales,* 62). Strong claims that Lord de la Warr was portrayed as a Protestant knight who would conquer the devil and propagate the gospel in the New World.

135. For further information on Crashaw, and particularly a discussion of his library, see P. J. Wallis, "Library of William Crashawe."

136. "The Relation of Cptne Gosnols Voyage to the North part of Virginia, begun the six and 20th of March, Anno 42 Elizabethae Reginae 1602 and delivered by Gabriel Archer, a Gentleman in the said Voyage" (Purchas 1625, Part II:1647–50); and "A letter of Master Gabriel Archer, touching the voyage of the Fleet of Ships, which arrived at Virginia, without Sir Thomas Gates and Sir George Summers (1609)" (Purchas 1625, Part II:1733).

137. Although Hakluyt groomed Purchas as his successor, the two were estranged before Hakluyt's death, and Purchas had to buy Hakluyt's manuscripts after Hakluyt's death (E. G. R. Taylor, *Late Tudor Geography,* 57–59). Purchas did entitle his 1625 book *Hakluytus Posthumus: or, Purchas His Pilgrimes,* in acknowledgment of the use of these manuscripts.

138. Abbot (Balliol, Oxford, B.A. 1582, M.A. 1585, B.Th. 1593, D.Th. 1597), besides having written a rather dull and ordinary geography textbook, *A Briefe Description of the Whole Worlde* (London, 1599), is most famous for having uttered those immortal words, "geographie is better than divinitie." H. Wallis used this tag in her article, "'Geographie is Better than Divinitie': Maps, Globes, and Geography in the Days of Samuel Pepys."

PVRCHAS
HIS
PILGRIMES.

IN FIVE BOOKES.

The first, Contayning Peregrinations and Discoueries. in the remotest North and East parts of *ASIA*; called TARTARIA *and* CHINA.

The second, *Peregrinations*, *Voyages*, *Discoueries*, *of* CHINA, TARTARIA, RVSSIA, *and other the North and* East parts of the World, by *English-men* and others.

The third, Voyages and Discoueries of the North parts of the *World, by Land and Sea, in* ASIA, EVROPE; *the* Polare Regions, and in the North-west of *AMERICA*.

The fourth, English *Northerne Nauigations, and Discoueries*: Relations of *Greenland*, *Groenland*, the North-west passage, and other Arctike Regions, with later *Russian* OCCVRRENTS.

The fifth, *Voyages and Trauels to and in the New World*, called AMERICA: Relations of their Pagan Antiquities *and of the Regions and Plantations in the North and* South parts thereof, and of the Seas *and Ilands adiacent*.

The Third Part.

Vnus Deus, Vna Veritas.

LONDON
Printed by *William Stansby* for *Henrie Fetherstone*, and are to be sold at his shop in *Pauls* Church-yard at the signe of the Rose.
1625.

FIGURE 11. Title page from Samuel Purchas, *Purchas His Pilgrimes* (London, 1625; reprint, no. 19 of March of America Facsimile Series, Ann Arbor, Mich.: University Microfilms, 1966).

often in the form of local history or antiquarian activity. These men included Thomas Nash (Lady Margaret Scholar 1584, B.A. 1586, expelled ca. 1590); Lord William Howard (attended ca. 1577), who assisted Camden; Sir Hugh Platt (B.A. 1572), an inventor who maintained gardens at Bethnal Green and in St. Martin's Lane, where he performed agricultural experiments; and the famous Dr. William Gilbert (B.A. 1561, M.A. 1564, M.D. 1569). Gilbert's work on magnetism was on a very different level from that of the other men mentioned, since he was a true natural philosopher in terms of his theoretical achievements in understanding the lodestone. He has been criticized for his impractical bent and the lack of any application of his magnetical knowledge to navigation,[139] but Gilbert nevertheless deserves to be ranked among the top investigators of geographically related phenomena in sixteenth-century Europe. His work influenced Briggs, who followed him at St. John's, as well as scores of other young mathematicians eager to solve the theoretical problem of navigation at sea.

St. John's made up in intensity what it may have lacked in numbers. There was an extremely strong program of geography at the college, encouraging all three branches of the study within their own discrete circles of interest. Mathematical geography was the strongest subdiscipline, centering around Dee in the early period and Gilbert and Briggs later in the century. Local history attracted a fair number of followers, especially those with gentry or clerical connections. Finally, St. John's showed a particularly strong attachment to the affairs of the Virginia settlement, both in actual support and in descriptive prose written by its members.

The Typical Academic Geographer

A portrait of the English geographical community now begins to take shape. Not all geographers attended the universities, of course, but a large proportion did. For those who wished to pursue the study of geography in a scholarly manner, while still recognizing and utilizing practical experience, the universities provided centers of opportunity and enthusiasm. Geography was investigated at a number of colleges at Oxford and Cambridge, to a far greater degree than could be explained by the presence of one or two interested tutors or coteries. The majority of men interested in geography pursued the full seven-year Arts degree, demonstrating that geography fit well within the guidelines of and was probably encouraged by the statutory Arts curriculum. Geography was practiced by serious

139. For example, Francis Bacon criticized Gilbert (in *Advancement of Learning,* 64) for creating a philosophy from a lodestone. See also William Gilbert, *On the Loadstone and Magnetic Bodies* (1893), p. xv. Julian Martin gave an interesting paper on this issue, "Bacon, Gilbert, and the Status of New Natural Knowledge," at "William Gilbert(d) and His World," a conference in Colchester, July 1993.

TABLE 5 Class Origins of Geography Students at Four Colleges (Christ Church and New College, Oxford; Trinity and St. John's College, Cambridge)

Class Origin	*Oxford*	*Cambridge*
Nobility	2	5
Gentry	21	19
Plebeian	17	4
Other	3	2
Unknown	15	34
Total	58	64

scholars, not simply by dilettantes, and this is reflected in the numbers of professors and academics within the ranks of these geographically inclined men. Centers of keen geographical interest developed, most noticeably, though not exclusively, at Christ Church and New College, Oxford, and at Trinity and St. John's, Cambridge. These were colleges that encouraged the collection of geographical books, as the book list analysis demonstrates, and provided a venue for geographical study and publication, shown by the many geographers who passed through their doors.

The social makeup of students interested in geography indicated that these young men were representative of the wider student population. Of the students at the four colleges we have examined in detail whose backgrounds are known, half of the Oxford associates and more than half of those from Cambridge came from gentle families (see table 5).[140]

More than 40 percent of the Oxford students recorded their origins as "plebeian"; three of these plus one who did not list family origin (9.3 percent) are known to have come from urban mercantile families.[141] Only two students at Oxford and one at Cambridge were known to have been sons of clergymen. This closely parallels Lawrence Stone's breakdown of student social origins as 50 percent gentle, 9 percent clerical, and 41 percent plebeian.[142] Given that most of the men involved in the study of geography completed the Arts requirements for both B.A. and M.A., geography must

140. Of course, this must be a tentative statement. The Oxford registers, as recorded by Clark and Boase in *Register of the University of Oxford* and J. Foster in *Alumni Oxonienses*, note the class of the student at matriculation, while Cambridge registers, as recorded by Venn in *Alumni Cantabrigienses*, do not. Thus, while a quarter of the Oxford students have unknown class origins, this is the case with more than half of those from Cambridge.

141. These is a danger in taking the matriculation record at face value. Since it was cheaper for the student to matriculate should he come from a plebeian background, it is quite possible that some students from higher families claimed this status. On the other hand, some students may have claimed a higher status in order to gain prestige. It is impossible to know to what proportion these caveats apply, so all that can be done is to recognize the limitations of any such analysis.

142. See note 11 for chapter 1 above. Stone, "Educational Revolution," 61.

be seen as part of that curriculum and not merely a pastime for those students interested in a smattering of culture rather than in serious study. This suggests that the new men invading the universities were not seeking easy polish but were as interested as their poorer clerical brothers in exploring the challenging Arts curriculum. Geography was a new subject, but it fit neatly into the older Arts curriculum.[143]

Geography was thus an important subject for many students at Oxford and Cambridge, the influence of which spread far beyond the men here examined. Of course, only those who produced or owned geographical works remain for posterity to judge. A much larger group of students of geography stands behind these 362 university men—students who have left few traces. Of course, the influence of these men on the geographical community as a whole was profound. The circle of men at Oxford and Cambridge and beyond who were interested in geographical topics was large and diverse; Oxford and Cambridge provided ideal environments for geographical study, allowing and perhaps encouraging an interplay between theoretical, scholarly analysis and practical needs and desires.

Many members of the geographical community, both within and outside the universities, traveled great distances around the globe. Several of them traveled to Russia a number of times, or around the world, or to America and back. The tale of Sir Anthony Sherley, with which this chapter began, although spectacular, was not without precedent. Of the men associated with the four colleges examined here in detail, twenty traveled abroad as part of their education or in pursuit of a career. Fourteen traveled to Europe, where most studied medicine, law, or mathematics at a European university. Two students traveled to the East: George Turberville sojourned to Russia, and Edward Terry went to Turkey as chaplain to the East India Company. Four went to the Americas—two to Virginia and two to the West Indies. For an age of comparatively primitive transportation, the ease with which these men traversed their world was astonishing and argues for a love of adventure and a need for excitement not met in an England which had been largely at peace (except for the invasion of the Armada) since the fall of Calais.[144] These remarkably peripatetic careers emphasize the interest these geography students felt in the expanding globe, seen in both scholarly and practical applications.

The attitudes of these travelers demonstrate their growing security in

143. McConica, "Elizabethan Oxford," 695, 701.

144. The theme of travel taking the place of warfare for these young bloods in search of adventure deserves to be explored more fully, a task quite beyond the scope of this study. Hale begins this discussion, arguing that travel, especially to Europe, became more commonplace in the sixteenth century, while travelers of more modest means became more restricted in their ability to travel (Hale, *Civilization of Europe*, ch. 4, esp. 182–83).

their own identity and their ability to experience at the same time the excitement of exotic and dangerous places and cultures and their sense of the superiority of English customs and beliefs. George Sandys, for example, was unimpressed by the pyramids of Egypt, "the barbarous monuments of prodigality and vain glory," and complained about the food in the Middle East. Typical was "a beastly kind of unpressed cheese that lieth in a lumpe."[145] William Lithgow, on his tour to the Holy Land and the Middle East, complained about the food, the climate, and the customs, claiming that "Although Arcadia in former times was pleasant, yet it is now for the most part wast and disinhabited," and on seeing Rome declared, "would God all Papists in *Brittaine* had the like eie-witnessing approbation as I have had, I am certainely perswaded . . . they would heavily bemone the terrible fall of that *Babylonian* whoore which . . . is their holy mother church."[146] William Parry, traveling with Sherley, compared antisocial foreign behavior with the race the English were more used to oppressing: "The people whereof [the Kurds] are altogether addicted to theaving, not much unlike the Wilde Irish."[147] The experience of travel was not, perhaps, the broadening influence we might now take it to be. These men did not travel to understand other cultures; they traveled for gain, either monetary, political, or spiritual, and they returned firmly convinced of their own superiority. (Perhaps this is the end of much travel!) Especially for those who began to understand geography at university, their experiences abroad often confirmed attitudes they had found in the discipline they had studied.[148]

The students who studied geography at Oxford and Cambridge absorbed a view of England and of the world outside that encouraged a mentality of pride in their own nation and of control and exploitation of the world outside its boundaries, as the next three chapters will demonstrate. This university education gave these men a common set of attitudes and beliefs from which to choose in their later careers. As we have seen, many became teachers or clerics, while a significant number served the state or their monarch in a more direct way. These were people who could affect the government, policy, and people of the English nation. The training they received from the study of geography was important in that enterprise.

145. Sandys, *Relation of a Journey,* 127, 65.

146. Lithgow, *Most Delectable, and true Discourse,* sigs. E1a, B3b.

147. William Parry, *New and large discourse,* 17.

148. Hale shows that the stereotypes used to describe the various peoples of Europe to one another died hard, used by learned men well into the seventeenth century (*Civilization of Europe,* 51–66).

CHAPTER THREE

Mathematical Geography: Theory at Practice

> To conclude, some, for one purpose: and some, for an other, liketh, loveth, getteth, and useth, Mappes, Chartes, and Geographicall Globes. Of whose use, to speake sufficiently, would require a booke peculier.[1]

Mathematical geography, essentially a branch of applied mathematics designed to explain the use and aid in the improvement of maps, charts, and globes, developed in late-sixteenth-century England as a study separate from both pure mathematics and descriptive geography. This subdiscipline had its roots in classical antiquity and sought a complete mathematical understanding of the globe that allowed English geographers to measure, categorize, and thus control its frontiers. With these textual and theoretical faces, it is not surprising to discover that mathematical geography was largely the concern of Oxford and Cambridge scholars. Yet mathematical geography was a study that was inexorably linked with practical application, even in its most arcane manifestations. While this mathematical study of the world should be seen as distinct from the more general political, descriptive geography and chorography popular with less mathematically inclined students, mathematical geographers did not exist in isolation from the rest of the world. Through a mathematical mapping of the earth's features, mathematical geographers were attempting to understand and control the world for England, to its economic, political, and religious betterment. The study of mathematical geography involved more theoretical manipulation and mathematical skill than did the other subdisciplines of geography, but equally it required an application of that mathematical understanding to the world of practical affairs. This rigorous subdiscipline attracted a relatively homogeneous, dedicated, and comparatively small group of scholars who developed a new methodology of experimentation, the systematic tabulation of facts, and mathematical manipulation. This methodology was vitally important in the development of a new mathematical science that used practical knowledge to advance its aspirations to

1. Dee, *Mathematical Preface,* sig. a4a. While maps, charts, and globes could also be employed in the other two branches of geography, Dee had a more strictly mathematical application in mind.

new and fruitful theories of the earth, allowing the application of scientific ideas to the practical world of human affairs.

The Development of Mathematical Geography

From antiquity, the study of geography had followed two different paths. Ptolemy, possibly the most famous ancient geographer, was interested in the mathematical mapping of the globe using the tools of astronomy. Ptolemy treated the celestial and terrestrial globes as equivalent, applying the same grid system to each and using the same spherical geometry to plot particular points thereon. He divided the globe into a series of parallel belts or "climates," separated by the equator, the tropics, and the arctic and antarctic circles. He developed a grid of longitude and latitude coordinates, creating a mapping system that has never been entirely superseded.[2] Ptolemy, then, was more concerned in *Geographia* with the organization of the globe than with the people inhabiting it, since such description could only be inexact and subjective.[3] To make this preference explicit, Ptolemy drew a clear distinction between geography, which he studied, and chorography, which he left to others.[4] The second ancient geographical model came from Strabo. In contrast to Ptolemy, Strabo concentrated on the description of different towns and regions, a study that could potentially descend merely to a series of entertaining anecdotes. As John Greaves, an Oxford mathematician and geographer of the early seventeenth century, expressed it, "[Strabo's] description of places ha[s] more of the historian and philosopher (both which he performed with singular gravity and judgement) then [sic] the exactness of a mathematician."[5]

Sixteenth-century geography took up these distinctions. With the rediscovery of Ptolemy's *Geographia* in the fifteenth century,[6] mathematical geography emerged as a subdiscipline differentiated by classical precedent from its descriptive cousin. Humanists and natural philosophers alike saw

2. Ptolemy, *Geographia.* Ptolemy discusses both the five climatic regions and the longitude-latitude grid (sigs. C2–5). See Waters, *Art of Navigation,* 43; and especially Harley and Woodward, eds., *The History of Cartography,* vol. 1. For a good description of various map projections, including those called Ptolemaic (though with little analysis), see Snyder, *Flattening the Earth.*

3. Bowen, *Empiricism and Geographical Thought,* 31.

4. Ptolemy, *Geographia,* sig. a1a.

5. Oxf. Bodl. MS Add. A 380, f. 153b.

6. Ptolemy's *Geographia* was not included in the Ptolemaic opera introduced into the West in the twelfth century. It was only rediscovered in the West around 1406, when it was translated into Latin by Jacobus Angelus in Florence. In addition to numerous manuscript copies, it appeared in six printed editions in the fifteenth century: Bologna, 1462 [1482?]; Vicenca, 1475; Rome, 1478; Ulm, 1482; Ulm, 1486; and Rome, 1490. It appeared in numerous editions in the sixteenth century in both folio and quarto: twenty in Latin, six in Italian, and two in Greek. For a discussion of the rediscovery of Ptolemy, see Bagrow, *History of Cartography,* 77.

the advantages of a world made navigable and mathematically exact, and mathematical geography developed within the tradition of the *quadrivium* as a form of applied mathematics, much akin to astronomy. This branch of geography, which combined mathematical rigor and practical curiosity, encouraged a reevaluation of the natural world as based on both number and action. This world could be measured, mapped, and thus owned and manipulated. Hence mathematical geographers helped to develop a new methodology that combined mathematics, inductive experimentation, and practical application, all affecting the development of seventeenth-century science.[7]

Mathematicians in late Elizabethan and early Stuart England found this quantitative and exact study of the earth compelling. At a time when the English hoped to gain a foothold in the European race for new trade routes and colonies in the New World and the East, they needed accurate methods of navigation. Tales of hitherto unknown lands, moreover, cast increasingly more doubt on the accuracy of Ptolemy's maps and called for new and improved charts and globes. Ptolemy's maps, after all, had never been intended for navigational purposes. While Ptolemy's *Geographia* might have inspired the study of mathematical geography, the very existence of this ancient source, especially with its later Byzantine maps, compelled geographers to compare it with current knowledge and to create a new picture of the globe that would supersede the one it offered.[8] This discrepancy between Ptolemy and contemporary knowledge soon led mathematical geographers to abandon any slavish devotion to Ptolemy in favor of the investigation of problems of navigation and map projection.[9]

Sixteenth-century English mathematical geographers developed three main areas of investigation, all centering on practical problems. While navigators and sailors did not always make use of the more theoretical solutions proposed by these geographers, the investigations were certainly prompted by questions of utility and service to the empire. The first and most pressing was the need to determine longitude at sea. As Brian Twine, a fellow at Corpus Christi College, Oxford, said, "The Chiefest worke of Navigation dependeth uppon ye Longitude and latitude of places."[10] This became essential as more transatlantic voyages were made, since it was de-

7. See Bennett, "Mechanics' Philosophy," for a discussion of geography and navigation as the foundation for seventeenth-century mechanical philosophy.

8. The original Greek maps, if indeed they existed, were soon separated from Ptolemy's text, and the map tradition that came to fifteenth-century cosmographers originated with the "A" manuscript of the *Geographia* from ninth-century Constantinople (Crone, *Maps and Their Makers,* 68). See Dilke, *Greek and Roman Maps,* esp. 80, for a discussion of whether Ptolemy did, in fact, draw maps.

9. This point is illustrated in Grafton et al., *New Worlds, Ancient Texts.*

10. Corpus Christi College, Oxford, MS 264, f. 86b.

sirable to be able to plot the shortest path to the New World rather than the longer course along a fixed latitude. It was even more desirable to avoid foundering off the Scilly Islands on the return trip, as was the fate of many ships. Second, English geographers faced the difficult question of how to navigate in northern waters. In northern climes, most common charts distorted directions and distances, and extreme magnetic variation caused compass needles to point to many different directions as north. The solution lay in the creation of a polar projection map and the charting of the isogonic lines of compass variation. Third, geographers needed to develop a mathematical map projection that would allow sailing and rhumb lines to be drawn in a straight line. This promised to make navigation easier and a more exact art.

The problem of determining longitude at sea plagued geographers from the first Renaissance voyages until the eighteenth century.[11] Latitude determination was a relatively simple matter, since a navigator could use the elevation of the stars or of the sun.[12] It was necessary to have accurate instruments, of course, and to understand the rudiments of astronomy, but by the mid-sixteenth century, any moderately skilled navigator could find latitude at sea. Longitude was not so simple. Although sixteenth-century geographers were able to determine longitude on land in Europe, it was quite a different matter at sea. On land, positions of discrete objects could be viewed from different angles, compass directions could be taken, the distances between points plotted could be measured, and simple triangulation could be calculated. This could involve relatively sophisticated equipment, including the theodolite introduced in the sixteenth century, but could also use the cruder surveying table and drawing compass.[13] At sea, the distance between points was unknown and unmeasurable, making differences in local times the crucial variable missing from the equation. All sixteenth-century geographers and navigators were aware that the essential problem was relative time, but it was extremely difficult to measure time on shipboard. Sand glasses, although used to change the watch and establish ship's time for most functions, were simply not accurate enough for calculations of longitude. Likewise, pendulum clocks, tried in the seventeenth century, were both inaccurate and impossibly cumbersome. Columbus tried to use lunar eclipses to judge the difference in time between a standard table (probably Regiomontanus's) and the observed eclipse;[14] oth-

11. This problem was solved only with the invention of an accurate chronometer, first tested by Captain Cook on his famous voyages. See Marshall and Williams, *Great Map of Mankind.*

12. Waters, *Art of Navigation,* 47.

13. See chapter 5 below.

14. Waters, *Art of Navigation,* 58.

ers also sought to perfect the method of "lunars," but with no great success. Galileo suggested observing the eclipses of Jupiter's moons, but this proved too difficult to observe from a pitching ship.[15]

One popular idea in England in the late sixteenth century was the possibility of using the variation of the compass to determine longitude. Compass needles, it was realized very early, did not point to true north. Furthermore, sixteenth-century mathematical geographers and navigators soon discovered that this compass variation was not uniform throughout the world but varied from place to place. It was commonly suggested in the sixteenth and early seventeenth centuries that this variation of compass bearing, if charted, could allow navigators to determine longitude, since compass variation was thought to vary consistently with geographical location. Some geographers believed that the compass needle was affected by large land masses, since, for example, it seemed to point to Africa throughout the circumnavigation of that continent.[16] Thus, it was believed that the determination of the degree of variation could be used to establish the degree of longitude, using mathematical calculations or following previously compiled charts.

As late as 1574, William Bourne, author of a popular textbook on navigation, admitted that he did not know how to determine longitude at sea.[17] Thomas Digges, soon after, although no closer to a solution, claimed that ignorant mariners (such as Sebastian Cabot) "could as easily solve the longitude as one could fly from one mountain top to another."[18] This he felt was a problem for the learned mathematician, not for the mere practitioner.

Neither the theoretician nor the more practical craftsman had much luck in cracking the longitude problem. Robert Norman, a self-taught instrument maker who likened himself to the naked (that is, unlettered) Archimedes, springing from his bath to tell the world of his magnetical dis-

15. Ibid., 300. The observation of the eclipses of Jupiter's moons was unaffected by the position on the earth, and so a table of the eclipses combined with observation would give the difference in time between the standard position and the observer. In practice, too many technical difficulties got in the way. The moons proved difficult to observe accurately, and there were differences of opinion regarding the exact moment of eclipse. In addition, telescopes on board ship that would be accurate enough to spot the Medician planets were cumbersome and often not properly understood by navigators and sailors.

16. For example, Gilbert, *De Magnete.* He used this idea to predict a northeast passage and a large Terra Australis. Dutch pilots claimed that the line of zero-degree variation ran through Java, while Gilbert favored the Peloponnisos. So stated an anonymous reporter of Guillaume de Nautonier's work, *Metrocomie,* in BL Burney MS 368, f. 32a–b. See E. G. R. Taylor, *Late Tudor Geography,* 69.

17. Bourne, *Regiment of the Sea;* quoted in Waters, *Art of Navigation,* 139.

18. L. Digges, *Prognosticon Everlastinge;* quoted in E. G. R. Taylor, *Mathematical Practitioners,* 36.

coveries,[19] thought the solution might lie in the dip of the compass needle, rather than its variation. Norman felt that variation was too erratic to be useful in finding longitude: "This Variation is iudged by divers travailers to bee by equall proportion, but herein they are muche deceived, and therefore it appeareth, that norwithstandyng their travaile, they have more followed their bookes then experience in that matter. True it is, that Martin Curtes doeth allowe it to be by proportion, but it is a moste false and erroneous rule. For there is neither proportion nor uniformitie in it, but in some places swift and sudden, and in some places slowe."[20]

He sought the answer in a related phenomenon, the dip of the compass needle, that is, its pull toward the center of the earth (see figure 12). Norman claimed that this carefully observed dip might be used to determine at least latitudinal position—he used it to calculate the latitude of London as 71°50″—and perhaps longitudinal position as well.[21]

Also in 1581, William Borough, another self-taught mathematician, published *A Discourse of the Variation of the Cumpas,* usually bound with Norman's *New Attractive.* In it, Borough explained the phenomenon of compass variation, "knowing the variation of the Cumpasse to bee the cause of many errours and imperfections in Navigation."[22] Recognizing that the only way to understand variation was by collecting a mass of inductive data, Borough charged mariners and navigators to make continual, accurate observations.[23] He blamed variation for wreaking havoc with chart construction, "For, either the partes in them contained, are framed to agree in their latitudes by the skale thereof, and so wrested from the true courses that one place beareth from an other by the Cumpas, or els in setting the parts to agree in their due courses, thei have placed them in false latitudes; or abridged, or overstretched the true distances betweene them."[24]

Borough was credited by Henry Coley, a seventeenth-century English mathematician, with the discovery of the variation of the compass,[25] although Borough himself recognized his predecessors. Borough acknowledged Mercator as the savior of navigators from this variational quagmire,

19. Norman, *New Attractive,* "Epistle," f. A2a. Norman identifies himself with Archimedes because Norman was himself naked; i.e., he had not the Latinate training of the universities. This was not the normal sixteenth-century interpretation of the nakedness of Archimedes, which was usually viewed as representing purity of thought and purpose. For the changing attitudes toward Archimedes, see Laird, "Archimedes among the Humanists."

20. Norman, *New Attractive,* 21.

21. Ibid., 10. The latitude of London is now usually given as 51° 30″.

22. Borough, *Discourse on the Variation of the Cumpas,* "Preface," sig. 2a. For biography of Borough, see *DNB,* 2:867; E. G. R. Taylor, *Mathematical Practitioners,* 68.

23. Borough, *Discourse on the Variation of the Cumpas,* "Preface," sig. *3a.

24. Ibid., sig. F2a.

25. BL Sloane MS 2279, f. 91b.

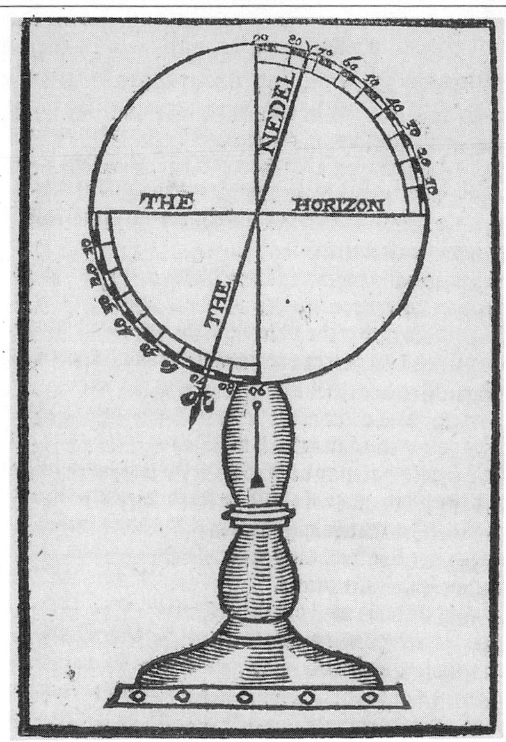

FIGURE 12. Dip of the compass needle, from Robert Norman, *The Newe Attractive* (London, 1581), p. 10. (BL c.31.d.2.(1). By permission of The British Library.)

with the publication of his "Universal Map" of 1569 containing his new projection.[26] Borough tentatively proposed variation as a longitude-finding tool, although he temporized that "if there might be had a portable Clocke that would continue true of space of forty or fifty houres together . . . then might the difference of longitude of any two places of knowen Latitudes . . . be also most exactly given."[27] He acknowledged that this was, for the time, at least, an impossible dream.

In the 1590s and 1600s, many people wrote treatises claiming they had unlocked the mystery of longitude but chose to keep these solutions secret.[28] E. G. R. Taylor suggests that this was the result of Neoplatonic secrecy, but I suspect it was more because of a vain hope of achieving prefer-

26. Borough, *Discourse on the Variation of the Cumpas,* sig. F3a. This map, however, remained largely unused by sailors.

27. Ibid., sigs. D1b–2b. E. G. R. Taylor (*Mathematical Practitioners,* no. 58) claims that this was "a technique for finding the longitude by carrying a number of spring driven watches," not quite what Borough had in mind.

28. E. G. R. Taylor, *Late Tudor Geography,* 72.

ment by the queen for the solution. Since Elizabeth was so notoriously parsimonious, it is perhaps not surprising—and probably shows good sense on her part—that none of these several solutions was made public.[29]

In 1599, Edward Wright, perhaps England's foremost mathematical geographer and an important member of the university-based community of mathematical geographers, translated Simon Stevin's *The Haven-finding Arte* from the Dutch.[30] In this work, Stevin claimed that magnetic variation could be used as an aid to navigation in lieu of the calculation of longitude.[31] He set down tables of variation, means of finding harbors with known variations, and methods of determining variations. In his translation, Wright called for systematic observations of compass variation to be conducted on a worldwide scale, "that at length we may come to the certaintie that they which take charge of ships may know in their navigations to what latitude and to what variation (which shal serve in stead of the longitude not yet found) they ought to bring themselves."[32]

The work of both Wright and Borough demonstrates the close connection between navigation and the promotion of a "proto-Baconian" tabulation of facts meant both for practical application and for scientific advancement. Here appears the foundation of an experimental science, grounded in both practical application and theoretical mathematics, quite separate from any more traditional Aristotelian natural philosophy or Neoplatonic mathematics. Unfortunately, Wright's scheme was not entirely successful. By 1610, in his second edition of *Certaine Errors in Navigation,* Wright had constructed a detailed chart of compass variation, but he had also become more hesitant in his claims concerning the use of variation to determine longitude.[33]

Belief in the equivalency of variation to longitude or the use of variation in determining longitude died hard. In about 1620, in a manuscript called "A Magnetical Problem," the university-trained divine and mathematician Thomas Lydiat, following William Gilbert's lead, claimed that a compass touched by a lodestone resembled the earth, itself a great lodestone, and thus that each part of the compass represented a different part of the earth. Compass variation, therefore, mirrored natural magnetic variation of the globe. "Nowe if this prove true," claimed Lydiat, "that thereby is given a

29. New work on secrecy in early modern science stresses the role played by patronage, older traditions of guild secrecy, and the development of concepts of copyright. See especially Long, "Invention, Authorship, 'Intellectual Property'"; Long, "Military Secrecy in Antiquity"; Eamon, *Science and the Secrets of Nature.*

30. Wright, *Haven finding art* (1599). E. G. R. Taylor, *Mathematical Practitioners,* no. 100.

31. Wright, *Haven finding art,* 3.

32. Ibid., "Preface," sig. B3a; Waters, *Art of Navigation,* 237.

33. Wright, *Certaine Errors in Navigation,* 2nd ed. (1610), sigs. 2P1a–8a; Waters, *Art of Navigation,* 316.

most certain and redy means of measuring the Longitude, or East and West distances, and withal a most easie way of sayling by a great Circle."[34] It was not until 1634 that John Pell, John Marr, and Henry Gellibrand discovered that magnetic variation itself varied over time. Their discovery that the variation had significantly diminished at two observational sites since Edmund Gunter's measurements in 1622 and 1624 effectively laid to rest the chimera of variation as a longitude-finding tool.[35] While a great deal of time had thus been spent on observations and calculations related to the problem of compass variation, to little avail, this investigation brought together theoretical and practical methods and ideas in a way that blended observational, mathematical, and indeed imperial concerns. Compass variation was interesting from the theoretical viewpoint of understanding the magnetical world, but both the data themselves and the motivation to collect them came from the practical world of navigation.

The difficulty of navigating in northern waters fueled this English attempt to link longitude and variation. Because the English entered the "Age of Discovery" later than other European nations, their share of the New World was a northern one, and their only route to Cathay was through a hypothetical northwest or northeast passage. Much ink was spilled, in fact, convincing the various English monarchs that such a passage existed and that the nascent English empire would be aided by its exploration and exploitation.[36] John Dee, for example, explicitly linked the idea of empire and navigational improvement in his 1577 work, *General and Rare Memorials.*[37] It soon became apparent to navigators and geographers, however, that the plane charts that were adequate for Mediterranean travel were useless for polar sailing. By extending the point of the North Pole into the top edge of the map, these charts made nonsense of compass navigation and sailing distance. Some northern charts at least acknowledged this problem by depicting two compass needles, one pointing north,

34. Oxf. Bodl. MS Bodl 313, f. 63b.

35. E. G. R. Taylor, *Mathematical Practitioners,* no. 158. This finding was announced in Gellibrand, *A Discourse Mathematicall.* For an important interpretation of Gellibrand's discovery, see Pumfrey, "'O tempora, O magnes.'"

36. For example, Anthony Jenkinson, "A proposal for a voyage of discover to Cathay, 1565," BL Cotton Galba D.9, ff. 4–5; "Advise of William Borrowe for the discerning of the sea and coast byyonde. Perhaps whather the way be open to Cathayia, or not," ca. 1568, BL Lansdowne 10, ff. 132–133b; Humphrey Gilbert, BL Add. 4159, f. 175b; Richard Hakluyt, "The chiefe places where sundry sorts of spices do growe in the East Indies, gathered out of sundry the best and latest authours," Oxf. Bodl. MS Arch. Selden 88, ff. 84–88; Bourne, *Regiment of the Sea,* f. 76; Barlow, *Navigator's Supply,* sig. b1a; D. Digges, *Of the circumference of the earth;* Henry Briggs, "A Treatise of the NW Passage to the South Sea, to the Continent of Virginia and by Fretum Hudson" (London, 1622), published in Purchas 1625, Part II:848–54. For a discussion of the imperial implications of these proposals, see Cormack, "The Fashioning of an Empire: Geography and the State in Elizabethan England."

37. Sherman, *John Dee,* 152 ff.

for example, along Labrador and a second pointing along a convergent path following the coast of Greenland (see figure 13).[38] The answer to this problem, which lay in stretching a point into a line consisting of an infinite number of points, and to the problem of conflicting sailing directions, was a stereographic polar projection showing the pole at the center and the latitudinal lines as concentric circles (see figure 14). John Dee apparently developed such a projection, and perhaps a companion calculational device, which he called the "Paradoxal Compass," in 1556.[39] Although this technique did not facilitate compass use, it did have the advantage of projecting correct spatial relations and allowed a navigator to use the radiating rhumb lines to plot an approximate great or lesser circle course.

Since John Dee was much given to secrecy, it is not clear how many people knew of his innovative map projection. By 1595, at least, Captain John Davis included this projection and instructions for navigating with it in *The Seaman's Secrets.* Davis may have relied directly on Dee's work in this, since he lived with Dee for a time while Dee taught him cosmography, and he is known to have stolen several books from Dee's library.[40] Likewise, in 1604, George Waymouth offers the following in *The Jewelle of Artes:* "The demonstration of an Instrument to finde out the degree and minute of the meridian of the paradoxicall chart whose degrees dothe increase from the pole towardes the equinoctiall as you may plaine finde in the demonstration next after this."[41]

Just as the plane chart proved totally inaccurate for polar sailing, it was soon found wanting for sailing at most latitudes away from the equator. The basic problem with a plane chart was that it was drawn as if the lines of latitude and longitude were a constant distance apart. Since this was approximately true for lines of longitude at only a very limited distance from the equator, in northern or southern waters distortion rapidly became a problem. Thus, a straight-line course on a plane chart would not have corresponded to a straight course on the globe, and a great circle course could only have been approximated by a series of steps across the map.

38. Borough describes such a map in *Discourse on the Variation of the Compass,* sig. F2b.

39. John Dee announces this compass in *General and Rare Memorials,* claiming that it will appear in an now-unknown manuscript, "The Brytish Complement, of the Perfect Arte of Navigation." Waters (*Art of Navigation,* 210) claims this paradoxical compass to have been the projection itself, while Sherman (*John Dee,* 154) sees this compass as an invention to compute longitudes and latitudes. Dee did draw a map of the polar regions for Burghley, now contained in Burghley's Ortelius of 1570. See Skelton, "Maps of a Tudor Statesman."

40. John Davis, *Seaman's Secrets.* See Waters, *Art of Navigation,* 201. Concerning Davis's sojourn at Dee's house, see Sherman, *John Dee,* 174 and n. 86.

41. George Waymouth, *The Jewelle of Artes.* BL Add. MS 19,889, f. 78b. Both Waymouth and Davis had been on northern expeditions. Waymouth's travels are recorded in Purchas 1625, Part II:809–13, Davis's at 463–50.

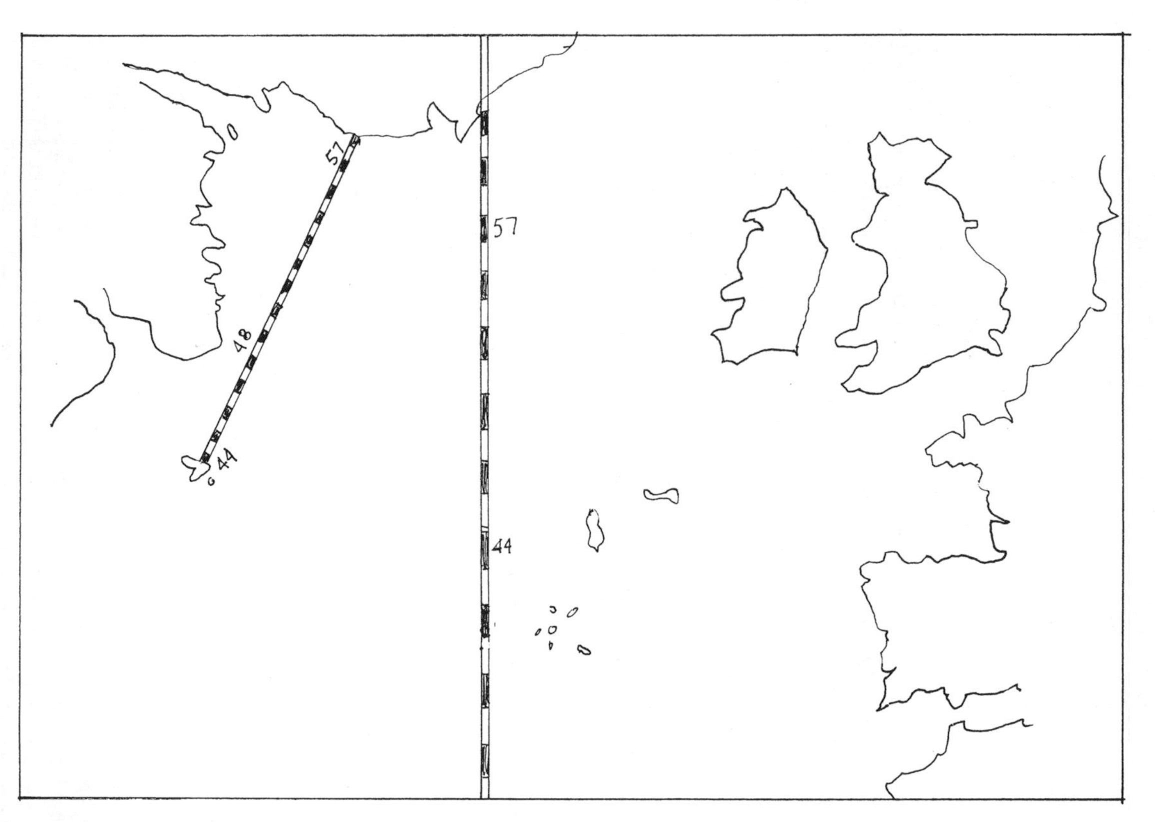

Figure 13. Oblique meridian, drawn after a chart of the Atlantic, ca. 1502, by Pedro Reinel.

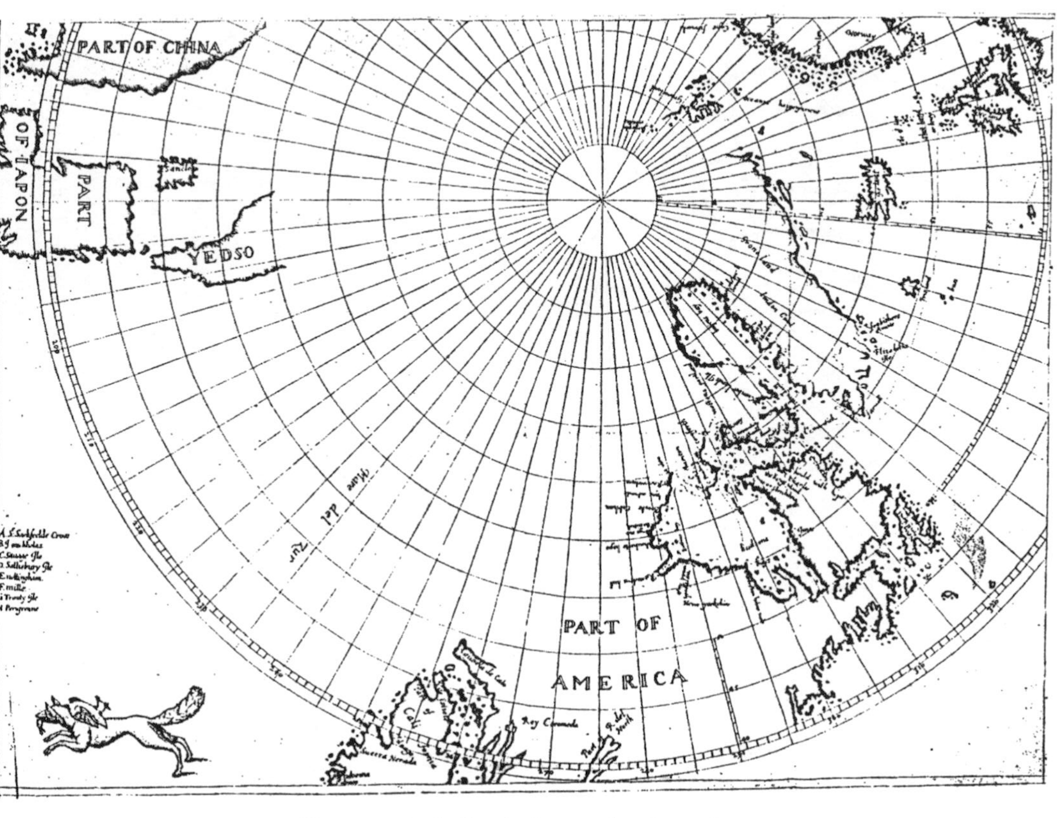

FIGURE 14. Luke Foxe's circumpolar chart, 1631.

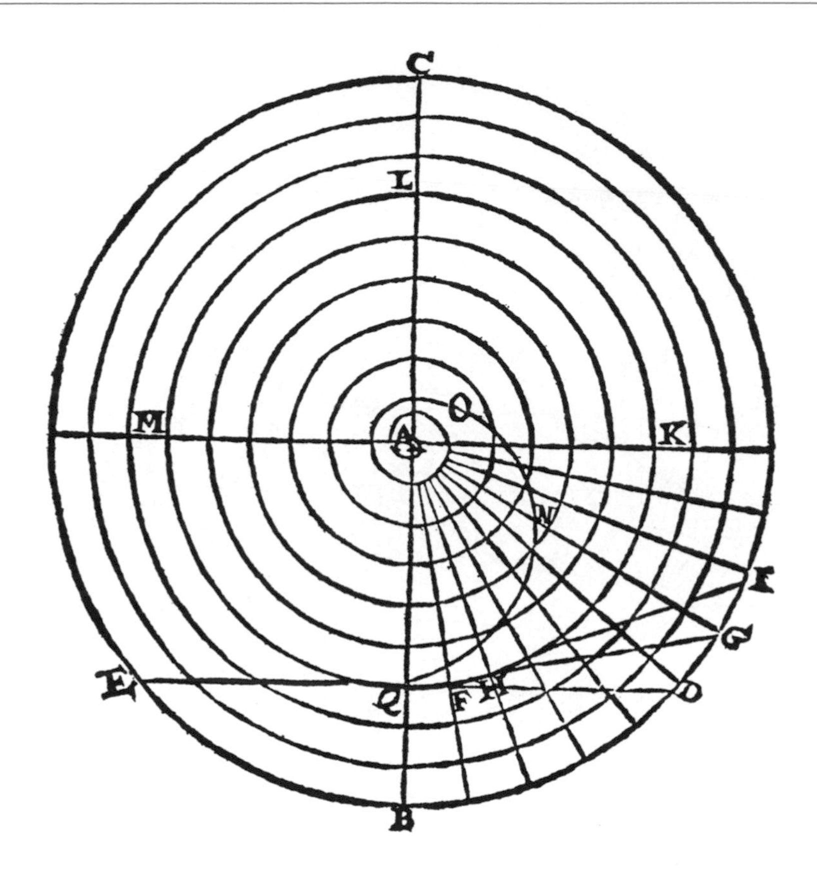

FIGURE 15. Loxodrome spiral, from Thomas Blundeville, *Blundeville his Exercises* (London, 1591; facsimile, Amsterdam: Da Capo Press, 1971), f. 326a.

Mathematical geographers throughout the late sixteenth century had been intrigued by the fact, first discovered in 1537 by Spanish geographer Pedro Nuñez, that loxodromes or rhumb lines formed a spiral on the globe.[42] Although this demonstration had no real practical application, the elegance of the proof appealed to the English mathematical geographers, and they delighted in repeating it. Thomas Blundeville, for example, in an often-reprinted mathematics and geography manual, *Blundeville his Exercises,* explained and illustrated this phenomenon (see figure 15).[43] This

42. "A loxodrome is a line that makes the same angle with all successive meridians. When a ship sails along a loxodrome it crosses successive meridians at the same angle and maintains steadily the same ratio of northing or southing and easting or westing. Thus on a sphere a loxodrome, except when it lies along a parallel or meridian, because of the convergence of the meridians, traces out a spiral" (Waters, *Art of Navigation,* 71–72).

43. Among others, Borough (*Discourse on the Variation of the Cumpas,* sig. G1b) talks about this proof, citing Pedro Nuñez de Medina; also Barlow, *Navigator's Supply,* sigs. I1b–

proof provided an intellectual challenge and aesthetic delight for these mathematical geographers, as did the creation of map projections, demonstrating that the theoretical dimension of their investigation was as compelling as the practical one. It would thus be misleading to view these new map projections as merely practical or technological improvements, since they also satisfied the aesthetic requirement of neatness of fit so important to the theoretician.

In 1569, Gerard Mercator published his now-famous map projection, which allowed sailing courses and rhumb lines to be drawn as straight lines. Mercator used a geometric progression to increase the distance between lines of latitude in proportion with the increasingly distorted lines of longitude. The result was the map we all recognize, where areas became increasingly enlarged and distorted toward the poles (see figure 16 and figure 4 in the introduction above).[44] Distances sailed became difficult mathematical calculations, but proportions between lines of longitude and latitude remained true and compass bearings corresponded to straight lines, making it possible to plot courses using a ruler and drawing compass.

Although Mercator drew and published this map in 1569, it was not until the end of the century that the precise technique of using geometric progression for mapmaking was enunciated mathematically in print. In 1597, William Barlow, in *The Navigator's Supply,* presented a graphical method for creating a Mercator projection, which encouraged navigators to draw their own charts.[45] Barlow described all the instruments that were useful and necessary for the navigator, beginning with the compass, whose invention he, like Bacon, greeted as the herald of the modern age, and ending with Mercator's map projection. Barlow's method for creating this projection did not entail any insight into the mathematics involved, merely an ability to use a quadrant and a ruler, and to follow his somewhat confusing directions.[46] The second, more significant explanation of Mercator's work came in Edward Wright's *Certaine Errors in Navigation* (1599). Wright provided an elegant Euclidean proof of the geometry involved in this map projection. He pictured the globe as a cylinder, in which every parallel was equal to the equinoctial, and proved that the rhumb lines "must likewise

---

2a. Blundeville, *His Exercises* (1636, 693) also refers to the proof but not specifically to Nuñez.

44. Edward Wright, who calculated and explained the mathematical structure of the Mercator projection, used such a projection to illustrate the title page of the second edition of his *Certaine Errors* (1610). Notice the astronomical instruments surrounding the title, reminding the reader that proper navigation must be informed by mathematics, proper understanding, and accurate observation. In this, the map was both a tool and a product. See Ede, "When Is a Tool Not a Tool?" for a more modern discussion of this question.

45. Barlow, *Navigator's Supply;* E. G. R. Taylor, *Mathematical Practitioners,* no. 95.

46. Barlow, *Navigator's Supply,* sigs. A1b, K4b.

FIGURE 16. Title page from Edward Wright, *Certaine Errors in Navigation,* 2nd ed. (London, 1619). (BL c.114.b.13. By permission of The British Library.)

by streight lines."[47] He also published a table of meridian parts for each degree, which enabled cartographers to construct accurate projections of the meridian network and offered straightforward instructions on map construction.[48] He also constructed his own map using this method (see figure 4 in introduction above). Its publication in Hakluyt's *Principal Navigations* shows the close relationship, as well as the distinction, between mathematical and descriptive geography. Wright's work was the first truly mathematical rendering of Mercator's projection and placed English mathematicians, for a time, in the vanguard of European mathematical geography.

English mathematical geographers were thus interested in problems that were current, sophisticated, and usually practical. Grounded in the mathematical framework of Ptolemy and attempting to manipulate their image of the world into a geometrically satisfying design, mathematical geographers developed a research program, unspoken though it might have been, that was driven by practical problems and an attempt to use mathematics to improve England's standing in the world. The questions of longitude determination, navigation in northern waters, and map projection were all of paramount importance to England in its drive for new trading routes and new colonies. In addition, problems relating to the determination of longitude and map projection had to be solved before geographers could understand the magnetic globe in a geometrical manner.[49] Mathematical geography was thus an integrated blend of the theoretical and the practical.

Mathematical Geography in the Universities

This blend of theory and practice can be seen by the attempts of various academic mathematical geographers to encourage and teach mariners the theory of their practice. Although these proposals seldom seem to have been completely successful, several lectureships were proposed or established, numerous books were written and translated, and much lobbying of government was carried out. Thomas Hood provided the most famous of these lecture series, first at Leaden Hall and later at the house of Thomas Smith. These lectures were described by both George Buck and John Stow, indicating that they were highly regarded as an asset to the metropolis.[50] Richard Hakluyt proposed another series of lecture for seamen in 1599:

47. Wright, *Certaine Errors in Navigation* (1599), sig. C4a.

48. Ibid., sigs. D3a–E4a. E. G. R. Taylor, *Late Tudor Geography,* 76; E. G. R. Taylor, *Mathematical Practitioners,* no. 99; Waters, *Art of Navigation,* 219.

49. See Edgerton, "From Mental Matrix to Mappamundi," for an interesting discussion of the changing mental construct necessary to accept Mercator's image of the world.

50. Stow, *Survey of London,* 57; Buck, *Third Universitie of England,* 981.

> When I call to minde, how many noble ships have bene lost, how many worthy persons have bene impoverished by losse of great Ordinance and other rich commodities through the ignorance of our Sea-men, I have greatly wished there were a Lecture of Navigation read in this Citie, for the banishing of our former grosse ignorance in Marine causes, and for the increase and general multiplying of the sea-knowledge in this age, wherein God hath raised so generall a desire in the youth of this Realme to discover all parts of the face of the earth, to this Realme in former ages not knowen.[51]

Thus, mathematical geographers attempted to put their knowledge to good use and to provide England with the mariners and navigators it so desperately needed. In order to do this, however, mathematical geography had to exist as well as an academic occupation, within the universities.

This development of mathematical geography as a new mathematical science, with its own agenda, its own membership, and its own methodology, was part of a larger movement of mathematical studies that mingled practical and theoretical methodology and ideology to form a new emphasis on mathematics and on application.[52] These mathematical sciences were to prove influential in changing the mental framework of scholars at the universities and beyond, especially since they were not necessarily tied to Aristotelian natural philosophy and so developed with some flexibility. Part of the burgeoning emphasis on the mathematical sciences, mathematical geography was part of the course of study at Oxford and Cambridge, as has been demonstrated to have been the case with geography more generally. The presence of mathematical geography at the English universities is shown both by the number of men interested in the subject who attended one of the universities and by the number and type of mathematical geography books owned by students. These books, especially, indicate a constant interest in mathematical geography and a university community connected with these important issues of the construction of the globe. While the problems that interested students were much less sophisticated than the program of England's more prominent mathematical geographers, they show a geographically literate group of students, for whom mathematical geography might provide a common base of knowledge and expectations, as they pursued their careers, engaged in a highly connected and political world.

An analysis of the university book lists reveals that mathematical geography was popular at a smaller number of colleges at Oxford and Cam-

51. Hakluyt, *Principal Navigations* (1598–1600), 1: sig. *3a.

52. Bennett, "Mechanic's Philosophy"; Shapin and Schaffer, *Leviathan and the Air Pump*. I discuss this in a different context in Cormack, "Flat Earth or Round Sphere."

bridge than was the case for geography generally.[53] There were, however, several colleges at both universities at which the acquiring of mathematical geography books indicates substantial interest in that subject. Specifically, the book lists of Corpus Christi and St. John's College, Oxford, and of Peterhouse and Corpus Christi College, Cambridge, show a real predilection for the study of mathematical geography in terms of numbers, while New College, Oxford, and Trinity College, Cambridge, seem to have housed a smaller though keener group of geography book owners. The lists of St. John's College and Gonville and Caius, Cambridge, and University, Christ Church, and All Souls colleges, Oxford, contained significant though smaller proportions of mathematical geography books.

The Corpus Christi College, Oxford, book list contained a significant collection of geography books in all three areas, including a respectable number of mathematical geography books.[54] Four copies of Ptolemy's *Geographia* represented the basic knowledge available for the study of the mathematical world, as did copies of Pierre d'Ailly's *Imago Mundi* (ca. 1410) and Apian's and Münster's *Cosmographia* (1529, 1544). The latter, though not primarily mathematical, contained Ptolemaic and other maps, as well as basic cosmographical mapping techniques.[55] There were several map collections, including three copies of Ortelius's *Theatrum Orbis Terrarum* (1570). John Gillies claims that Ortelius's popular collection, the first to personify the continents as female, represented the world as a stage and showed the relationship between the vast cosmography of the universe and the more contained world of human experience.[56] William Gilbert's *De Magnete* (1600) also appeared, indicating some acquaintance with controversial theories of the physical world. The book list also contained fourteen books of mathematics, which are relatively innovative and quite interesting. These include Proclus's commentary on Euclid, two copies of Vitruvius's *De Architectura,* and Archimedes' *Opera.* Although all ancient sources, these were newly rediscovered and thus indicate a small mathematically literate group of Corpus men who were keen to keep up with recent developments. Although this seems more true for mathematics generally than

53. See chapter 2 above.

54. For a list of geography books at Corpus Christi, see appendix B.

55. See chapter 4 below for a discussion of the descriptive aspect of Münster's work. This is one of the many cases of close connections among the different branches of geography.

56. Gillies, *Shakespeare and the Geography,* ch. 3. Gillies argues that Ortelius used the term *theatrum* advisedly, explicitly pointing the reader to the Ciceronian cautionary tale of the artificiality of world affairs (71). Maps, of course, provided a clear crossover between mathematical and descriptive interest, but they are included here since they clearly describe the world in a geometrical and controlled way and deal with space and emplacement rather than human interaction.

for mathematical geography, Corpus Christi was clearly a college where mathematical geography was read and reflected upon.

The St. John's College, Oxford, book list displayed a distribution of geography books in all three areas, including a strong showing in mathematical geography and mathematics generally. The list contained all the standard sources one would expect, including Ptolemy's *Geographia* and the *Cosmographia* of both Peter Apian and Sebastian Münster, as well as a copy of the *De Geographia* of Henricis Loritus (Glareanus) (1534)[57] and Robert Norman's *Safegard of Sailers* (1584). The latter volume indicates an interest in native productions and ideas, as well as reflecting the emphasis on practical problems so important to mathematical geography. Norman's work was a practical book of navigational problems and implies that St. John's readers of mathematical geography shared in this practical inclination, substantiated by the presence of a book entitled *The Merchants' New-Royall Exchange* (1604). The mathematical section of the St. John's book list indeed was predominantly English as well, containing the rather elementary arithmetics of Robert Recorde, Thomas Hill, and Humphrey Baker and exhibiting an unusual emphasis on English mathematics and applied mathematics. The presence of these mathematical books, as well as the standard geographies owned at this college, argues for a rather low level of sophistication in dealing with mathematical ideas and problems, since all had long been surpassed on the Continent.

New College, Oxford, proved an important center for mathematical and astronomical book ownership. In contrast with St. John's, the mathematics, astronomy, and mathematical geography books contained imply a real awareness of new theories and problems. Several of the astronomy texts introduced the Copernican system or proposed alternatives based on observation.[58] The mathematical works of Oronce Finé, Petrus Ramus, and Gemma Frisius laid a sophisticated foundation of solid sixteenth-century mathematical thought, while the presence of mathematical works by Christopher Clavius represented the contemporary research of the innovative Collegio Romano.[59] The mathematical geography section contained a number of classical works, which seem to have been gaining in importance as the century progressed, including Ptolemy, Theodosius, and Pomponius Mela. The editions of Theodosius were edited by Clavius, while modern cosmographical concepts were included in works by Mercator and especially by Ortelius. Thus, we see a real emphasis on both classi-

57. Listed in John Glover, Inventory 1578, Oxford VCC, as "Gloriani Geographia" (and so transcribed in the nineteenth century) but must refer to Loritus (Glareanus).

58. These include Tycho Brahe, *Astronomias Instauratae Mechanica* (1st edition Wandesburg, 1598), and Thomas Lydiat, *Epistolam Astronomicam de Anni Solaris,* MS.

59. Lattis, *Between Copernicus and Galileo.* Also Wallace, *Galileo and His Sources.*

cal and innovative solutions to mathematical topics at New College, both in the area of traditional mathematics and astronomy and in the applied field of mathematical geography.

Peterhouse students at Cambridge also acquired mathematical geography books and mathematical and astronomy books more generally, in comparatively large numbers. The volumes owned were not as innovative as those at New College, Oxford, tending to be early-sixteenth-century standard texts. The mathematical geography books consisted of relatively traditional Continental sources, such as Peter Apian, *Cosmographia* (1529); Gemma Frisius, *De Principiis Astronomiae & Cosmographiae* (1530); and Oronce Finé, *De Mundi Sphera sive Cosmographia* (1542). There were several copies of Ptolemy and of Pomponius Mela, demonstrating an interest in classical sources, and one representative of contemporary English thought, Leonard and Thomas Digges's *A Prognostication Everlasting* (1555). The latter is particularly interesting, since it contained both an older description of the Ptolemaic universe by Leonard and a newer translation or English paraphrase of Copernicus (the first such translation) by his son Thomas, as well as two navigational treatises. This book reminds us that we should not see the "Copernican Revolution" as a clear transition and that early modern thinkers could develop alternative theories at one and the same time.[60] The mathematical books included Euclid's *Elements,* as might be expected, as well as Finé and Gemma Frisius, showing an awareness of Continental trends. There were also several books on perspective, including perhaps a copy of Alhazan's *Perspectiva,* an unusual book for English consumption. The most interesting part of the Peterhouse book list lies in the astronomy section, which included, besides the earlier work of Sacrobosco, several important writers from the first half of the sixteenth century, chiefly Peurbach, Peucer, Regiomontanus, Stöffler, and Mizauld. It also included Copernicus's controversial book, indicating the Copernican debate might well have been waged at Peterhouse, especially given the presence of Digges's book as well. The appearance of these astronomy texts indicates a healthy interest in astronomy and cosmography going hand-in-hand with a similar interest in mathematical geography. Thus, Peterhouse represents a focus of interest in mathematics and mathematical geography, though one that was not particularly innovative.

Corpus Christi, Cambridge, had a smaller number of mathematical

60. See Heninger, *Cosmographical Glass,* for illustrations of this double vision. Gillies, in *Shakespeare and the Geography,* makes a similar claim about the geography of Shakespeare, contending that Shakespeare employed both "poetic" geography (derived from classical and medieval sources) and "scientific" geography (from modern sources). Although I would disagree with Gillies's clear differentiation here, he is correct to see the early modern geographical mind as an interesting amalgamation of ideologies and concepts.

geography books than Peterhouse but exhibited a similar collection of mathematical materials. In the mathematical geography section, the students and library owned a collection of late-fifteenth- and early-sixteenth-century books, as well as Ptolemy's *Geographia.* The one modern entry was Georg Henisch's *Principles of geometrie, astronomie, and geographie* (1591), an English translation. This book was translated by Francis Cook, to be used by his teacher Thomas Hood in his mathematical lectures.[61] As we have seen, this was part of the impetus on the part of mathematical geographers to educate seafaring men, in order to improve England's imperial position. Henisch's book provided a very basic grounding in all three subjects of geometry, astronomy, and geography, though navigation would clearly require a much wider knowledge. The mathematical section is small and included Euclid, Recorde, and Tunstall, indicating a basic mathematical understanding, if little in the way of innovative ideas. The exception to this might be Archimedes' *Opera* in Greek and Latin—the inspiration of Galileo, among others. Corpus, then, was a college where mathematics and mathematical geography were read and a basic understanding of established wisdom was available from the books at hand.

Trinity College, Cambridge, although not at the top of the mathematical geography colleges in terms of numbers alone, contained a much more interesting list than that of Corpus Christi. The mathematics books were remarkable for their English bent, including Robert Recorde's *Pathway to Knowledge* (1551) on geometry and Tunstall's classic *De Arte Supputandi* (1522) on arithmetic. There was also a copy of Euclid's *Elements,* of course, and another of Ramus's *Arithmetica et Geometria* (1569), linking Continental and English thought, as Ramus himself had done on his visits to England. The astronomy texts showed an interesting mixture of standard sources and revolutionary tracts, since the list included the likes of Peurbach, Münster, Sacrobosco, and Regiomontanus, as well as Copernicus, Kepler, and Thomas Digges's *Scala Mathematica* (1573), a Copernican treatise. The mathematical geography section was less exciting than its astronomical counterpart. There were two copies each of Ptolemy's *Geographia* and Apian's *Cosmographia,* as well as single exemplars of Münster's *Cosmographia* and Kaspar Peucer's *De Dimensione Terrae* (1587). Significantly, there were a number of innovative texts that come under the rubric of maps and navigational treatises. Mercator's *Atlas* was included, a work that popularized this type of map collection. Of course, it did not contain Mercator's new map projection, but it provided a real updating of especially European maps often based on quite new surveys. There were also two

61. Henisch, *Principles of Geometrie, Astronomie, and Geographie;* sig. A3a is a letter from Cook to Hood.

navigational works, which were relatively rare in this book list as a whole: Martin Cortes's *Art of Navigation* (translated by Richard Eden, 1561) and William Bourne's *A Regiment for the Sea* (1574). These two works are indicative of the English bent of Trinity's mathematics and mathematical geography. Both Bourne and Eden saw these works as part of the English enterprise to lay claim to new markets and lands, in part by educating those who would sail forth. Both works were consciously useful and aimed at practical men: Bourne dedicated his book to the Lord High Admiral, and Eden dedicated his translation to the Merchant Adventurers of London.[62] These books thus indicate that Trinity students tended to be interested in practical mathematical geography, perhaps more than in theoretical and classical works. In other words, in mathematics and mathematical geography, the Trinity book list tended to be strong in some revolutionary and in many English sources, providing an important mixture of pride in English ideas and the desire to learn new ideas and techniques.

St. John's and Gonville and Caius colleges, Cambridge, and All Souls and Christ Church colleges, Oxford, all had members who showed a serious interest in mathematical geography and mathematical topics through their book ownership, though not to the extent of the colleges already discussed.[63] The study of mathematical geography was thus more widespread than just at the colleges described above. All Souls' students apparently had a distinct leaning toward classical mathematical authors, with two copies of Ptolemy's *Almagest,* three of Euclid's *Elementa* (two in Greek), and Ptolemy's *Geographia.* They also possessed a number of maps, especially Ortelius's collection. The St. John's, Cambridge, book list is interesting for its English emphasis. The mathematical geography section is especially striking for its loyalty to English authors. The only copy of *Blundeville his Exercises* (1594) to appear on the whole list, for example, materialized here, as did Leonard Digges's *Tectonicon* (1562), on surveying, and his astronomically interesting *Prognostication* (1555). The Gonville and Caius book list contained many of the expected favorites, as well as Edward Wright's *Certaine Errors in Navigation* (1599), an interesting practical source. Christ Church, such a strong center for geography generally, had fifteen mathematical geography books associated with the college, including Ptolemy's *Geographia* and Münster's *Cosmographia* (1544), but also such innovative books as Paulus Merula's *Cosmographiae generalis libri 3; Geographiae partic-*

62. Bourne, *Regiment of the Sea,* sig. A2a; Cortes, *Arte of Navigation,* sig. c2a.

63. See appendix A for all book list sources and appendix B for St. John's College, Cambridge, and Christ Church, Oxford, geography book lists. University College, Oxford, had sixteen maps in the possession of its masters. This does not necessarily indicate the members' intellectual development, although it shows the growth of what P. Barber (in "England I" and "England II") calls "map-consciousness."

ularis libri 4 (1605), Robert Hues's *Tractatus Globis Caelestis et Terrestris* (1617), describing the construction and use of Molyneux's globes, and Ortelius's *Theatrum Orbus Terrarum* (1575). Thus, these colleges tended to emphasize both the classics and several modern (especially English) and important sources. It is clear that these colleges exhibited a moderate interest in mathematical geography, implying a relatively widespread development of this mathematical subdiscipline at Oxford and Cambridge.[64]

There was thus a widespread interest in mathematical geography at a large number of foundations at both Oxford and Cambridge. This interest seems to have gone hand-in-hand with an interest in mathematical and astronomical works, indicating the presence of a mathematical subculture at the universities—a group of scholars investigating mathematical geography as part of a broader commitment to mathematical topics.[65] The enthusiasm for things mathematical clearly varied from college to college, running the gamut from a fairly superficial treatment of basic mathematical concepts to a relatively sophisticated analysis of modern theories. Still, Oxford and Cambridge both provided the opportunity for students and fellows to investigate mathematical geography in a serious and detailed manner.

Evolution of Mathematical Geography as a University Subdiscipline

The geographical ideas available to these mathematically inclined students changed over time, allowing the subdiscipline to grow and mature. From the examination of those mathematical geography books owned by Oxford and Cambridge associates through time, a pattern of changing emphases emerges. Within the five decades under consideration, there appears to have been a stable readership for some important texts, combined with a shifting emphasis from early-sixteenth-century standard works in the early decades to rediscovered classics and modern texts by the 1620s (see table 6).

Ptolemy's *Geographia* and Mela's *De Situ Orbis* appear in multiple copies in the list for each decade. Münster's *Cosmographia,* more a descriptive geography book but with a significant mathematical geography section,

64. As well as these colleges with large numbers of mathematical geography books in their book lists, a significant number of other Oxford and Cambridge college and hall lists contained mathematical geography texts. These included Sidney Sussex, Emmanuel, Christ's, Queens', Kings, and Jesus colleges and Clare, Pembroke, and Trinity halls, Cambridge; and Merton, Balliol, Magdalen, Brasenose, Trinity, Oriel, Queen's, and Lincoln colleges and Broadgates and Gloucester halls, Oxford (see tables 2 and 4, above). This suggests that mathematical geography, far from being an extracurricular pursuit, was an integral part of the university curriculum.

65. Feingold, in *Mathematicians' Apprenticeship,* has proven the existence of the larger mathematical community.

TABLE 6 Mathematical Geography Books Owned, by Decade

	1580	*1590*	*1600*	*1610*	*1620*
Ptolemy	5	7	7	6	7
Apian	5	5	8	5	—
Mela	5	4	3	3	4
Münster	5	5	4	5	5
Ortelius	4	5	5	5	4
Mercator	3	—	—	4	4
Honter	2	4	—	1	—
Theodosius	—	—	1	2	5

was also owned in multiple copies throughout the period.[66] The steady ownership of classical geography sources, as well as Apian's *Cosmographia* in most years and the appropriate sections of Münster, indicates a foundation for mathematical geography laid in the pre-Copernican framework of the early sixteenth century. Apian's work combined cosmographical theory, basic instruction in mapping techniques, and maps of European and later New World countries and regions. The continued presence of Ptolemy's *Geographia,* though acting as a firm classical foundation for the subdiscipline of mathematical geography, does not imply a slavish devotion to ancient or outmoded ideas. Unlike Apian's or Münster's book, which remained essentially the same through numerous editions, Ptolemy's work was revised with each edition, usually with up-to-date maps of newly discovered and surveyed parts of the globe. It is probable that the continuing interest in Ptolemy through all five decades demonstrates both the establishment of a firm foundation in the work of that most excellent creator of the science of geodesy and the satisfaction of curiosity concerning newly revealed information on all parts of the world.

Maps and atlases were consistently popular throughout the period, demonstrated in part by the continuing popularity of Ptolemy and Münster. New information could even more effectively be sought in such innovative atlases as those of Ortelius and Mercator. The continuing and significant presence of Ortelius, combined with the increase in Mercator's *Atlas* ownership, indicates a slight shift in focus from Ptolemy's maps, probably with modern interpolations, to genuinely new and innovative atlases. This is particularly so since both Ortelius's and Mercator's atlases changed significantly in later editions. Although it is usually impossible to know which editions were owned on these lists, it is at least possible that later ownership of these map collections was of current rather than older

66. I include Münster in this list for comparative purposes, since this book did contain much mathematical geography information. Its primary location, however, is with descriptive geography, and it is therefore included as well in table 7, below.

editions. An interest in maps is clear from the five copies of Ortelius's *Theatrum Orbis Terrarum* owned in 1600 and 1610, four copies of Mercator's *Atlas* appearing on the list for 1610, and multiple copies of atlases and globes by Mercator, Ortelius, and Hondius owned in 1620.

Although the mathematical geography titles varied far less than those of other sections of geography, there was a shift of emphasis from 1580 to 1620. In 1580, one of the most popular books (in terms of numbers of copies owned) was Apian's *Cosmographia.* Apian continued to hold an important position in this book list until 1610, after which his name disappeared entirely, to be superseded by Theodosius's *De Sphaera,* rediscovered and translated in 1518, though gaining popularity in the late sixteenth century only in its original Greek,[67] and by the atlases of modern Dutch cosmographers. This demonstrates a subtle shift from the rigorous but conservative cosmography of German geographers in the early period to the newly discovered classics and, more important, the innovative geography of the Dutch school in the later period.

This shift is even more evident upon examination of the nationality of the authors of the mathematical geography books owned by Oxford and Cambridge students (see figure 17).[68]

In 1580, German authors (among whom Apian and Münster are conspicuous) accounted for 15 percent of all mathematical geography books, while fewer than 5 percent of the titles were written by Dutch authors. This is not surprising, since Dutch publishing at this period was still very much in its infancy. By 1600, the German share had increased to almost 27 percent, while Dutch authors accounted for 12 percent of the mathematical geography books. By 1620, this situation had changed dramatically. Fewer than 14 percent of mathematical geography books had German authors, while almost one-quarter originated from Dutch sources. This change corresponded to a large increase in the number of mathematical geography books published in the Low Countries, as Dutch knowledge and hegemony in matters geographical and especially navigational began to emerge. The Netherlands had not yet begun to challenge England's carrying trade, as it was to do in the First Dutch War of 1652–54, but the Dutch were a force to be reckoned with, and it is not therefore surprising that those young English politicians, merchants, and warriors in training would be interested in increasing their knowledge of Dutch scholarship in mathematical geography. Here was an English audience increasingly familiar with Continental trends and eager to learn the newly developed theories

67. Bulmer-Thomas, "Theodosius of Bithynia," 320.

68. Figure 17 charts the relationship, expressed as a percentage per decade, of the countries of origin of the authors of these mathematical geography texts. These books include maps and navigational treatises, as well as more general mathematical geography books.

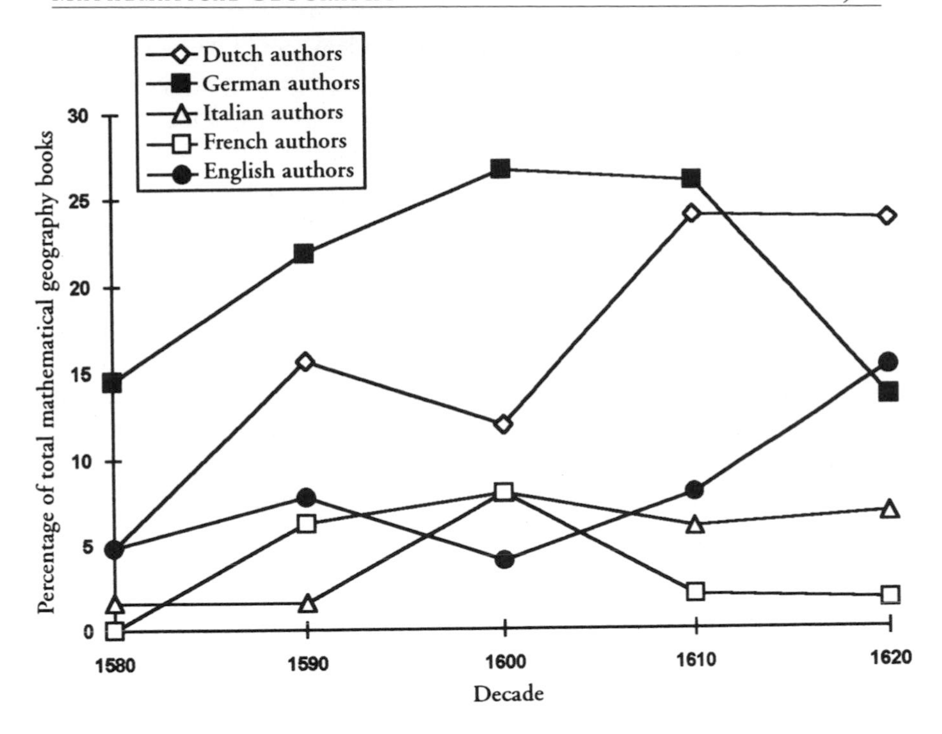

FIGURE 17. Countries of origin of mathematical geography book authors.

and discoveries of the Dutch nation, which was both their political and religious ally and increasingly their economic competitor.

Mathematical geography maintained a relatively constant audience in terms of numbers of books owned, although this is somewhat obscured by the meteoric rise of interest in descriptive geography. Comparing the data for mathematical geography book ownership with that for mathematics and astronomy (see figure 8, above) reveals that the three curves follow one another closely, suggesting a similar ownership pattern for all three disciplines. In addition, these three mathematical curves decrease only slightly throughout the period under discussion, indicating a readership, if relatively small, of constant size. From an examination of contemporary definitions of geography and cosmography,[69] scholars interested in mathematical geography appear to have viewed it as a branch of applied or mixed mathematics or astronomy. They were fascinated with many diverse mathematical problems, some of which were geographical in nature.

69. For example, Dee, *Mathematicall Preface;* Blundeville, *His Exercises* (1594, esp. 2). Carpenter defines geography as "lesse regarding their qualities, inquires rather of the Quantities, measures, distances, which places have as well in regard one of the other, as of the whole Globe of the Earth . . . requires necessary helps of the Sciences Mathematicall, chiefly of Arithmeticke, Geometry, and Astronomie" (*Geography delineated forth in two Books,* 3).

These mathematical and geographical problems could be viewed from several different vantage points. Mathematical geography can be divided into a number of related categories: cosmography, navigation, and maps and atlases. The theoretical study of the earth, cosmography, was the most widely read of these branches. Within this category, however, there was a change in emphasis during the period under investigation, from an analysis of the world as an integral part of the wider cosmos, a true cosmography, to a conceptualization of the earth as an entity complete unto itself, or even consisting of smaller subdivisions.[70] Compared with cosmography texts, books on navigation, such as Norman's, consistently accounted for fewer than 10 percent of all mathematical geography books, suggesting that this predominantly practical aspect of mathematical geography was not of such paramount importance to those at the universities as the more general cosmography. Mathematical geographers studied geography with theoretical as well as practical interests in mind, wishing to understand the globe and its irregularities, rather than merely how to sail from one point to another.

The third mathematical subcategory, maps and atlases, had a fairly stable ownership, accounting in 1580 for half of all mathematical geography books, decreasing to 28 percent in 1590, and remaining relatively stable over the next three decades. This pattern was a result of an early investment in individual maps and globes (the latter, especially, when once donated to a college, likely satiating demand for some time) and a later inclination toward the larger and more complete atlases, often of Dutch manufacture.

Thus, the book list shows a relatively constant interest in mathematical geography, especially at Corpus Christi College, St. John's, and New College, Oxford, and at Peterhouse, Corpus Christi, and Trinity College, Cambridge, corresponding to a similar interest in mathematics and astronomy. The books owned change over time, from the staid and conservative German authors in the early period to the innovative and exciting work of the Dutch in the later years. There is a consistently firm foundation in the work of Ptolemy, but these students and masters became better acquainted with new Continental trends and problems as the seventeenth century progressed.

70. R. Porter, "Terraqueous Globe," 285–324; Lestringant, in *Mapping the Renaissance World,* although his transformation is from a more poetic, subjective, and complete description of the world to the later development of observable, "objective" knowledge. This change could be exemplified by the prevalence of Münster's *Cosmographia* (1544) and Apian's *Cosmographia* (1529) in the earlier period and works by Mercator, *Atlas sive Cosmographia* (1595), and Hondius, *Globus Terrestris* (1600), in the later.

Students of Mathematical Geography

The students interested in these books of mathematical geography formed a compact but cohesive body, interested in mathematical subjects in general and mathematical geography specifically. They often maintained their university connections throughout their lives and were concerned with establishing and encouraging England to take its place among the nascent empires of Europe.

Of the 178 men with some known interest in mathematical geography between 1580 and 1620, 117, or 66 percent, attended university.[71] This total includes those men whose interest in mathematical geography was part of the requirement of their employment, since as navigators, surveyors, and instrument makers they were expected to construct the globe in a mathematical manner. While some of these men, including several already mentioned, contributed in an integral way to the development of geographical knowledge, by supplying the geographical community with information and instruments, by writing vernacular treatises, and on occasion by acting as liaisons between their more theoretical brethren, others lived out their lives on board ship or in their instrument shops and did not contribute to the mathematical dialogue I am charting.[72] If we remove those practitioners who did not engage in the lar geographical community from the total of mathematical geographers, the proportion of university men increases to 87 percent. Thus, mathematical geography, as a subject that made use of practical information and applied knowledge to practical problems, but maintained its theoretical framework and motivation, was studied and practiced almost exclusively within or in connection with the universities. As will be seen, this was not the case with descriptive geography and thus suggests the more academic nature of mathematical geography.

Equally, mathematical geography seems to have attracted the attention of the new groups of students entering the universities. Of the mathematical geographers whose class origins are known, thirty-four (65 percent) were from the gentry or merchant class. More than half of these were of gentle birth. Among the other mathematical geographers, five were known to have been sons of clergy, eleven were of plebeian stock, and one was a

71. Biographical information for this section was drawn from the *DNB;* Venn, *Alumni Cantabrigienses;* J. Foster, *Alumni Oxonienses;* and Clark and Boase, *Register of the University of Oxford.* Other sources are cited individually.

72. This is an extremely difficult distinction to make, and I am conscious of the layers of theoretical baggage it carries. We need to investigate further the interactions between instrument makers and navigators and the more academically inclined thinkers, since I believe we will discover fundamental sources for scientific change in just such interactions.

noble scion. Although simply tracing family status cannot be conclusive, these class origins indicate that mathematical geography did attract the "rising classes" and especially the gentry out of proportion to their presence at the universities.

Given the clear majority of mathematical geographers who attended the universities and who were members of the new classes at Oxford and Cambridge, the degrees achieved by these acolytes and the careers that they chose help confirm the central role of the Arts curriculum as well as the universities more generally in developing the subdiscipline of mathematical geography. Seventy-one percent of the university-trained men achieved the degree of Bachelor of Arts, and most of the university-trained men (61 percent) remained to take the Master's degree. A large proportion of these mathematical geography students thus remained at Oxford or Cambridge for the seven years required for the completion of the Arts curriculum, or at least maintained some relatively strong connections with their university and college. A fair proportion, moreover, remained to study for various graduate degrees. Eleven achieved an M.D., seven a D.Th., and five a D.C.L. This amounts to 20 percent of the total of university-trained mathematical geographers, a significant number when the length of time for these degrees is considered.

The careers of these mathematical geographers confirm the close ties between this subdiscipline and the universities. Thirty-five of those with mathematical geography interests (30 percent of the university men) pursued careers as teachers of mathematics and geography, either inside or outside the universities. Many remained within their colleges as fellows or professors. Thomas Allen (B.A. 1563, M.A. 1567, Trinity, Oxford), for example, was a fellow at Trinity College until 1570, when he retired to Gloucester Hall, living the rest of his life as an independent academic fellow, teaching several generations of mathematicians and mathematical geographers, including Thomas Harriot, Kenelm Digby, Sir Thomas Aylesbury, and perhaps Richard Hakluyt. Allen, in fact, was the center of a lively geographical research circle at Oxford, although one with no formalized structure.[73] Similarly, Robert Hegge (B.A. 1618, M.A. 1620, Corpus Christi, Oxford), Samuel Ward (B.A. 1593, M.A. 1596, D.Th. 1610, Christ's, Cambridge), and William Bedwell (Trinity, Cambridge) remained at their colleges as fellows until their deaths. The latter two, as translators of the Authorized Version (with Ward as part of the company responsible for the Apocrypha),[74] combined literary and mathematical studies in a way only possible for Renaissance men. Ward produced a last-

73. See McConica, "Elizabethan Oxford," 717–18.
74. Opfell, *King James Bible Translators,* 67.

ing memorial to his interest in geography through the donation of a petrified human skull, acquired from a sea captain on his return from Crete, to Sidney Sussex College in 1627.[75] John Bainbridge (B.A. 1603, M.A. 1607, M.D. 1614, Emmanuel, Cambridge) spent most of his life at various academic posts, as the first Savilian Professor of Astronomy at Oxford from 1621 to 1643, and as Linacre Lecturer from 1631 to 1635. From Bainbridge's notebooks, it is clear that he taught up-to-date astronomical discoveries, as well as working on his own observations. Bainbridge was very interested in organizing a trip to South America to perform astronomical observations and announced Captain Roger Fry's observations from Brazil to his class in 1633.[76] He was one of the English mathematical geographers who attempted to solve the longitude problem through studying the change of compass variation over time.[77]

Several other mathematical geographers became Gresham professors, most notably Henry Briggs, Edmund Gunter, and Henry Gellibrand (B.A. 1619, M.A. 1623, Trinity, Oxford). Briggs was Gresham Professor of Geometry from 1596 to 1620 and Savilian Professor of Astronomy at Oxford from 1619 to 1630. Gunter was Gresham Professor of Astronomy from 1619 until his death in 1626, when he was succeeded by Gellibrand. Gunter is best remembered for his work with triangulation, both theoretical and practical. At the age of twenty-two, he wrote a theoretical manuscript entitled "A new Projection of the Sphere," which gained him the notice of Henry Briggs and William Oughtred.[78] Later, his *Canon Triangulorum,* a table of sines and tangents, revolutionized trigonometry and therefore map theory.[79] But his greatest contribution to mathematical geography lay in his invention of a number of observational and calculational instruments: a sector, "Gunter's Quadrant," and a type of prototypical slide rule known as "Gunter's Rule."[80] Gellibrand was still a young man as our period ended,

75. The skull, which was borrowed by William Harvey in the 1630s to show to Charles I, still exists and has sparked some anthropological interest. Personal communication with Nicholas Rogers, Keeper of Muniment Room, Sidney Sussex College, Cambridge, 1990.

76. Feingold, "Universities and the Scientific Revolution," 43; N. Tyacke, "Science and Religion at Oxford," 83. Both taken from Bainbridge's class notes, in the Trinity College Archives, Dublin.

77. John Bainbridge, "De Meridianorum sive Longitudinum differentiis inveniendis, Dissertatio," Oxf. Bodl. MS Add. A 380, ff. 194–197b. A second copy of this is located in Oxf. Bodl. MS Smith 92, ff. 9–14, which also contains "Roger Fry's observations at Maranhano in Brazil," f. 26.

78. E. G. R. Taylor, *Mathematical Practitioners,* no. 107.

79. Gunter, *Canon Triangulorum.* This book, whose printing history is very muddled, with combinations of Latin and English versions produced in a rather haphazard manner, contains the first use of the terms *cosine* and *cotangent* (*DNB,* 8:794). STC 12519 ff.

80. Edmund Gunter, *De Sectore. Descriptio et Usus,* MS 1607. *Description and Use of a Portable Instrument [known by the name of Gunter's Quadrant],* MS 1618. *The Description and Use of the Sector, the Crosse-Staffe and other Instruments* (London, 1623). Adamson claims

only coming into his own geographically with the publication of *An Appendix concerning Longitude* and *An Epitome of Navigation* in 1633 and *A Discourse Mathematicall of the variation of the magneticall needle* in 1635.[81] The *Appendix* records the simultaneous observations of a lunar eclipse by Thomas James off the coast of North America and by himself at Gresham, leading to Gellibrand's corrections for lunar parallax.[82] In the latter, he announced the change of magnetic variation over time, based on observations made by himself, John Pell, and John Marr.

Of those who pursued an academic career, several became schoolmasters or private tutors. Nathanael Carpenter (B.A. 1610, M.A. 1613, D.Th. 1626, Exeter, Oxford), author of the first theoretical geography book in English, earned his living as Schoolmaster of the King's Wards in Dublin. His major work, *Geography delineated forth in two books* (Oxford, 1625), set down the study of geography as "a *Science* which teacheth the *Measure* and *Description* of the whole *Earth*."[83] He divided the study of geography into "Sphaericall" and "Topicall," corresponding to my categories of mathematical and descriptive geography: the former dealt with the mathematical form of the globe, magnetism, the measure of the earth, the zones, climates, and parallels, longitudes and latitudes; the latter with descriptions of regions and especially local history.[84] In the first section, he identified his astronomy as Tychonic and the best map projection as coming from Edward Wright. In the second section, Carpenter explained the masculine nature of the northern hemisphere and the feminine nature of the southern.[85]

Among the private tutors, the best known are Thomas Harriot, Edward Wright, and William Oughtred. Harriot (B.A. 1580, St. Mary's Hall, Ox-

---

that "Gunter is one of the better examples of the catalytic effect which Gresham College had upon aspiring scientists freed from the strictures of the universities" ("Foundation and Early History of Gresham College," 142). E. G. R. Taylor, on the other hand, believes that Gunter was far from being an innovative and influential thinker: "Gunter's writings had only an indirect influence on navigation and on the other mathematical arts that he dealt with . . . for they were too mathematical for most people and needed the interpretation of a teacher as intermediary" (*Mathematical Practitioners,* 64).

81. Gellibrand, *An Appendix concerning Longitude; An Epitome of Navigation* (London, 1633). E. G. R. Taylor infers this date for the latter, since no edition of the *Epitome* is known before 1674 (*Mathematical Practitioners,* no. 158). *A Discourse Mathematicall of the variation of the magneticall needle: together with its admirable Diminution lately discovered* (London, 1635).

82. See E. G. R. Taylor, *Mathematical Practitioners,* 72.

83. Carpenter, *Geography delineated forth in two Books,* 4. This is discussed in Bowen, *Empiricism and Geographical Thought,* 73.

84. See Carpenter's schema, *Geography delineated forth in two Bookes,* a foldout frontispiece facing sig. 2¶ 1a; Watson, *Beginnings of the Teaching,* 118–24; Baker, "Nathaniel Carpenter and English Geography," 262–67.

85. Carpenter, *Geography delineated forth in two Books,* Book 1:111, 170, and Book 2:40.

ford) was a mathematical tutor to Sir Walter Raleigh and gave personal instruction to Henry, Prince of Wales, before being pensioned by Henry Percy, ninth Earl of Northumberland. As Richard Hakluyt said of Harriot, in a dedication to Raleigh: "By your experience in navigation you saw clearly that our highest glory as an insular kingdom would be built up to its greatest splendor on the firm foundation of the mathematical sciences, and so for a long time you have nourished in your household, with a most liberal salary, a young man well trained in those studies, Thomas Hariot, so that under his guidance you might in spare hours learn those noble sciences."[86]

Edward Wright, the most famous English geographer of the period, was educated at Gonville and Caius College, Cambridge, receiving his B.A. in 1581 and his M.A. in 1584. He remained at Cambridge until the end of the century, with a brief sojourn to the Azores with the Earl of Cumberland in 1589.[87] In the early 1600s, he became tutor and librarian to Henry, Prince of Wales. Finally, in 1614, Wright was appointed by Sir Thomas Smith, governor of the East India Company, to lecture to the company on mathematics and navigation, being paid fifty pounds per annum by the company.[88] William Oughtred (B.A. 1597, M.A. 1600, King's, Cambridge), an eminent mathematician, earned his keep as tutor to Lord William Howard, son of the Earl of Arundel. He began his career with an analysis of the geometry of dialing and soon introduced the new mathematical study of algebra to an English audience.[89] His greatest contribution to geography was in the realm of measurement, creating better instruments for astronomical observation and for measuring the globe and surveying smaller plots of land.[90]

86. Richard Hakluyt, introduction to Peter Martyr, as quoted in J. W. Shirley, "Science and Navigation," 80. See chapter 6 below for further information about Harriot. For a complete biography of Thomas Harriot, see J. W. Shirley, *Thomas Harriot. A Biography.*

87. As a result of this voyage, Wright wrote *The Voiage of the right honorable George Erl of Cumberland to the Azores* (1589), which was later printed by Hakluyt, "written by the excellent Mathematician and Enginier master Edward Wright" (Hakluyt, *Prinicipal Navigations* 1598–1600, 2.2:155 [misnumbered as 143]–68). M. B. Hall (*Scientific Renaissance,* 204), Waters (*Art of Navigation,* 220), and J. W. Shirley ("Science and Navigation," 81) all cite this trip to the Azores as the turning point in Wright's career, his road to Damascus, since it convinced him in graphic terms of the need to revise completely the whole navigational theory and procedure.

88. Waters, *Art of Navigation,* 320–21. There is some speculation about whether or not Wright actually gave these lectures, since he died in the following year.

89. William Oughtred, *Horologiographia Geometrica: an Easy Way of Delineating Sundials by Geometry* (MS 1597); *Clavis Mathematica* (London, 1631).

90. William Oughtred, *The Circles of Proportion and the Horizontall Instrument,* translated from the Latin by William Forster (London, 1632). This book described two instruments that Oughtred had invented, the "Circles of Proportion," which was a type of slide rule, and the "Horizontall Instrument," a sort of astrolabe. He also wrote *An Addition unto the use of the Circles of Proportion for the working of Nauticall Questions* (London, 1633). In

Some mathematical geographers set themselves up as mathematics and navigation teachers in the "Third University," the city of London.[91] These included William Pratt, Richard Norwood, and Robert Hartwell, all self-taught.[92] They also included Thomas Hood, whose mathematics and navigational lectures have already been discussed. Hood also established a reputation as a chartmaker and a manufacturer of astrological tables, demonstrating that his practical interests were taken very seriously. Hood was replaced as lecturer for the Virginia Company at his death by Edward Wright, who later moved on to lecture for the East India Company, also under the governorship of Smith.[93]

Within the ranks of the academic geographers should be included those supported directly by patronage. Thomas Harriot in part falls into this category, having been supported by Northumberland after Raleigh's fall from grace. Also patronized by Northumberland were Harriot's friends Robert Hues (B.A. 1578, Magdalen Hall, Oxford), author of *De Usu Globis* (1594),[94] and Nathaniel Torporley, Harriot's executor and designer of a new instrument to be used instead of a quadrant. Hues was the more interesting of the two, sailing around the world with Thomas Cavendish in 1586 to 1588 as his mathematician, tutoring Northumberland's son Algernon, and maintaining his ties with the theoretical world of Oxford as well as the practical world of London.[95] Since Cavendish's global circumnavigation must be seen as an important boost to English imperial aspirations, Hues's presence shows once again a strong link between geography and an imperial vision.

While teaching was the most common career choice among mathematical geographers, several opted for a career in medicine. Eleven of the men

---

this work, Oughtred rejected the use of magnetic variation as an aid to finding the longitude, recommending water clocks as the way to establish standard time. See E. G. R. Taylor, *Mathematical Practitioners,* nos. 152, 156.

91. Buck, *Third Universitie of England.*

92. Robert Hartwell may have attended Oxford for a time. Clark and Boase, in *Register of the University of Oxford,* list a student of that name who matriculated at the age of seventeen at University College in 1587, but it is impossible to know whether this is the same man who edited *Blundeville His Exercises* in 1636, adding on the last page, "Arts and Sciences Mathematicall taught in Fetter Lane . . . by Robert Hartwell, Teacher of the Mathematicks, viz. Arithmetick, Geometry, Astronomy, Cosmography, Geography, Navigation, Architecture, Fortification, Horologography, etc." Watson, *Beginnings of the Teaching,* 324.

93. See Hood, "The Humble petition of Thomas Hood Mathematicall Lecturer in the Citie of London," BL Lansdowne MS 101, ff. 57b–58a. *Copie of the Speache,* BL Lansdowne MS 101, ff. 56–57a. See also E. G. R. Taylor, *Mathematical Practitioners,* 40–41; Waters, *Art of Navigation,* 185–89.

94. Hues, *Tractatus de Globis.* This work was dedicated to Sir Walter Raleigh and contained instructions on the use of Molyneux's globes.

95. J. W. Shirley, "Science and Navigation," 80; M. B. Hall, *Scientific Renaissance,* 202.

on the mathematical geography list were physicians, most practicing medicine for at least a portion of their lives. These men included Gabriel Harvey (B.A. 1570, M.A. 1573, Christ's, Cambridge; D.C.L. Oxon. 1585), William Gilbert, Matthew Gwinne (B.A. 1578, M.A. 1582, M.D. 1593, St. John's, Oxford), Mark Ridley (B.A. 1581, M.A. 1584, M.D. pre-1592, Clare Hall, Cambridge), and Robert Fludd. These men all had interesting and unusual careers. Ridley was physician to the English merchants in Russia, became physician to Ivan the Terrible, and finally returned to England to practice medicine and teach mathematics and navigation.[96] One suspects that a little descriptive geography was thrown in as well. Fludd, a student of Thomas Allen, practiced medicine in London and was famous for his studies of chemistry and alchemy.[97] Matthew Gwinne was the first Gresham Professor of Physic, practiced medicine in London, and was in Prince Henry's entourage. He was also a friend of Sir Philip Sidney and Sir Fulke Greville, as well as an acquaintance of Giordano Bruno. Gabriel Harvey, father of William, the discoverer of the circulation of the blood, earned his living as an author and critic, only turning to medicine late in life. He was probably the least serious mathematical geographer of these physicians but provided a literary bent to the discussion sometimes lacking in the drier, more scholarly works.[98] Finally, William Gilbert, author of *De Magnete* (1600), practiced as a physician in London and was a fellow of the College of Physicians (and its president in 1600). He also maintained court associations as physician to Elizabeth and briefly to James, and he may have dedicated his great work informally to Henry, Prince of Wales.[99] Unfortunately, we know relatively little about Gilbert, because of the almost complete lack of personal papers. Gilbert's house in Colchester was destroyed during the Civil War, and his papers deposited with the College of Physicians were lost when the college burned down.

A significant number of mathematical geographers thus adopted a scholarly way of life, maintaining their level of interest in mathematical geography while pursuing careers in mathematical investigation, teaching,

96. E. G. R. Taylor, *Late Tudor Geography,* 73. Bennett sees Ridley as typical of the new practical follower of the mathematical sciences ("Mechanic's Philosophy," 18).

97. For a more thorough treatment of Fludd and his circle, see Huffman, *Robert Fludd,* now superseding Yates, *Rosicrucian Enlightenment.*

98. The only modern biography is Sterne, *Gabriel Harvey,* unfortunately inadequate. For an important analysis of Harvey's methodical and active reading of the classics, see Jardine and Grafton, "'Studies for Action.'"

99. Strong, *Henry, Prince of Wales,* 212. There is no major biography of William Gilbert. Perhaps the most important works are Pumfrey, "William Gilbert's Magnetic Philosophy 1580–1684"; Hesse, "Gilbert and the Historians"; and Dibner, *Doctor William Gilbert.* Editions of Gilbert's work have appeared, such as Kelly, *The De Mundo of William Gilbert;* and Roller, *De Magnete of William Gilbert.*

or medicine. Many of these mathematical geographers knew and corresponded with one another. These acquaintances, often based on mutual interests and a sense of cooperation and sharing of geographical information, helped mathematical geography develop as a public science, involving the display of practical application as well as the definition of a specific community equipped to deal with the geographical problems identified as important. In the self-definition of these mathematical geographers, the development of groups of like-minded individuals was to prove vital. Geographical coteries, identifiable by the exchange of letters, by dedications, by mention in one another's books and manuscripts, as well as by known associations of place and interest, developed in four specific locations. Particularly, groups sprang up around Thomas Allen at Oxford, Henry Briggs at Oxford and Gresham, John Dee in London and Mortlake, and Henry Percy, Earl of Northumberland, in London and abroad. These coteries differed in emphasis, with Allen's group most concerned with the teaching and study of geography at Oxford, Briggs's circle interested in rigorously applying astronomical and cosmographical laws to practical problems, Dee's group most interested in practical applications, and Percy's coterie linking both theory and practice in a loose but fruitful manner. All four groups, with the possible exception of Briggs's circle, maintained a degree of flexibility of membership within the enclosed ranks of mathematical geography, since these scholars were often known in two or more circles. This helps to confirm the closed and almost recondite nature of mathematical geography as a whole, since it attracted a small, cohesive group of men, from similar backgrounds, with similar education, and often with parallel career patterns. Most of these mathematical geographers knew one another but gravitated, generally, to one or the other of these different coteries.

The mathematically minded scholars surrounding Thomas Allen at Oxford were attracted by his special interest in mathematics and geography.[100] Allen taught Thomas Harriot, Kenelm Digby, Sir Thomas Aylesbury, Sir John Davies, Robert Fludd, and perhaps Richard Hakluyt. He was also a friend of John Dee, with whom he acted as consultant to Robert, Earl of Leicester, and who gave Allen his manuscripts and his inverting mirror.[101] Allen also corresponded with the Earl of Northumberland, thus providing a link between these two groups of mathematical geographers. He befriended Thomas Gent, a somewhat obscure character who was at Gloucester Hall from the 1580s until his death in 1613, who donated more than four hundred scientific and medical books to the Bodleian Library

100. McConica, "Elizabethan Oxford," 717–18.

101. Feingold, "Occult Tradition," 85. For a fuller account of Allen's life, see M. Foster, "Thomas Allen, Gloucester Hall."

and who was part of the circle of scholars surrounding William Gilbert, a subset of Percy's coterie.[102]

The circle of scholars surrounding Henry Briggs seems to have been connected by ideas rather than proximity. Three of its members taught at Gresham College: Matthew Gwinne, Edmund Gunter, and Henry Gellibrand.[103] Two more hailed from Cambridge: Samuel Ward and William Oughtred. Two others were physicians in London: Richard Forster and, for a time, John Bainbridge. What they all shared was an interest in the most abstract mathematical geography studies. Closely linked with these mathematicians was Bishop Ussher's circle of chronologers, including Thomas Bodley, Sir Robert Cotton, William Camden, John Selden, William Crashaw, and Thomas Lydiat, who was at the same time an astronomer, a cosmographer, and a chronologer.[104] This is a most compelling connection, indicating a close parallel between mathematical descriptions of the earth and mathematical calculations of its age. This chronological group was centered about the Inns of Court, as is seen by the numerous connections both they and Briggs's circle developed with the Inns.[105] There were few links between Briggs's coterie and the other mathematical groups, with the exception of John Dee, who was ubiquitous.

Dee's personal circle seemed to have been peopled with mathematical geographers more concerned with practical problems. No explorer or navigator would think of beginning his voyage without a consultation at Mortlake. Martin Frobisher, Richard Chancellor, Pet, Jackman, Humphrey Gilbert, and Sir Walter Raleigh all took Dee's advice about navigation and strategy.[106] Indeed, Dee advised the Privy Council on occasion concerning such imperial issues as the creation and maintenance of sovereignty of the

102. Feingold, "Occult Tradition," 85. Gent does not appear in either Clark and Boase, *Register of the University of Oxford,* or J. Foster, *Alumni Oxonienses.*

103. In *Canon Triangulorum,* Gunter adds logarithms compiled by his "old colleague and worthy Friend Mr. Henry Briggs" (quoted in E. G. R. Taylor, *Mathematical Practitioners,* no. 135). Gellibrand completed Briggs's book *Trigonometria Britannia* in 1633. Gunter is mentioned in Gellibrand's book, *An Appendix Touching Longitude,* sig. Qa–b. See Adamson, "Foundation and Early History of Gresham College," 143.

104. Briggs and Lydiat correspondence in Oxf. Bodl. MS Bodl. 313. Reprinted in Halliwell, *Collection of Letters,* 46–47, 58. Ussher and Lydiat correspondence in Oxf. Bodl. MS Add. C297, ff. 9–10b. Bodley to Cotton, 31; Camden to Cotton, 35, 93, 117; Ussher to Cotton, 133; Selden to Cotton, 141, in Oxf. Bodl. MS Smith 71.

105. Fisher, "William Crashawe's Library," 120. Crashawe was a preacher at the Inner and Middle Temples. Camden was admitted to the Inner Temple in 1571, though he never achieved the status of a barrister (J. Foster, *Alumni Oxonienses).* Selden became a barrister, studying first at Clifford Inn and then at the Inner Temple.

106. For example: "Jan. 23rd [1582], the Ryght Honorable Mr. Secretary Walsingham came to my house, where by good lok he found Mr. Awdrian Gilbert [Humphrey's brother], and so talk was begonne of North-west Straights discovery . . . we made Mr. Secretarie privie of the N.W. passage, and all charts and rutters were agreed uppon in generall" (Dee,

seas and the expansion of naval control and settlements.[107] He numbered among his friends Robert Recorde, the mathematician; Leonard and Thomas Digges; Cyprian Lucar, the surveyor; Thomas Blundeville, the popular mathematical and equestrian writer; Thomas Harriot; and William Camden. Anyone who was anyone in Elizabethan geography knew Dee. Dee's connection with Camden's circle of correspondents, which included Mercator and Ortelius, shows an important link between mathematical geography and chorography, as well as reminding us of the close interconnections of many people interested in geographical subjects.[108] His coterie should be characterized by its interest in practical questions, as opposed to Allen's, which was pedagogical, or Briggs's, which was theoretical. Proposals for new English paths to the Orient also came from this group, emphasizing its interest in governmental policy and power and in the development of the English empire.[109]

The final group of mathematical geographers developed around two foci: Henry Percy, ninth Earl of Northumberland, and Edward Wright, the eminent geographer. Northumberland's group included Thomas Harriot (also linked to Dee and to Allen), Robert Hues, and Walter Warner, the three men who were later to be associated with Percy's supposed necromancy and potential treason.[110] All three had attended Oxford at approximately the same time, as had Nathaniel Torporley, and all received pensions from Northumberland to enable them to work on mathematical problems. Both Harriot and Hues spent time on voyages, Hues with Cavendish and Harriot with Raleigh, demonstrating that productive tension between the theoretical and practical that was to characterize this coterie as well as the study of geography more generally. Raleigh, indeed, might be seen as part of this group, until his imprisonment made participation difficult. Northumberland had been a youthful devotee of the older Ra-

Private Diary, 18); "May 18, 1590, the two gentlemen, the unckle Mr. Richard Candish, and his nephew the most famous Mr. Thomas Candish, who had sayld round about the world did visit me at Mortlak" (33).

107. Sherman (*John Dee,* ch. 6) argues persuasively that "Brytannicae Reipublicae Synopsis" (1570) must be read as an advisory document.

108. Ibid., 119.

109. See Cormack, "Fashioning of an Empire."

110. Part of this accusation was a result of Northumberland's implication in the gunpowder plot in 1605. He was imprisoned from 1605 to 1621, and these three men, though never arrested or only briefly detained, were all suspect (Public Record Office [hereafter PRO], SP 14/216, Part 1, 110, 111; Part 2, 112, 113, Nov. 23–26, 1605). For example, Nov. 27, 1605, "Examination of Nathaniel Torporley, about his casting the King's nativity for Mr. Heriot, who lived at Essex House, the Earl of Northumberland's" (PRO, SP 14/216, Part 2, 122). This examination, demonstrating Torporley's and Harriot's presence in Northumberland's house, was part of the proceedings in the investigation of the Gunpowder Plot. J. W. Shirley, *Thomas Harriot: A Biography,* 327–57.

leigh from the 1580s and was perhaps inspired by him to pursue science and geography.[111] Through him, Emery Molyneux, the globe and instrument maker; Humphrey Gilbert, Raleigh's stepbrother and unsuccessful adventurer; and Thomas Hood, the preeminent mathematician and teacher all gained admittance into this group. This assemblage, more than the other three groups, was most interested in the real expansion of England's dominion. Several members sailed on expeditions to establish colonies or new trading routes, and this, combined with their close connections with Elizabeth's court, enabled them to use the lessons of academic mathematical geography to help forge a new conception of England's place in the world.

Connected with this practical yet theoretical circle was a group of scholars surrounding Edward Wright and William Gilbert.[112] This group included Mark Ridley, the physician to the czar, who also knew Thomas Gent, thus connecting this group with Allen's.[113] Gent also knew William Gilbert, and, of course, John Dee knew members of both circles. Molyneux could be considered a member of Wright's circle, since he worked with Wright in creating Wright's projection, while his ties with Hues, who wrote the published description of Molyneux's globes, link him more directly with Percy's circle. Blundeville, as well, was a friend of Wright's and Gilbert's, while his friendship with Dee marks yet another point of contact.

These four related coteries of mathematical geographers developing in the late sixteenth and early seventeenth centuries show the small but intense world of English mathematical investigation. The group around Thomas Allen at Gloucester Hall was largely pedagogical in nature, uniting its members by educational training and intellectual excitement and thus a shared ideology. These men were often in contact with the other groups, especially as they left Oxford behind them. Briggs's circle, on the other hand, was linked by academic rigor and mathematical ideas, rather than by time and place, and therefore, not surprisingly, tolerated little interaction between its ranks and those of the other geographical coteries. The links it formed were rather with chronologers, both connected with Ussher in Ireland and with the Inns of Court in London. This helps to confirm the interest felt by these mathematical practitioners in the different applications of their mathematical ideas and theories. Dee's circle of associates tended to be extremely practical and less concerned with theory, though Dee himself provided the bridge between his group and the others

111. J. W. Shirley, *Thomas Harriot: A Biography,* 168–70.

112. Gilbert obtained help from Wright in writing his book on magnetism (introductory letter by Wright, *De Magnete,* sigs. 3b–5b). See E. G. R. Taylor, *Mathematical Practitioners,* no. 101. This point was recently made much more provocatively by Stephen Pumfrey in "The Unknown William Gilbert," public lecture at "William Gilbert(d) and His World," a conference in Colchester, July 1993.

113. E. G. R. Taylor, *Late Tudor Geography,* 73.

by his interaction between theory and practice. Indeed, Dee's group seems to have been most concerned with the expansion of the English empire and geography's role in that movement. Finally, the somewhat amorphous group centered around Northumberland and Wright provided a clear link between theory and practice, on the one hand, and between court and college, on the other, again suggesting that mathematical geography was a tool both of government and of imperialism.

❦

MATHEMATICAL GEOGRAPHY should be seen as an identifiable and discrete subdiscipline. Related to geography generally by its focus on the globe and its constructs, mathematical geography distanced itself from description by its relationship with the mathematical sciences more generally. The men who were interested in a mathematical geography program were a small and relatively homogeneous group. They pursued mathematical geography because of an intrinsic interest in mathematical problems in any form, although motivated as well by a perceived duty to use their knowledge for the good of the state. This dynamic tension between utility and theoretical sophistication was essential to the study of mathematical geography. It drew its inspiration from the classical sources beginning to be in vogue at the universities, which encouraged some emphasis on elegant mathematical proofs, but it thrived on practical and patriotic problems. Mathematical geographers, concerned with England's role in Continental and worldwide affairs, worked hard to develop better and more accurate means of exploring and explaining the world. In the process, they helped to develop better mapping techniques, new magnetic theories, and a more sophisticated structure for navigational theory. In fact, the very substance under geographical investigation required this twofold approach of theoretical and practical; it was through such a study as mathematical geography that the "new science" began to emerge, separating itself from the more traditional natural philosophy but using the same institutions and educational facilities. The English universities encouraged this venture into practical mathematics, and it is thus evident that the universities, far from remaining bastions of fading scholasticism, were deeply engaged, through students and associates who pursued mathematical and geographical studies, with a discipline and a methodology that were both theoretical and practical, that addressed academic and political concerns alike. While the universities generally encouraged this pursuit of geography, within the Arts curriculum and beyond it, the coteries of geographical men that grew up at and through the universities provided foci for this study and allowed mathematical scholars at Oxford and Cambridge to tackle new ideas and to make their contribution to the budding English empire.

CHAPTER FOUR

Descriptive Geography: Tales of Prester John and of the Palace of Edo

> dar'st thou lay
> Thee in ships woodden Sepulchers, a prey
> To leaders rage, to stormes, to shot, to dearth?
> Dar'st thou dive seas, and dungeons of the earth?
> Hast thou couragious fire to thaw the ice
> Of frozen North discoveries?
>
> (John Donne, *Satyre III*)

Lurid tales of torture and adventure, descriptions of grotesque races, stories of arcane customs and peoples—sixteenth-century undergraduates' taste in literature seems remarkably close to that of their twentieth-century counterparts. The popularity of Münster's *Cosmographia,* providing fantastic descriptions and illustrations of far-off places and people as well as its more exacting maps and mathematical geography, demonstrates this well. Most spectacular is Münster's inclusion of the Mandevillian races, clearly based on the medieval travel tales in vogue since the fourteenth century (see figure 18). These bizarre people, according to Münster living both in Asia and in Africa (the illustration is included in both sections), are presented as living specimens, equally as plausible as the leopard, the unicorn, and men riding on elephants. There was, however, much debate in late-sixteenth-century England about the truth of these fantastic tales, which Thomas Blundeville, for one, categorized in 1594 as "meere lyes, invented by vaine men to bring fooles into admiration, for monsters are as well borne in Europe, as in other partes of the world."[1] Abraham Hartwell, however, believed these people would be found with more exploration, expressing the hope in 1597 that "in good time, some good *Guianian* will make good proofe to our *England,* that there are at this day both *Amazons,* and *Headlesse Men.*"[2]

It is not surprising that the geography that most interested undergraduates at Oxford and Cambridge between 1580 and 1620 was not the dry

1. Blundeville, *His Exercises* (1594), f. 262b.
2. Abraham Hartwell, *Report of the Kingdom of Congo,* sig. 4a.

FIGURE 18. Mandevillian races, from Sebastian Münster, *Cosmographiae Universalis* (Basle, 1550), book 5, p. 1080. (BL 566.i.14. By permission of The British Library.)

mathematical tract but rather fanciful descriptions of faraway places and peoples, liberally sprinkled with tales of adventure and intrigue. These were sometimes supplied by native English writers as an English tradition of descriptive geography developed. More often, however, this interest was met by Continental sources, either in the original language or in translation. Descriptive geography was quite different from the mathematical geography discussed in chapter 3 above. Both originated from classical roots, but descriptive geography, being less rigorous, less exacting, less arcane, attracted more casual investigators. Undergraduate interest and serious scholarly concern were not nearly as closely linked in descriptive geography as they were in the investigation of mathematical geography. The result was a study that was less theoretical than mathematical geography but fundamentally useful, socially, politically, and economically. Part of that utility lay in the fact that descriptive geography was closely linked to the politics of the new governing classes. This subdiscipline supplied descriptions of European nations, peoples, and cultures, important in the growing rivalry between England and European powers. It also included tales of new markets and parts of the world where England could establish contact and control. Descriptive geography, more than its more arcane mathematical counterpart, encouraged its English students to develop a sense of superiority over other countries and peoples, as well as an interest in the exploitation of the rest of the world.

The Development of Descriptive Geography

Descriptive geography owed its origins to Strabo's *De Situ Orbis.* Strabo, a Greek geographer and historian born around 64 B.C., wrote his eight-volume geography in approximately 7 B.C.[3] He set out to describe every detail of the known world, based on both his own extensive travels and also on authoritative sources and other travelers' accounts. Strabo was a historian before he was a geographer, and this historical bent is clear in many of his descriptions of foreign lands. As Thomas Elyot was to remark in the sixteenth century, "In the part of cosmographie wherwith historie is mingled Strabo reigneth."[4] Strabo defined geography as a science that was the concern of the philosopher, while maintaining that it must be first and foremost a *useful* enterprise. For Strabo, the utility of geography "is manifold, not only as regards the activities of statesmen and commanders but also as regards knowledge both of the heavens and of things on land and sea, animals, plants, fruits, and everything else to be seen in various regions—the utility of geography, I say, pre-supposes in the geographer the same philosopher, the man who busies himself with the investigation of the art of life, that is, of happiness."[5]

Strabo did not exclude mathematical studies from geography but viewed mathematics as less useful or applicable to the world of affairs.[6] Strabo cast his net as far afield as possible in his search for geographical information, eschewing the specialized investigation inherent in mathematical studies. The geographical "science" he developed was an inductive investigation of interesting detail, with utility of information (in the widest sense) as the final arbiter of what was to be examined and reported.

Many sixteenth-century geographers followed Strabo's lead, and those who did not devote themselves to that more specialized study of mathematical and astronomical geography, seen by Strabo as of lesser importance, developed a descriptive subdiscipline that owed much to this anecdotal and rambling source.[7] Strabo's geographical path soon divided into

3. There is some controversy about this date, with Strabo scholars placing it anywhere from 7 B.C. to A.D. 18–19. See J. R. S. Stennett, "Introduction," in Strabo, *Geography,* xxiv.
4. Elyot, *The Governour,* f. 38a.
5. Strabo, *Geography,* 1.1.1:3.
6. Ibid., 1.1.21:45.
7. Strabo's *De Situ Orbis* was published in five Latin folio editions in the fifteenth century: Rome, trans. G. Veronensis and G. Tifernas, 1469; Venice, 1472; Treviso, 1480; Venice, January 1494; Venice, April 1494. In the sixteenth century, one Greek (Venice, Aldus, 1516), three Greek and Latin (Basel, 1549, 1571 [ed. G. Xylander], and Geneva, 1587 [ed. Isaac

descriptive or "special" geography, as Varenius was to call it,[8] and chorography or local history, the subject of chapter 5 below. Descriptive geography, following this distinction, became a more or less rigorous investigation of the physical features of the globe, including mountains, rivers, cities, and climates, as well as political institutions, different peoples and cultures, and the ever-varying flora and fauna. The most popular aspects of this descriptive branch in the sixteenth and seventeenth centuries were descriptions of the physical and political aspects of foreign countries, though practitioners took occasional forays into investigations of culture and primitive peoples. While earlier works such as Münster's attempted to include all of this and mathematical geography as well in a single work, creating a "cosmography" as defined by Lestringant,[9] by the end of the sixteenth century, descriptive geographers were concentrating on smaller geographical areas and more limited topics of explanation.

Early modern students of geography also took to heart Strabo's blend of descriptive geography and history. Sixteenth-century scholars viewed geography and history as complementary, and it is often difficult to separate descriptive history, which included descriptions of the scenes of action as a backdrop to historical incidents, from historical description, which, though mainly concerned with the geographical lay of the land, used history to place this land in perspective.[10] Renaissance historians felt compelled to place history in space, in a geographical setting, and, in a like manner, Renaissance geographers wished to impart to geography a historical perspective.

For most students and scholars at the English universities, the study of geography tended to concentrate on these descriptions of other countries, rather than on the more restricted mathematical geography studied by the small, though significant, group discussed in chapter 3 above. At a time when England was attempting to gain a foothold in the European race for

Casaubon]), three Latin (Venice, 1510; Basel, 1523; and Lyon, 1559), and one Italian (Venice, Ferrara 1562–65) versions appeared. It was published only once in the seventeenth century (Greek and Latin, Paris, 1620) and thereafter not until the nineteenth century. See Kristeller and Cranz, *Catalogus translationum et commentariorum,* 2:225–33.

8. Bernardus Varenius, *Geographia Generalis, in qua affectiones generales telluris explicantur* (Amsterdam, 1650); published in Cambridge, edited by Newton, in 1672. This became the standard geographical text for the next century, establishing a framework that could incorporate new geographical knowledge and systematize old. See *Dictionary of Scientific Biography* 13; Baker, "Geography of Bernhard Varenius." Livingstone, in *Geographical Tradition,* provides an important introduction to Varenius and eighteenth-century geography.

9. Lestringant, *Mapping the Renaissance World,* ch. 1.

10. Erasmus, for example, in 1511 discussed the value of these two studies. Likewise, Vives and Elyot both advocated the importance of knowing where things were so that history could be understood (Watson, *Beginnings of the Teaching,* 91–92).

new trade routes and colonies in the New World and the East, interest increased in new lands and peoples, as well as in the more abstract problems of how to navigate unknown waters. Descriptive geographers began to compile books and manuscripts concerning both old and new discoveries, in an attempt to catalog what was known and to rationalize that knowledge through inductive tabulation. They were also interested in the exotic "other," as well as the fate of those who transgressed the boundaries between the known and the unknown.[11] English geographers initiated this compilation of information by translating Continental works, but they soon started to write their own accounts, either based on personal experience as privateers, navigators, politicians, and merchants or relying on the transmission by such men of tales of discovery and adventure. These descriptions proved extremely popular at English universities, where students who sought cautionary tales for future sermons, or useful information for future political advancement, eagerly read and recorded a wide range of geographical notions.

The descriptive geography they read, as well as that written or translated by more committed descriptive geographers, fit into two different categories: those books primarily concerned with political relations and organizations and those whose main focus was natural history—the people, physical characteristics, flora, and fauna of the region. The first group of books tended to be drawn from historical and political accounts, involving scholarly research and erudition rather than firsthand experience on the part of the author. An excellent example of this genre is Richard Knolles's *The General Historie of the Turks* (1603), which he composed from historical sources while a fellow of Lincoln College.[12] This book was organized by the reigns of emperors and sultans, using historical detail to compare Turkish politics and customs with those of Europe and England, usually to the disadvantage of the former. The sources for the second descriptive category were more likely to include firsthand travel accounts and so had an immediacy lacking in the first group. Abraham Hartwell's translation of Odoardo Lopez's work, *A Report of the Kingdom of Congo* (1597), demonstrates the eclectic and exciting character of this latter classification. Hartwell recorded several theoretical speculations about Africa, including the question of the cause of the blackness of the inhabitants of the Congo. Mustering as evidence the fact that Portuguese children born in the Congo were not black and that people living in the West Indies at the same degree of latitude were much lighter in color, Hartwell concluded that "*Signor*

11. Gillies demonstrates this fascination—as well as the cost of transgression—through examining Othello's doomed marriage and Marc Antony's bizarre behavior in Egypt. *Shakespeare and the Geography*, esp. 25–36.

12. Knolles, *The Generall Historie of the Turks* (London, 1603), sig. A6b.

FIGURE 19. Zebra, from Abraham Hartwell, trans., *A Report of the Kingdom of Congo, a Region of Africa . . . Drawn out of the writings and discourses of Odoardo Lopez, a Portingall, by Philippo Pigafetta* (London, 1597), p. 72. (BL 279.e.36. By permission of The British Library.)

Odoardo was of the opinion, that the blacke colour did not spring from the heate of the Sunne, but from the nature of the seede."[13] He also described many curious creatures, including the zebra (see figure 19) and the noble mother elephant, whose child fell into a pit and was unable to escape; she buried him alive in the pit, "to the ende that the hunteres shoulde not enjoy her calfe, choosing rather to kill it her selfe, then to leave it to the mercie of the cruell huntsemen."[14]

The most significant contribution made by English descriptive geographers was the collection of tales of exploration and adventure. Although this genre was first seriously developed by Europeans such as Giovanni Battista Ramusio, whose *Delle Navigationi et Viaggi* (3 volumes, Venice, 1554–59) recorded, inter alia, Marco Polo's adventures and Columbus's discoveries, it might be said to bring to mind the earlier, if more fanciful, accounts of the supposed Englishman Sir John Mandeville (see figure 18, above).[15] In a genre dominated by such giants as Ramusio, Peter Martyr,

13. Hartwell, *Report of the Kingdom of Congo,* 19.

14. Ibid., 67.

15. *The Voyages and Travels of Sir John Mandeville* was written in about the middle of the fourteenth century, probably by Jean d'Outremeuse. See Mandeville, *Travels,* introduc-

Joseph de Acosta, Jan Huygen van Linschoten, and Theodore and Johann Theodore de Bry,[16] Richard Hakluyt and Samuel Purchas take their places as significant and popular contributors to the European field of collections of voyages and new lands.

When Thomas More encountered Raphael Hythlodaeus and questioned him concerning that hitherto unknown island of Utopia, he was describing the vocation later taken up by Richard Hakluyt. Hakluyt, a Master of Arts from Christ Church, Oxford, and a diplomat, spy, and churchman, spent his life interviewing mariners, navigators, and travelers and collecting stories of new countries, hair-raising adventures, and sea dramas.[17] His great work, *The Principal Navigations, Voyages, Traffiques and Discoveries of the English Nation* (1598–1600), enumerated the voyages and discoveries of the English in the Americas and the East.[18] Hakluyt's collection, seen with Sidney's *Arcadia* and translations of Camden's *Britannia* as the cornerstone of both the mature English language and nascent English patriotism, encouraged Britons to see themselves as leaders in the explora-

---

tion by Jules Bramont, vii. It is largely an imaginary account of travels through the Middle East to India and Africa, deriving from Pliny and the Romance of Alexander. Its putative author claims to have made the pilgrimage to Jerusalem and then an extended journey eastward. From this book come tales of dog-faced men, men with heads in their chests (cited by Othello and in *The Tempest)*, and people with one large umbrella foot. Mandeville was very popular in the late Middle Ages and preserved that popularity well into sixteenth-century Europe. See Moseley, "Availability of *Mandeville's Travels*." The *Travels* went through numerous editions in the sixteenth and seventeenth centuries, as well as appearing in epitome in Purchas 1625, Part 2:128–57; interestingly enough, however, the book only appears on one university book list, that of St. John's College, Oxford (St. John's Library Benefaction Book). This could imply that the book was only popular in nonuniversity circles, or simply that it was read to the point of disintegration.

16. Petrus Martyr Anglerius, *De Orbe Novo Decades. 8 Decades* (Alcalà de Henares, 1530); Joseph de Acosta, S.J., *De Natura Novi Orbis, libri duo, et de promulgatione Evangelii, apud barbaros . . . sive de procuranda Indorum salute libri sex* (Salmantica, 1589); Jan Huygen van Linschoten, *Itinerario. Voyage ofte Schipvaert, van J.H.v.L. naer Ooost ofte Portugaels Indien* (Amsterdam, 1595–96), in Latin as *Navigatio ac itinerarium J. H. Linscotani in Orientalem sive Lusitanorum Indiam* (The Hague, 1599), or the English version by Wolfe (1598); T. de Bry, *America, partes 1–13* (Frankfurt, 1590–1634); Johann Theodore de Bry (son), *India Orientalis, partes 1–10* (Frankfurt, 1598–1613). For a discussion of Hakluyt's place in the ranks of these great compilations, see Parks, "Tudor Travel Literature."

17. The authoritative biography of Hakluyt, though now old, continues to be Parks, *Richard Hakluyt and the English Voyages*. For more modern treatments of this important geographer, see E. G. R. Taylor, "Richard Hakluyt" and *Original Writings and Correspondence;* Lynam, ed., *Richard Hakluyt and His Successors;* and Quinn, ed., *Hakluyt Handbook*.

18. Hakluyt, *Principal Navigations, Voiages and Discoveries of the English Nation* (1589) was largely concerned with explorations of America. His later work, too much enlarged and revised to be seen as an edition of the first, dealt with exploration of the whole world: *Principal Navigations, Voyages, Traffiques, and Discoveries of the English Nation* (vol. 1, 1598; vol. 2, 1599; vol. 3, 1600). The addition of the word *Traffiques* in the later title indicates the greater stress on trade and commerce. Subsequent references, unless otherwise stated, are to the latter work.

tion of and trade with the wider world.[19] Rather than offering tales of the valor of the Spanish or Dutch, *Principal Navigations* supplied the English with reflections of themselves. As Hakluyt claimed, in his dedication to Lord Charles Howard, Lord of the Admiralty, "I began at length to conceive, that with diligent observation, some thing might be gathered which might commend our nation for their high courage and singular activitie in the Search and Discoverie of the most unknowen quarters of the world."[20]

Hakluyt perused historical sources for long-forgotten English voyages, and in contemporary reportage he used the words of people who had been there, a style of reporting that lent great verisimilitude to his stories and allowed his readers to see the real passion and poetry, as well as the hard-nosed business sense, of England's travelers.[21] His book let the English mariner or merchant develop a self-consciousness of his role in the world and encouraged him to risk life and limb for the glory of queen, country, and purse. Thus, Hakluyt's book provided an ideological foundation for a descriptive study of the wider world. It combined an energetic and often dramatic literary style, and patriotic and pragmatic pride, with a huge portion of fascinating descriptive and navigational information.

Hakluyt's great work encouraged the English to see themselves as separate from the Continent and the rest of the world in two different ways. First, he stressed the primacy of English exploits and contacts, beginning with Arthur's voyage to Britain and including the trade of Britons in the Mediterranean "before the incarnation of Christ" and the "ancient trade of English marchants to the Canarie Isles, Anno 1526," among others.[22] Second, Hakluyt stressed the dissimilarities between the English and other peoples, by describing strange customs and practices. For example, the Lappians and "Scrickfinnes" "are a wilde people who neither know God, nor yet good order . . . they are a people of small stature, and are clothed in Deares skinnes, and drinke nothing but water, and eate no bread but flesh all raw."

From another part of the world: "The king of Persia (whom here we call the great Sophy) is not there so called, but is called the Shaugh. It were there dangerous to cal him by the name of Sophy, because that Sophy in the Persian tongue, is a begger, & it were as much as to call him, The great begger."

Finally, concerning China: "so great a multitude is there of ancient and

19. For a discussion of the development of the English language, see C. Barber, *Early Modern English;* and Jones, *Triumph of the English Language.* Helgerson deals with the relationship between language and patriotism, the creation, as he calls it, of "The Kingdom of Our Own Language" (*Forms of Nationhood,* 1 and esp. ch. 1).

20. Hakluyt, *Principal Navigations* (1598–1600), I: sig. *2a.

21. Ibid., I: sigs. *4a, *5a.

22. Ibid., I:1; II, part I:1, taken from Camden's work; II, part II:3. See also II, part II:96.

grave personages: neither doe they use so many confections and medicines, nor so manifold and sundry wayes of curing diseases, as wee saw accustomed in Europe. For amongst them they have no Phlebotomie or letting of blood: but all their cures, as ours also in Japon, are atchieved by fasting, decoctions of herbes, & light or gentle potions."[23]

By stressing odd and foreign attributes, he drew a distinction between the "other" and the familiar English reader. Although Hakluyt deals with these strange customs with a light hand, not explicitly condemning them as inferior, the result is a ranking of different cultures beneath the English. Eating raw flesh is clearly, in Aristotelian terms, a category error (meat is, by definition, cooked) and condemns the "wilde" Lappians to deeply inferior status. Insulting the Shah of Persia through an incorrect appellation shows Europeans to be making a joke at his expense, dangerous though this may have been. And the Chinese lack of modern medical techniques, though passed over lightly, indicates to Hakluyt and his readers that this great civilization had not yet attained the technological and scientific heights of England.

Hakluyt's book also reveals an interesting movement toward a "Baconian" or collecting methodology in the human and descriptive sciences. The collection of useful facts—insignificant when taken individually but amalgamated into a complete worldview that was greater than the sum of its parts—corresponds to Francis Bacon's later methodology of tabulation.[24] The many lists of foreign phrases, useful for the traveler and trader, demonstrated an early consciousness of comparative languages (see figure 20), while descriptions of natives and their customs introduced concepts of anthropology.[25] Botany, zoology, and natural history were not forgotten, with descriptions of native plants and animals, and Hakluyt could even be said to have been contributing to economic theory with his faithful rendering of merchants' reports and commodities pricing.[26] These human

23. Ibid., for Lappians, I:233; for Grand Sophy, I:397–98; for China, II, part II:88.

24. Francis Bacon, in the *Novum Organum* (1620), set up a method of tables of presence, absence, and comparison by which one could distance discovery from the idols that distorted them.

25. For example, on languages: "Divers words of the language spoken in New France, with the interpretation thereof" (Hakluyt, *Principal Navigations* 1598–1600, 3:211, 231); "The interpretation of certeine words of the language of Trinidad annexed to the voyage of Sir Robert Duddeley" (3:577–78). For anthropological material, e.g., "Certaine letters in verse, written out of Moscovia, by M. George Turbervile, Secretary to M. Randolfe, touching the state of the Countrey, and maners of the people" (1:384); "Observations of the Sophy of Persia and of the Religion of the Persians" (1:397).

26. On native plants and animals, e.g., "A testimony of Francis Lopez de Gomara, concerning the strange crook-backed oxen, the great sheepe, and the might dogs of Quivira" (ibid., 3:308); "A Notable description of Russia—The native commodities of the Countrey" (1:477–79). And on topics of economic interest, e.g., "The letters of the Queenes Majestie

Here followeth the language of the countrey, and kingdomes of *Hochelaga* and *Canada*, of vs called *New France*: But first the names of their numbers.

Secada	1	*Indahir*	6
Tigneni	2	*Aiaga*	7
Hasche	3	*Addigue*	8
Hannaion	4	*Madellon*	9
Ouiscon	5	*Assem*	10

Here follow the names of the chiefest partes of man, and other words necessary to be knowen.

the Head	*aggonzi*	a Womans member	*castaigne*
the Browe	*hegueniascon*	an Eele	*esgueny*
the Eyes	*higata*	a Snaile	*vndeguezi*
the Eares	*abontascon*	a Tortois	*heuleuxima*
the Mouth	*esahe*	Woods	*conda*
the Teeth	*esgongay*	leaues of Trees	*hoga*
the Tongue	*osnache*	God	*cudragny*
the Throate	*agonhon*	giue me some drink	*quazahoaquea*
the Beard	*hebelim*	giue me to breakfast	*quaso hoa quascaboa*
the Face	*hegouascon*	giue me my supper	*quaza hoa quatfriam*
the Haires	*aganiscon*	let vs goe to bed	*casigno agnydahoa*
the Armes	*aiayascon*	a Man	*aguehum*
the Flanckes	*aissonne*	a woman	*agruaste*
the Stomacke	*aggruascon*	a Boy	*addegesta*
the Bellie	*eschehenda*	a Wench	*agniaquesta*
the Thighes	*hetnegradascon*	a Child	*exiasta*
the Knees	*agochinegodascon*	a Gowne	*cabata*
the Legges	*agouguenehonde*	a Dublet	*caioza*
the Feete	*onchidascon*	Hosen	*hemondoha*
the Hands	*aignoascon*	Shooes	*atha*
the Fingers	*agenoga*	a Shirt	*amgoua*
the Nailes	*agedascon*	a Cappe	*castrua*
a Mans member	*ainoascon*	Corne	*osizi*

Bread

FIGURE 20. Table of Indian words by Jacques Cartier, from Richard Hakluyt, *Principal Navigations*, vol. 2 (London, 1599), p. 231. (BL 212.d.3. By permission of The British Library.)

sciences are found in nascent form in this huge tabulation of geographical description. Hakluyt's massive work can thus be seen as helping to develop a methodology of data collection, as well as an ideology that helped establish this technique of collection and its product as morally neutral. In other words, by collecting information in a putatively objective manner (simply recording what others reported), Hakluyt presents the information itself

written to the Emperour of Russia, requesting licence and safe-conduct for Anthonie Jenkinson" (1:338); "A note of all the necessary instruments and appurtenances belonging to the killing of the Whale" (1:413).

as neutral.[27] This was clearly not the case, since the very essence of Hakluyt's work was imbued with values of the supremacy of England and Protestantism, and of the power of the Old World over the New. *Principal Navigations* provides an early example of, and perhaps an inspiration for, the methodology of inductive tabulation and the underlying message of hierarchical power that went hand-in-hand with such a technique.

Samuel Purchas, Hakluyt's literary and spiritual successor, developed a very different emphasis in his great collection of English voyages. Purchas worked as Hakluyt's assistant during the latter years of Hakluyt's life and purchased Hakluyt's unpublished manuscripts after his death.[28] He used these manuscripts and other geographical research in two great works: *Purchas His Pilgrimage* (1613) and *Hakluytus Posthumus: or, Purchas His Pilgrimes* (1625).[29] Both were highly moralistic, almost Puritan in tone, and were aimed at the armchair traveler and island-bound country gentleman rather than at the practical navigator or merchant.[30] Indeed, Purchas claimed that travel might be injurious to the naive Protestant youth, while he could read about other countries with no danger. This change in intended audience reflects the growing popular interest in travel literature, as well as increased stress on religious controversies as the century progressed. Purchas told tales of new discoveries and described peoples, customs, and natural settings in great detail; he tended, however, to dwell on "Mans diversified Dominion in Microcosmicall, Cosmopoliticall, and that spirituall or heavenly right, over himselfe and all things, which the Christian hath in and by Christ" and "the diversities of Christian Rites and Tenents in the divers parts of the world," rather than dealing with rates of exchange or the feasibility of a northwest passage, as Hakluyt had done. He also stressed the treachery of other races toward the English and the victories over these lesser people by his superior Protestant countrymen.[31] He criti-

27. Ibid., I: preface, xxiv. This question of the development of "objectivity" in science has been discussed in the context of experimentation both by Shapin and Schaffer in *Leviathan and the Air Pump;* and by Daston and Galison in "The Image of Objectivity." My argument takes this into the realm of observations "from nature."

28. For an analysis of the relationship between Hakluyt and Purchas, see Steele, "From Hakluyt to Purchas."

29. *Purchas His Pilgrimage: or, Relations of the world* went through four editions: 1613, 1614, 1617, and 1626. The 1626 edition is often seen as vol. 5 of *Hakluytus Posthumus: or, Purchas His Pilgrimes* (1625) but is actually a completely separate work. *Hakluytus Posthumus* was published in four volumes and has been printed as twenty volumes in the definitive Glasgow 1905 edition. It is here cited as Purchas 1625. His third work, *Purchas His Pilgrim or Microcosmos: or, The Historie of Man* (1619), a moral rather than a geographical treatise, was highly condemnatory of the degeneracy of man and was never reprinted.

30. Purchas 1625, 1: preface, sig. ¶5a.

31. Ibid., 1.1:6, 147 (quotations). On treachery and triumph, see, e.g., "A true and briefe discourse of many dangers by fire, and other perfidious treacheries of the Iavans" (1.3:167–70); also 1.3:156, 179, 206, 251–53; 1.10:1853.

cized the Dutch as well, not surprising in an increasingly competitive economic atmosphere.[32] Purchas's pedantic prose is far less captivating than the more lively renditions contained in Hakluyt's compilation, and he succeeds best where he follows Hakluyt in allowing the explorers and travelers to speak for themselves. While Purchas continued to encourage the self-promotion of the English nation, this movement relied less on his original work than on the impetus already supplied by Hakluyt. What Purchas contributed to English descriptive geography was a growing Protestant bias and an increasing belief in the ability and need of the English to achieve a Protestant hegemony over the pagan and Catholic world.

The other main scholarly activity of English descriptive geographers involved the translation of modern Continental works. Between 1580 and 1620, twenty-eight books of descriptive geography were translated into English. Hakluyt sponsored or encouraged many of these efforts and himself translated a large number of tales, both for separate publication and for inclusion in *Principal Navigations.*[33] These translations were from Spanish, French, Italian, and Dutch sources and included descriptions of the West Indies and America, the East Indies and China, and Europe itself.[34] Most were translations of firsthand accounts of discovery, exploration, trade, and battle, although some European scholars of more theoretical or synthetic political descriptive geography were also included. This interest in European accounts indicates an awareness of the supremacy of other nations in exploration and trade and a desire to learn about new lands, both for their intrinsic interest and in order to better these *conquistadores* at their own game.

Most original English descriptive geography, aside from these collections and translations, consisted of travel literature. The vast majority of such English travel texts described the New World, especially Virginia. Many were promotional, attempting to attract colonists and investors to

32. Andrews, *Trade, Plunder and Settlement,* 364.

33. Hakluyt translated, inter alia, Peter Martyr, *De Orbe Novo* (Paris, 1587); René de Laudonnière, *A notable historie containing foure voyages made by certayne French captaynes unto Florida* (London, 1587); T. de Bry, *The true Pictures and Fashions of the People of Virginia* (London, 1590); Don Ferdinando de Soto, *Virginia richly valued, By the description of the maine land of Florida, her next neighbour* (London, 1609).

34. These included John Pory's translation (encouraged by Hakluyt) of Leo Africanus, *A Geographical Historie of Africa* (London, 1600); Martin Basanier's translation of de Espeio, *Historie des terres nommées le nouveau Mexico* (London, 1586); Thomas Danett's translation of *Description of the Low Countries by Ludovic Guiccardini gathered into an epitome* (London, 1591); Edward Grimestone's translations of José de Acosta, *The naturall & morall historie of the East & West Indies* (London, 1604) and *The Low Country Commonwealth, by Jean François le Petit* (London, 1609); Thomas Washington's translation of Nicholas de Nicolay, *The navigations, peregrinations, and voyages made into Turkey* (London, 1585); and William Phillip's translation (at the urging of Hakluyt, procured by John Wolfe) of Linschoten, *Discourse of voyages,* described in the introduction above.

the new settlement;[35] others were apologias for the strife-ridden colony.[36] Thomas Harriot's contribution, *A Briefe . . . Reporte of . . . Virginia,* was the most descriptive, although it was also imperial in tone, and rapidly became widely known beyond England by its inclusion in de Bry's great compilation.[37] (See figure 21.) The second and smaller group of travelers' tales described Europe. These were largely political in nature, commenting on government and mores in the various countries visited. The most frequently described country was Russia, significant for English mercantilism at this time. After the first disastrous voyage of Sir Hugh Willoughby and the somewhat more successful one of Sir Richard Chancellor, the English maintained significant contacts with the court of Ivan the Terrible, and through him were often able to circumvent the Mediterranean trade embargo of Suleiman the Magnificent and his successors. These adventures were frequently described in highly evocative prose.[38] As the examples in chapter 2 above demonstrated, the Middle East also held out lures of rich trade, exotic culture, and traditional peregrination routes. This region provided a third area of emphasis for original descriptive geography. Many a traveler encountered significant hardship as well as stunning beauty, as reported by such men as Sir Anthony Sherley, George Sandys, and that unlucky Scot, William Lithgow. Travel literature, by and for English readers and writers, thus stressed the New World, Europe (including Russia), and the Middle East, areas where England had significant financial involvement.

35. These included John Brereton, *A briefe and true relation of the Discoveries of the North Part of Virginia . . . made this present year 1602. With annexed: A treatise containing inducements for planting,* by Edward Hayes (London, 1602); Robert Johnson, *Nova Britannia. Offering most excellent fruites by planting in Virginia* (London, 1609); and Captain John Smith, *A description of New England* (London, 1616).

36. These included John Smith, *A True Relation of such occurrences as hath happened in Virginia* (London, 1608); William Symonds, ed., *The proceedings of the English colonie in Virginia* (Oxford, 1612); Rev. Alexander Whitaker, *Good newes from Virginia* (London, 1613).

37. De Bry, *America. Pars I.* See also M. Campbell, "Illustrated travel book."

38. Accounts of these Russian exploits included Jerome Horsey, *Coronation of the Emperor of Russia* (MS, printed by Hakluyt, *Prinicipal Navigations* 1598–1600, 1:466–70); Purchas 1625, II:70–74; Edward Webbe (servant at Moscow to Anthony Jenkinson), *The rare and most wonderful thinges which Edward Webbe an Englishman borne, hath seene and passed in his troublesome travailes* (London, 1590); Jenkinson's Map of Muscovy (1562) (unique copy, Wroclaw University Library; reissued in reduced form by Ortelius, *Theatrum Orbis Terrarum,* 1570), which could be considered a graphic form of descriptive geography; and Giles Fletcher, *Of the Russe Commonwealth,* BL Lansdowne MS 60 (London, 1591). There is some doubt about this date. According to E. G. R. Taylor (1934, 367), "Written 1583 and temporarily suppressed" until 1591. According to the *DNB,* the book was "suppressed and partially printed only in Hakluyt and Purchas." Both Horsey and Fletcher were reprinted in Bond, ed., *Russia at the Close;* a more recent facsimile of Fletcher was printed in G. Fletcher, *Of the Russe Commonwealth. Facsimile.*

FIGURE 21. Map of Virginia, from Thomas Harriot, *A Briefe and true Report of the New found land of Virginia,* in Theodore de Bry, *America Par I* (Frankfurt, 1590), facing sig. A1a. (BL G.6834. By permission of The British Library.)

Descriptive geography, as these examples show, was based largely on personal experience, an emphasis in which it differed from mathematical geography. Descriptive geography was also much more accessible than mathematical geography, lending itself to the enthusiastic amateur rather than only the highly trained specialist. In fact, descriptive geography, in line with its emergence as a "collecting science," rejected specialization in favor of the work of generalists. In seeking the place of descriptive geography in the universities, then, we should not expect to find small coteries of specialized scholars but rather a more general dissemination of descriptive geographical ideas. Indeed, the practitioners of descriptive geography wandered far afield, figuratively and literally, from the university atmosphere, and therefore an examination of descriptive geography at the universities must concentrate on the reception of geographical ideas and the instillation of basic geographical beliefs, instead of seeking there the mature development of descriptive geography.

Commonplace Books and Notebooks

Commonplace books supply much useful information about the types of geographical information more interesting or useful to university students. Sir Julius Caesar's commonplace book, whose title page has already been examined (figure 5, above), provides some indication in its printed running heads of an expected emphasis on both mathematical and descriptive geography topics. Among the subjects included in the running heads are both mathematics and geography, clearly as much a part of gentlemanly erudition as topics such as rhetoric and grammar. Indeed, there are four pages allotted to "Geographia, Cosmographia, Cosmographi Locorum situs. Distantiae Dimensiones."[39] A further two pages contain space for material on "Chronologia. Geographia. Historia. Historiographi Exempla,"[40] and one more for "Naves. Navigia. Nautica ars. Navigatio. Instrumenta navium, huca pertinentia."[41] This division of geography into cosmography, descriptive geography, and navigation is extremely instructive, demonstrating the distinction between mathematical and descriptive geography that has already been established in this analysis. The generous allotment of pages for these two branches of geography, and for the applied branch of navigation, demonstrates the arrival of this study as a legitimate concern of the student and scholar.

Caesar appears to have kept this commonplace book throughout his life; his first entry was made while at Magdalen Hall, Oxford, in 1577,

39. BL Add. MS 6038, ff. 250a–251b.
40. Ibid., ff. 267a, b.
41. Ibid., f. 408.

when he was nineteen years old, and the last entry is dated 1636, shortly before his death.[42] In it, he recorded a lifetime of citations, quotations, and ideas. He seems to have had relatively little to say on the pages devoted to theology and mathematics, but the sections of the notebook devoted to geography and navigation are closely filled. Indeed, he added several manuscript pages with the running heads "Cosmographia, Geographia."[43] He cited all the important geographical authors, including Ptolemy, Mercator, Strabo, and Pliny. He discussed navigation in terms of the care and design of ships, and included chorography in such entries as "The Singularities of England."[44] His descriptions of other countries were usually drawn from or referred to published authorities, although sometimes he recorded his own observations: "I was ownce in Italie my selfe: but I thanke god, my abode there was but 9. daies; and yet I sawe in that little time in the citie of Venice, more libertie to sinne, than ever I heard tell of in our noble citie of London in 9 yeare. I sawe, it was there as free to sinne, not onely without all punishment, but also without anie man's marking, as it is free in the citie of London, to chouse without all blame, whether a man hast to weare shoe or pantoche."[45]

This compilation of geographical information would have been extremely useful to Caesar in his long public career, especially as Judge of the Admiralty. It is striking, however, that he began this book while at university, indicating that part of his interest in geographical topics may well have begun there.

The geographical material written in other student notebooks falls into two main categories. The first was basically chorographical. This included local history, genealogies of important local families, and lists of place names, sometimes gleaned from ancient authors. The second category, by far the more extensive, was tales of strange lands and wondrous sights. These included diaries of early Grand Tours to Europe and some political reporting of European countries but by and large consisted of sensational descriptions of unusual places and customs.

The most common geographical descriptions that captured the imagination of undergraduates were tales (tall or otherwise) of distant lands. These were sometimes stories of discoveries of new lands and people, often the search for elusive semimythical characters, or even, closer to home, tales of European and Middle Eastern atrocities.

Good Protestants, as most of these undergraduates were, were naturally interested in tales of the persecution and martyrdom of other Protestants

42. L. M. Hill, *Bench and Bureaucracy,* 6.
43. BL Add. MS 6038, f. 348a.
44. Ibid., ff. 409b, 250a.
45. Ibid., f. 250a.

in near and distant lands. Lionel Day, a fellow of Balliol and son of John Day the printer, compiled a commonplace book in about 1603 in which he copied descriptions of a most grisly nature. "A poore naile maker," wrote Day, copying from *The History of France,* "because he would not blasphem God and give himself to ye Devill they pluckt him by ye eares about his shopp at last putt his head upon ye anvile and beat out his braines with a hammer."[46] In a similar vein, an unknown Cambridge student in the early seventeenth century copied extensively from William Lithgow's *The Horrid Barbarity of the Spanish Inquisition exemplify'd: In the Inhuman usage of Mr. William Lithgow a Scotch Gentleman by the Governour and Inquisitors of Malaga.*[47] In this excerpt, the description of the rack, partial hanging, and the breaking of Lithgow's limbs was extremely graphic. Even today, this work is chilling; seventeenth-century readers, with all of their prejudices and predilections, would have been horrified and yet found it strangely compelling.

Many undergraduates were interested in the wonders of Britain and of other lands. William Smith (B.A. Oxon.)[48] wrote *The Particular Description of England* in 1588. In it, he listed the seven wonders of England, which included the baths of Bath, "Stonehedge," a dripping well in Yorkshire that turned things to stone, Saltpits in Cheshire, and London Bridge.[49] Edward Pudsey (B.A., Oxon.),[50] in his commonplace book, described the palace of the "Great Chan" in China[51] and, closer to home, remarked that in Scotland there were "three things worthy note. 1. Theere in Lennox is a lake called Lowmond are finlesse fishes of a good taste. 2. Theere are Sheeting glas yt with the wynd are driven to and fro. 3. And in Agravia (as is reported) groweth a stone which beeing put to straw or stubble will kindle and set the same on fyre."[52] In Greece, "The river Melas makes whyte sheep blacke and ye river Cephis blacke sheep whyte: The sea called Euripus did in 24 houres seven tymes ebb and flow, which when Aristotle could

46. Oxf. Bodl. MS Add. A 115, f. 31b.

47. Cambridge University Library (CUL), Add. MS 3308. Probably from Lithgow, *Most Delectable and true Discourse,* reprinted in Purchas 1625 as "Relation of Travels of W. Lithgow Scot in Candie, Greece, the Holy Land, Egypt and other parts of the East" (I:1831–47).

48. BL Sloane MS 2596, frontispiece, claims this degree. Three men in J. Foster, *Alumni Oxonienses,* might correspond to this individual, who was later created "Rouge Dragon," namely: (1) B.A. 1567, n.s.; (2) Corpus Christi, scholar 1564, B.A. 1568, M.A. 1572; (3) St. Mary Hall, B.A. 1572, M.A. 1575. William Smith went on to produce a series of survey maps of England near the end of Elizabeth's reign. See P. Barber, "England II," 64.

49. BL Sloane MS 2596, ff. 12b–13.

50. Not found in J. Foster, *Alumni Oxonienses;* or Clark and Boase, *Register of the University of Oxford.*

51. Oxf. Bodl. MS Eng. poet. d.3, f. 62b.

52. Ibid., f. 69b.

not understand ye cause therof for shame and anger cast him self into ye said sea, whereuppon it was said yt becaus Aristotle cold not comprehend Euripus, Euripus had comprehended Aristotle."[53]

As well as demonstrating that undergraduate humor was as heavy-handed then as now, this shows that the unusual or exciting was what caught his attention.

Pudsey's interest was also piqued by an elusive personage often sought in the sixteenth and seventeenth centuries: Prester John. As Pudsey tells us, "he deemeth his pedegree from Miluch ye soone of Solomon and Saba. In his letters to ye king of Portingall uppon [condition] that hee wold wage warre against ye infidells hee offered him a million of gold and a million of men with provision accordingly."[54]

Prester John was that legendary Christian ruler, living somewhere in the East, who had promised aid to the Crusaders against the infidel.[55] Although such help never materialized, he and his Christian subjects were long sought, first in the Middle East, later in Africa (usually in Ethiopia), and later still in India. Abraham Hartwell, for instance, expressed the earnest "wish with all my hart, that not onely *Papists* and *Protestants,* but also all *sectaries* and *Presbyter-Johns men* would joyne all together both by word and good example of life to convert the *Turkes,* the *Iewes,* the *Heathens,* the *Pagans,* and the *Infidels.*"[56] Richard Cocks, in 1612, claimed that the people in the East Indies "are naked, Moors and Mahometans of religion, yet subject to Prester John, as we were given to understand."[57] Even the eminent scholar Richard Verstegen wrote to Robert Cotton in 1617 that "I have found something exciting, that is, a Catalogue of the books contained in the Library of the Emperor of Abissinia, vulgarly and corruptly called Prester John."[58] In 1625, Purchas continued to report stories concerning Prester John.[59] Although the later reports showed an element of skepticism, for Oxford and Cambridge undergraduates there was none, and Prester John continued to live in their minds well into the seventeenth century.

Another source of rumor and speculation came from the palace and country of the Grand Sultan, especially Suleiman the Magnificent. European nations, including England, had ambiguous relationships with "The

53. Ibid., f. 70.

54. Ibid., f. 62b.

55. Mandeville claimed to have visited Prester John in his city of "Suse" (*Travels,* Books 97, 98). This began a tradition that persisted with very little factual support for several centuries. See Gumilev, *Searches for an Imaginary Kingdom.*

56. Hartwell, *Report of the Kingdom of Congo,* sig. **1b.

57. "Richard Cocks to the Right Worshipful Sir Thomas Smith, Knight, Governor of the East India Company, . . . per Cptn Gabriel Towerson, in the Hector, whom God preserve. Bantam 12 of Jan, 1612" (Danvers, ed., *Letters received by the East India Company,* 217).

58. Oxf. Bodl. MS Smith 41, 105.

59. "Relations of Ethiopia, sub Ægypto, and Prester John" (Purchas 1625, I:1127–31).

Turk," since he provided, alternatively, a common enemy as the Infidel and an ally for European nations against one another. In either case, the lure of Constantinople was great, and many undergraduates whetted their appetites with such books as Richard Knolles's *Generall Historie of the Turks* (1603). The most spectacular English diplomatic journey to the Ottoman sultan, now Mohammed III, took place in 1598–99, when Thomas Dallam, an Oxford organ maker, delivered to the sultan a specially constructed organ as a gift from Elizabeth. Dallam wrote a manuscript diary of this trip,[60] which might have been read by students when Dallam returned to Oxford. Dallam described his voyage, which began adventurously with pirate attacks and included witnessing a Greek wedding. Its most exciting episode was Dallam's visit to the Grand Seraglio. His was an unusual experience, because men, aside from the eunuch attendants, were never admitted into the Grand Seraglio's innermost chambers. Dallam detailed the deaf-mutes in attendance, the dwarfs, and the concubines.[61] This story was doubtless very popular among undergraduates.

The newly discovered empire of Japan was another source of delight to English students. The East India Company had set up a factory in Japan in the 1590s, and the factors sent back letters telling of the wonders of that island kingdom.[62] In many ways, Japan was an extremely difficult country for Europeans to accept; it was a military empire with a culture at least as advanced as that of the Europeans, and the Europeans were forced to submit to "pagan" rule in ways that would not have been tolerated elsewhere. The Japanese were ruthless in their punishment of wrongdoing, and it was this brutality, viewed by the English with horror, but with a certain grudging respect, that was reported home. The letters home, officially to company representatives and less officially to friends in Cambridge and London, told tales of hands cut off for small infractions, crucifixions for theft, boiling in oil for greater crimes.[63] Even in an age of brutality, these punishments were seen as draconian. But, as the English factor Richard Cocks pointed out, the streets of Edo were safe to walk on at night, something that could not be said for London in the same period.[64]

60. BL Add. MS 17,480, published by the Hakluyt Society in 1893 in Bent, ed., *Early Voyages and Travels.*

61. Ibid. (Bent 1893), 70.

62. Many of these letters appear in Danvers, *Letters Received by the East India Company;* as well as in East India Company Papers 1611–18, BL Eg. MS 2086; India Office Marine Records, L/MAR/A/ I,II, etc.; Kent Record Office, Sackville MS ON 6014; PRO SP 14/96, 96.

63. E.g., William Adams, letter of Oct. 23, 1611, to English Merchants in Java (Danvers, *Letters Received by the East India Company,* 151).

64. Richard Cocks, 1616 diary entry (Cooper, *They Came to Japan,* 14). Adams also makes this point in "William Adams from Japan. To my assured good friend Augustin Spalding Laus Deo, written in Japan, in the island of Firando, the 12th of Jan 1613" (Dan-

Besides being impressed by the Japanese judicial system, Richard Cocks and the pilot Will Adams[65] were struck by the size of the palaces and armies of the emperor and especially the shogun of Japan. Cocks reported to Sir Thomas Wilson in 1617 that "Shongo Samme," the new shogun, had an entourage of more than a hundred thousand men,[66] and wrote to James I in 1619 that the "King's pallace [by which he meant the Shogun's palace in Edo] . . . [is] farr bigger then your Ma'tie's citty of York," to which the King added the marginal note, "Impossible!"[67]

For many undergraduates, these lurid tales of foreign wonders mingled with their study of history, and it is often difficult to distinguish where geography ends and history begins. Humphrey Lhuyd, a Welsh antiquarian and geographer probably from Jesus College, Oxford, blended geographical description with historical in *Mona, the Isle of the Druids, restored to its Antiquity*.[68] Lhuyd wished to establish the identity of the island known in antiquity as Mona; he demonstrated that the island described by Caesar and Tacitus could have been one of two islands of this name.[69] Polydor Vergil, Lhuyd explained, had chosen the wrong location, seen clearly if Ptolemy were read with some understanding of the Welsh tongue.[70] Thus, Lhuyd combined a historical analysis of the texts with linguistic and physical geographical arguments. With a similar blending of topics, Stephen Batman, a student at Oxford, copied portions of John Dee's *Generall and Rare Memorials pertayning to the Perfect Art of Navigation* into the same notebook in which he had recorded "Thomas Wolsey late Cardinall, his lyffe and deathe" by George Cavendish.[71] Here again is that

vers, *Letters Received by the East India Company*, 212). See also Needham, who comments at length on mutual perceptions of atrocities in East-West relations in *Science and Civilisation in China*.

65. James Clavell fashioned the character Blackthorn on Will Adams in his novel *Shogun* and clearly read Adams's log book, preserved on rice paper at the Bodleian (Oxf. Bodl. MS Savile 48); transcribed with notes by A. J. Farrington, 411 in Japan Series, India Office Marine Records; printed by Purnell, ed., *Log-Book of William Adams*.

66. Kent Record Office, Sackville MS ON 6014, transcribed by A. J. Farrington, 227 in Japan Series, India Office Marine Records, BL, 2.

67. PRO SP 14/96, 96, 1; transcribed by India Office, BL. My thanks to A. J. Farrington for pointing this letter out to me.

68. BL Add. MS 14,925, ff. 137–150b. Dated 1568. First printed in John Price, *Historia Britannia Defensio* (London, 1573), but best known from Ortelius, *Theatrum Orbis Terrarum*, sigs. a^a–b^b.

69. BL Add. MS 14,925, f. 139b.

70. Ibid., f. 140b.

71. Oxf. Bodl. MS Douce 363. "Internal evidence shows that his 'Life of Wolsey' was written in 1557; but it was not published, for the accession of Elizabeth brought forth changes, and it was dangerous to publish a work which necessarily spoke of disputed questions and reflected on persons who were still alive" (*DNB*, 9:346). First published in somewhat garbled form, for political purposes, as *The Negotiations of Thomas Woolsey, the Great*

close relationship between geographical and historical sources, enriching both disciplines, though making clear-cut distinctions virtually impossible.

Finally, some students filled pages of their notebooks enumerating world place names. For example, Brian Twine, the Oxford historian, noted down all the English geographical names found in Ptolemy.[72] Likewise, William Patten, who may have been a Cambridge undergraduate, supplied an alphabetical listing of world place names and their etymologies.[73] Patten also referred to Ptolemy, as well as to Pliny and Münster. This small subsection of descriptive geography was, in fact, motivated by classical rather than modern models of geographical methodology.

Student notebooks were often replete with geographical information, indicating a number of university students interested in the world displayed through descriptive geography. Students were interested in foreign and exotic locales, as well as unusual and bizarre customs. What attracted these students were aspects of the world that most demonstrated the contrast between their relatively secure, if dull, world and the exciting and dangerous world of the other.

Descriptive Geography Books at the Universities

A clearer picture of descriptive geography emerges with an examination of the books owned by Oxford and Cambridge students in the period. Descriptive geography book titles were present in the book lists of a large number of Oxford and Cambridge colleges. More accessible to the average undergraduate than mathematical geography, the descriptive branch of the discipline appears to have attracted especially wide attention of students at Corpus Christi, Christ Church, and New College, Oxford, and at Peterhouse and St. John's College, Cambridge. Significant numbers of descriptive geography books also appeared on book lists connected with St. John's, Oriel, All Souls, and Merton colleges, Oxford, and with Corpus Christi, Trinity, Gonville and Caius, and Sidney Sussex colleges, Cambridge, indicating the relatively broad attraction of descriptions of the expanding world (see tables 2 and 4, above).

Corpus Christi, Christ Church, and New College, Oxford, were all strongholds for the study of descriptive geography. All three book lists had a particularly large selection of books describing exploration of the New

---

Cardinall of England, containing his life and death. Composed by one of his owne servants, being his gentleman-usher [i.e., George Cavendish] (London, 1641). This commonplace book clearly used the manuscript version of this work, and its inclusion indicates a somewhat radical stance on the part of the student.

72. Oxf. Bodl. MS Wood D.32, ff. 163–73.

73. CUL Dd.11.40, ff. 2–43.

World, often gleaned from Continental sources.[74] The Christ Church list included several copies of *Purchas His Pilgrimage* (1613), and both Corpus Christi and New College boasted Hakluyt's *Principal Navigations, Voyages, Traffiques and Discoveries of the English Nation* (1598), but all other exploration accounts in these lists came from French, Dutch, and Spanish authors. Corpus students and library owned three copies of Linschoten's *Discourse of voyages into ye East & West Indies* (1598; see figure 2, above) and two of Theodore de Bry's *America* (1590), indicating a strong interest in tales of exploration and adventure. The Corpus Christi book list was particularly strong in descriptions of the Middle East, with several books on Persia, including Pietro Bizzari's *Rerum Persicarum Historia* (1583), as well as others on Turkey and the Holy Land. In European descriptive geography, most countries of Europe were represented, especially Germany (including Hungary, Bohemia, and Poland), Muscovy, and the Low Countries. Indeed, eastern Europe and the Low Countries represented significant financial and strategic areas of focus for the English, and so it is not surprising to find university men eager to learn more about these regions.

This trend is confirmed in the collection of books from Christ Church students. In these more mundane descriptions of the Old World, Christ Church students seem to have had a special interest in eastern Europe and the Middle East, including in their book lists such works as Knolles's *Generall Historie of the Turks* (1610), Martin Broniowski's *Tartariae Descriptio* (1595), Joannes Leunclavius's *Historia Musulmana Turcorum* (1591), Pietro Bizzari's *Rerum Persicarum Historia* (1601), and Antonio Possevino's *Moscovia* (1595). The last mentioned takes a largely theological view of Muscovy, while Knolles is more historical and Broniowski concentrates on the natural features around the Black Sea. In confirmation of this trend, the more strictly historical books from Christ Church also reflected this fascination with that part of the world, with such works as *Historia Byzantina Scriptores* (1615) and *Scriptores varii de emperio Ottomanico evertendo* (1601). This emphasis on eastern Europe (north and south) accords well with the general interest in Russia and the East already described. This stress on an area of Europe that was virtually unknown, as well as of such importance to England economically, demonstrates the twofold fascination—a blend of curiosity and economic promotion—that motivated university students as well as descriptive geographers more generally. It is striking, however, that Italy, the goal of the later Grand Tourists, is very poorly represented in these book lists.

New College students owned descriptive geography books that covered a wider spectrum of countries and areas than those at Christ Church, al-

74. See appendix B for geography books from Corpus Christi College and Christ Church, Oxford.

though in a smaller number of books. Strabo was included, providing a classical foundation for this descriptive and sometimes anecdotal study. There were also many modern descriptions of the Old World and of the New. There were descriptions of northern and eastern Europe, Asia, England, the Netherlands, Turkey, and Italy. Exploration accounts included those by Hakluyt, Maffei, Acosta, Trigault, Lescarbot, and Peter Martyr.[75] There were also books with a greater historical emphasis, including Francesco Guicciardini's *Historia d'Italia* (1561) and a number of English histories and descriptions, namely Sir Walter Raleigh's *History of the World* (1614), Raphael Holinshed's *Chronicles of England* (1577), and John Stow's *Chronicles of England* (completed by Edmund Howes in 1615). Thus, at New College, there existed a wide and more diffuse interest in the various parts of the world, from both a purely descriptive and a historical point of view.

As was the case with mathematical geography, the center for descriptive geography at Cambridge appears to have been Peterhouse.[76] More than half of this college's geography book list was made up of descriptive geography books. The selection of books of exploration was somewhat more limited than those of the Oxford colleges already mentioned, although there were copies of Ramusio's *Delle Navigatione* (1550), Geronymo Osorio's *De Rebus; Emmanuelis regis Lusitaniae virtute gestis libri xii* (1574), and Fynes Moryson's *An Itinerary (containing his ten yeeres travell through the 12 dominions of Germany, Bohmerland . . . and Ireland)* (1617). By contrast, the Middle East was as well covered as in the previous two collections, including several discussions of Islam, such as Joannis Cantacuzeni, *Contra Mahometicam fidem Christiana et Orthodoxa Assertio* (1543), and one copy of the Koran. Western Europe received more complete coverage, with descriptions of Italy, Greece, Germany, France, Poland, the Low Countries, and the northern peoples, as in Olaus Magnus, *Historia de Gentibus Septentrionibus* (1555).[77] Thus, for students at Peterhouse, the Old World held more immediate interest than the New, and the Middle East and western Europe seemed the most alluring.

Students and the library at St. John's College, Cambridge, also collected a large number of descriptive geography books.[78] This list has a stronger

75. Hakluyt, *Principal Navigations* (1598–1600); Giovanni Petri Maffei, *Historiarum Indicarum libri xvi* (Florence, 1588); Joseph de Acosta, *The Natural & Morall historie of the East & West Indies,* trans. E. Grimstone (London, 1604); Nicolas Trigault, *De Christiana Expeditione apud Sinas Suscepta ab Societate Iesu ex P. Matthaei Ricii* (Cologne, 1617); Marc Lescarbot, *Histoire de la Nouvelle France* (Paris, 1609); Peter Martyr Anglerius, *De Orbo Novo* (Cologne, 1574).

76. See appendix B for Peterhouse geography books.

77. For a discussion of both Johannes and Olaus Magnus, see Johannesson, *Renaissance of the Goths.*

78. See appendix B for St. John's geography books.

representation of exploration tracts than that of Peterhouse, including Purchas *His Pilgrimage* and two copies of Petrus Martyr's *De Novo Orbe* (1574). As well, the Low Countries seemed an area of special concern to St. John's scholars, counting books by Schrijver, Meteran, Groot, Does, and Meyers among their collections. Other than these two strengths, there were books describing the Middle East, the New World (including Harriot on Virginia), India, Muscovy, Hungary, Denmark, France, and Germany. It is clear that St. John's scholars were interested in many parts of the world, especially that area directly across the channel.

These five colleges were by no means the limit of interest in descriptive geography at Oxford and Cambridge. St. John's College, Oxford, students had access to Strabo and Dionysius's *De Situ Orbis,* as well as descriptions of Asia by Joseph de Acosta and Girolamo Benzoni and letters from the Jesuits. They also owned descriptions of the Middle East and Europe, including Samuel Lewkenor's *Discourse of Forraine Cities* (1600). The latter described all the major cities of Europe, dwelling particularly on the universities and on odd customs or local wonders. The English universities were clearly superior to all others, but Lewkenor was willing to give all foreign schools (with the exception of the Scottish ones) their due. Oriel students collected descriptions of most of Europe and explorations, including a disproportionate number in English. On this list were Camden's *Annales, The true historie of Elizabeth* (1625), Emmanuel de Meteran's *Generall History of the Netherlands* (1608), Louis Turquet de Mayerne's *Generall Historie of Spain* (1612), and Jean de Serres's *General Inventorie of the History of France* (1607), the latter three translated by Edward Grimstone. Grimstone's stated purpose in his translation of Serres was to show how the English had several times defeated the French and thus to demonstrate the continual supremacy of the English nation.[79] The All Souls College, Oxford, book list contained three copies of Strabo, as well as descriptions of the Old World by Martin Kromer, Paolo Giovio, George Buchanan, Avity, and Bolaius, and of the New World by André Thevet. Thevet's description of Brazil supplied tales of Protestant endurance, as well as reports of Amazons and cannibals.[80] Merton students could voyage to Siam with Saris, or enjoy descriptions of India by Giovanni Petri Maffei or the Jesuit explorers in their *Epistolae Indicae et Japanicae* (1570). At Cambridge, the Corpus Christi book list contained descriptions of Spain and the Spanish Inquisition, France, Muscovy, Poland, and Flanders, as well as Ramusio's exploration compendium. Trinity Cambridge students collected descriptive geog-

79. Grimstone, introduction to Serres, *General Inventorie,* sig. ¶ 4a.

80. Lestringant, in *Mapping the Renaissance World,*, ch. 3, argues persuasively that Thevet really *invented* the Brazil he described, especially with his reliance on much hearsay evidence.

Table 7 Descriptive Geography Books Owned, by Decade

	1580	1590	1600	1610	1620
Strabo	5	6	6	5	8
Münster	5	5	4	5	5
Olaus Magnus	3	3	2	4	2
Pius II (*Cosmographia*)	2	2	1	4	3
Philippson	3	3	2	—	—
Solinus	3	—	1	1	—
Giovio (*Descriptio Britanniae*)	—	2	2	—	—
Kromer (*Polonia*)	—	1	1	4	—
Herberstein	—	1	—	4	—
Zosimus	—	1	—	4	—
Lonicer	1	2	2	4	2
Buchanan	1	1	2	3	4
Pistorius	—	1	1	3	2
Theodore de Bry	—	—	2	3	—
Heuter	—	—	—	5	1
Camden	—	—	—	1	5
Hakluyt	—	—	1	1	4
Meteran	—	—	1	3	4
Possevino	—	—	—	5	—
Grimstone	—	—	—	1	4
Purchas	—	1	—	1	5
Serres	1	—	—	1	5
Cluvier	—	—	—	—	4
Knolles	—	—	—	2	4

raphy books on Belgium, France, Germany, Hungary, Poland, the Turks, and Spain. They also owned books on the newly discovered world, including those by Hakluyt, Linschoten, Purchas, and Lopez de Gomara. The lists from Gonville and Caius and from Sidney Sussex both contained books describing both the Old World and the New, again with a strong representation of Middle Eastern and Low Country destinations.[81] Thus, interest in descriptive geography was widespread throughout both universities, with a more significant activity in at least seven colleges at Oxford and six at Cambridge.

Upon examining the most frequently owned descriptive geography books through time, a clear evolution emerges between those books owned in the early part of the period and those owned in the later decades (see table 7).[82]

81. Descriptive geography books are also known to have appeared on book lists for the following colleges: Queen's, Balliol, Magdalen, University, Trinity, Brasenose, Exeter, Lincoln, and Hart Hall at Oxford; and King's, Emmanuel, Trinity Hall, Pembroke Hall, Christ's, Jesus, and Queens' at Cambridge.

82. Münster appears in table 7 and is also included in the mathematical geography list (see table 6, above).

A number of titles run consistently through all five decades: Strabo's *De Situ Orbis,* Münster's *Cosmographia,* Olaus Magnus's *Historia de Gentibus Septentrionibus* (1555), Pius II's *Cosmographia* (1475), George Buchanan's *Historia Scoticarum* (1582), and Philip Lonicer's *Cronica Turcica* (1578). Most of these were standard descriptions, especially of Europe. Most of the multiply owned descriptive geography books, however, do not fall into this category, and many were not purchased until the last two decades. By 1610, a number of important exploration accounts appear in multiple copies, including Theodore de Bry's *America* and Hakluyt's *Principle Navigations.* By 1620, these are joined by Camden's *Annales Rerum Anglicarum et Hibernicarum regnante Elizabetha* (1596) and *Purchas His Pilgrimage.* Likewise in 1610, several new descriptions of Europe appear on this list in multiple copies: five copies each of Possevino's *Muscovia* (1587) and Heuter's *Rerum Burgundicarum* (1584); four copies each of Kromer's *Polonia* (1555) and Herberstein's *Rerum Moscoviticarum* (1571); and three copies of Meteran's *Historia Belgica* (1598). There are some transitional texts here, but on the whole, the books owned after 1610 are a completely different collection from those owned earlier. The post-1610 descriptive geography texts were more modern and tended to concentrate on specific countries, peoples, and customs, or to tell exciting tales of new explorations and discoveries. These books were also increasingly English in origin, seven of the fifteen books owned in multiple copies in 1620 being by English authors. These included Hakluyt, Camden, Purchas, and Knolles, as well as Edward Grimstone's translations. This English flavor is borne out by an examination of the countries of origin of works of descriptive geography through time (see figure 22), which shows that descriptive geography by English authors was more frequently owned after 1600 than had previously been the case.

The type of descriptive geography that was read at Oxford and Cambridge thus changed dramatically after 1600 to reflect a readership that valued modern, vernacular (especially English) information, now concentrating on specific topics rather than general description. This corresponds to the influx of the new group of students into Oxford and Cambridge, students who were potentially less concerned with Latin erudition and style than with relevant, practical information and an attractive ideology of exploitation and the moral rectitude of English expansion.

We have already examined, in chapter 1 above, the remarkable increase in interest in descriptive geography during the period under investigation. Given the stable, serious group of mathematical scholars plodding away on mathematical geography subjects, this rapid rise of interest in descriptive geography must be explained by the changing character brought about by the influx of new students at Oxford and Cambridge. The study of descrip-

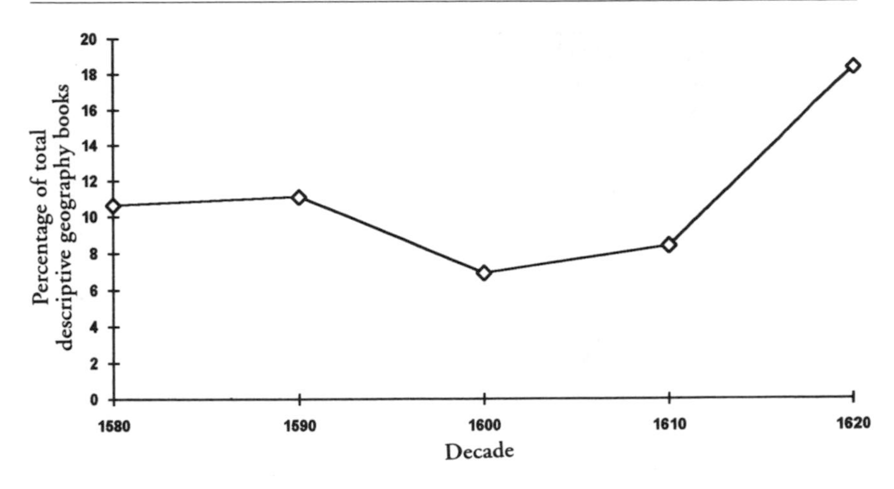

Figure 22. English descriptive geography book authors.

tive geography suited the career aspirations of these classes, since it provided them with up-to-date information about the countries with which they were likely to have business and about exploration ventures in which they might be called upon to invest. It is no surprise, then, that students of descriptive geography rapidly lost interest in the medieval and early-sixteenth-century authors by 1600 and instead concentrated on the new information found in the works of living authors. In addition, the proportion of vernacular accounts began to rise rapidly in the seventeenth century, accounting for almost a third of all descriptive geography books by 1620. This reflects an audience more interested in the practicalities of European exchange than in the international Latinate culture of scholars.

If we divide descriptive geography into subcategories of exploration accounts—physical, historical, medical/botanical, antiquarian description, and itineraries—and examine these categories through time, we find that the various areas occupied a relatively stable proportion of the total descriptive geography books owned (see figure 23).

Exploration accounts became steadily more prolific throughout the period, rising from 10 percent of all descriptive geography books owned in 1590 to more than 18 percent by 1620. This was, no doubt, partly because of the increased availability of such accounts, now beginning to appear in anthologies of English and Dutch authors. It may also have been caused by an accelerated level of investment in exploration, trade, and settlement schemes,[83] which led to an increased general interest on the part of those families investing, as well as requiring more specific research by the investors.

83. See R. Davis, *English Overseas Trade;* and Thirsk, *Economic Policy and Projects.*

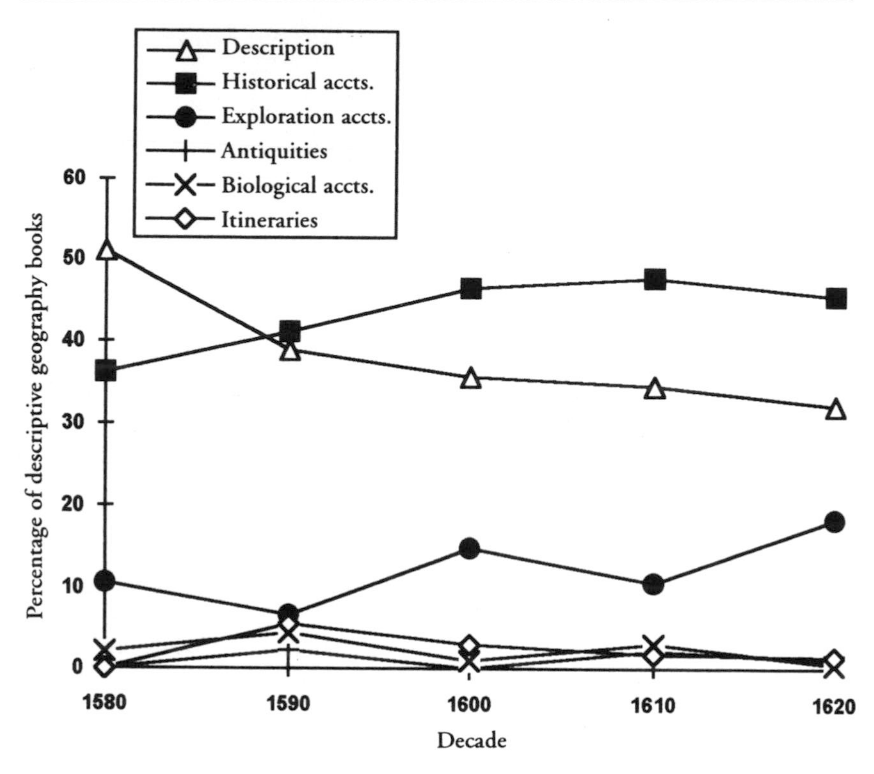

FIGURE 23. Descriptive geography books owned, subcategories.

Historical accounts and physical descriptive ones varied in an inverse proportion, indicating that the two need to be seen as complementary. Since both description and historical background were seen as necessary to a real understanding of the place being studied, this complementarity of numbers is not surprising. Descriptive and historical geography each accounted for approximately 40 percent of all descriptive geography books throughout the period, indicating that both the strictly descriptive and the historical accounts were given equal weight in the investigation of foreign lands.

Other descriptive geography books, which dealt with plants, medicine, or antiquities of other countries, accounted for a relatively insignificant portion of descriptive geography books. The basic division of descriptive, historical, and exploration accounts was a relatively equitable one, with more interest in the first two categories but a growing interest in exploration accounts over time.

This examination of the book list indicates that descriptive geography was an increasingly important branch of geography in the late sixteenth and early seventeenth centuries. It was a significant field of study at both

Oxford and Cambridge, especially at Corpus Christi, Christ Church, and New College, Oxford, and at Peterhouse and St. John's College, Cambridge. The book ownership in this subdiscipline changed significantly over time, both with a tremendous increase in volume by the seventeenth century and in subject matter. Descriptive geography began to emerge by the 1620s as a predominantly vernacular, English subdiscipline, the concern of those students engaged with the world, aware of their nationality and intent on developing their university training into a practical and economically or politically viable resource.

Students of Descriptive Geography

If we examine the men interested in descriptive geography, this emphasis on the utility of geographical knowledge becomes even clearer. Within the group interested in descriptive geography, there is a distinction to be made between the practical traveler, who might have had some university training but who generally spent his life in adventure and trade, and the armchair virtuoso, who developed at least the rudiments of his interest in descriptive geography at university. Of the 418 Englishmen with some known interest in descriptive geography, 60 percent (251) were travelers, colonizers, or adventurers, involved in practical rather than abstract aspects of the study. Most were deeply interested in the description of the globe and should be seen as part of the geographical community. Of these practitioners of descriptive geography, only 68 (27 percent) attended university, showing the relative disjunction between the adventurous life and that of scholarship. Of the 113 students of descriptive geography who were more concerned with developing a theoretical or at least a synthetic view of the world, 50 (44 percent) attended Oxford or Cambridge. Further, 44 university students were known to have owned descriptive geography books, though none became an active geographer, traveler, or adventurer. There was, then, a relative distinction between those who did not attend university but had adventures and those who went to university but only read about adventures. Partly because of the experiential nature of its research, descriptive geography was not by any means an exclusively university-based subdiscipline, as compared with mathematical geography, which owed its foundation to the university curriculum. Rather, the study of descriptive geography embraced both the hardy adventurer and the eager student.

The typical student of descriptive geography was a very different creature from his mathematical counterpart. Only half of the university-based descriptive geographers, not including those who only owned books, attained the degree of Bachelor of Arts, as compared with 71 percent of mathematical geographers; 48 percent received an M.A.; and only 14 percent remained at university long enough to earn a Doctor of Medicine, Theology, or Civil Law. Students interested in descriptive geography were far less committed to a long-term program of study, more often being content with a few years of Arts preparation, or with the attainment of the first Arts degree. While this does not necessarily imply that descriptive geography was a study external to the Arts curriculum, it does indicate that it appealed to the student less interested in completing his degree, whose career ambitions ran in other directions than that of church or academe.

Indeed, students of descriptive geography followed careers centered on the political world, rather than those focused on academic or religious aspirations. While twenty-three men did enter teaching professions of one sort or another (thirty including those who simply owned books of descriptive geography) and eighteen of the descriptive geography book owners entered orders, almost half of the descriptive geography students became diplomats, courtiers, soldiers, colonists, or adventurers. Giles Fletcher (B.A. 1570, M.A. 1573, King's College, Cambridge), for example, was an envoy to Ivan the Terrible's court in Russia for Elizabeth in 1588.[84] Likewise, Sir Robert Sherley was envoy for the Shah of Persia in the early seventeenth century and embarked on many diplomatic missions for Persian and other European heads of state. Sir Thomas Roe (commoner 1593, Magdalen College, Oxford), besides embarking on a voyage of discovery to the West Indies in 1610 at the behest of Henry, Prince of Wales, acted from 1615 to 1618 as ambassador for James to the court of Jahangir, Mogul emperor of India.[85] James Howell, Sir Thomas Wilson, and the renowned Sir Philip Sidney all performed diplomatic missions as well as demonstrating a keen interest in descriptive geography.[86] Other people who studied descriptive geography also took up public positions. Sir Robert Heath (pensioner 1587, St. John's, Cambridge), who wrote concerning trade with Europe,[87] became first a Member of Parliament and Attorney-General and later Chief Justice of Common Pleas and of the King's Bench. Several courtiers expressed an interest in geography, including Sir Robert Gordon (M.A. hon. caus. 1615, Cambridge), Gentleman of the Privy Chamber to James VI and

84. BL Lansdowne MS 60, ff. 157–60, deals with negotiations performed by Fletcher for the queen.

85. *DNB*, 17:89.

86. James Howell, *A letter from Scotland to a friend in England*, MS. Printed in 1647; *Instructions for Forreine Travel* (London, 1642), and *Lustia Ludovici, or the History of Lewis XII* (London, 1643). Sir Thomas Wilson translated *Diana* from the Spanish in 1596, a romance that formed the basis for *Two Gentlemen of Verona*. He wrote *Relation of Ormuz* in 1602, which was printed by Purchas.

87. Sir Robert Heath, *Several Grievances concerning Trade*, MS; cited by E. G. R. Taylor in *Late Tudor Geography*, 261.

I and geographer of Scotland and New Scotland; Sir Robert Dallington (M.A. 1601, St. John's, Oxford), Gentleman of the Privy Chamber in Ordinary to Prince Henry and translator of Italian geographical treatises; and Sir Thomas Overbury (B.A. 1598, Queen's, Oxford), traveler, adviser to Rochester at court, and victim of a famous poisoning plot.[88] Finally, a large group of students of descriptive geography became involved with the Virginia enterprise. Included in their ranks was the famous mathematical geographer Thomas Harriot. Also included were John Pory (B.A. 1592, M.A. 1595, Gonville and Caius, Cambridge), translator of Leo Africanus; Alexander Whitaker (B.A. 1605, M.A. 1608, Trinity College, Cambridge), preacher in Virginia and author of *Good Newes from Virginia* (1613); and George Percy, seventh son of Henry, Earl of Northumberland, and younger brother of the "Wizard Earl," who traveled to Virginia in 1620. This sample of the public and adventurous careers of such a large segment of the students of descriptive geography strongly suggests the close relationship between such a study and an active rather than contemplative life. Indeed, it speaks of a life devoted to the expansion of English power and influence, as well as personal status and political success. These publicly oriented occupations demanded a knowledge of the Old and New Worlds, which these men began to gain while at university. It could well be argued that the enthusiasm for global description that these men caught while at Oxford and Cambridge led them to venture beyond the safe confines of their native island, in search of knowledge but especially in search of glory and wealth. These forays into diplomatic and mercantile adventures were the mechanisms for the early development of imperial thinking and action; in this development, descriptive geography played a role.

Indeed, more than half the geographically minded men in this group spent some time abroad, either in Europe (Italy being the most popular country to visit), in Virginia and the New World, or, less often, in Russia, Persia, or Turkey. Given the conditions of transportation, the globe trotting of some of these individuals, such as Sir Anthony Sherley and his entourage, must have provided thrilling reading material for the less fortunate students left at home. Most did not venture so far, but a trip to the Continent was becoming more often part of a young man's education. Of

88. Sir Robert Gordon, *Genealogie of the History of the Earls of Sutherland,* written ca. 1630; *Encouragements of such as shall have intention to be undertakers of the New Plantation of Cape Breton, now New Galloway, In America, by mee, Lochinvar* (Edinburgh, 1625). Sir Robert Dallington, *A Survey of the Great Duke's State of Tuscany in the year of our Lord 1596* (London, 1605); trans. *Aphorismis Civile, & Militare . . . out of Guicciardine, written in 1609* (London, 1613). For Overbury, see *DNB.* Sir Thomas Overbury, *Observations in His Travailes Upon the State of the XVII Provinces as they stood Anno Dom. 1609* (London, 1623). Discussed in Glacken, *Traces on the Rhodian Shore,* 451.

these geographically inclined students, 18 percent visited Europe at some point in their careers.[89] Sir Fulke Greville (M.A. 1588, Jesus College, Cambridge), the famous courtier, friend of Camden and Giordano Bruno, patron of John Speed, and author of a manuscript on travel and a life of Sidney, studied with Sir Philip Sidney in Heidelberg in 1577. Thomas Winston (B.A. 1599, M.A. 1602, Clare Hall, Cambridge), a prominent member of the Virginia Company and on the committee to consider the marketing of Virginia tobacco,[90] studied at Padua and Basel in the 1600s, earning his M.D. from Padua around 1608. Such notables as John Pory, Sir Robert Dallington, Thomas West, Baron de la Warr, John Harington, second Baron Harington of Exton, and George Sandys traveled to Italy at about the turn of the century.[91] Although this flood of Continent-bound young men should not be confused with the de rigueur Grand Tour of eighteenth-century young gentlemen, it represents the beginning of that movement.[92] As Englishmen began to see themselves as separate from the Continent in terms of culture, religion, economics, and government, they could afford the luxury of observing these foreign countries and gaining information and culture from them, in a discriminating and attenuated fashion. If we combine this collection of overseas experiences with those of students of descriptive geography who went to the New World, it becomes clear that a firsthand examination of other parts of the globe was possible and desirable for a large number of students. The result of this was an English population that was becoming aware of its separateness from the rest of the world while at the same time gaining a more complete understanding of that world.

In addition to experiencing the wider world through travel, these students of descriptive geography were also engaged with the political life of England. A full 30 percent of these students attended the Inns of Court, and one-quarter became Members of Parliament. Descriptive geographers were involved in the world of affairs much more than the world of ideas. The Inns of Court trained common lawyers while also acting as a finishing school for aspiring gentlemen or merchants eager to gain polish and enough legal knowledge to protect themselves in an increasingly litigious society.[93] The study of geography filled a similar function. It provided polish for courtly conversation and information for investment and adventur-

89. See also Stoyre, *English Travellers Abroad.*

90. Adamson, "Foundation and Early History of Gresham College," 150.

91. See chapter 6 below for a discussion of the Italianate influence.

92. Hibbert, in *Grand Tour,* discusses this early prehistory of the later passion for European travel, claiming that the sixteenth- and seventeenth-century travelers were more concerned with diplomatic, political, and educational pursuits, while their eighteenth-century counterparts traveled for cultural betterment.

93. See Charlton, "Liberal Education and the Inns of Court"; Prest, *Inns of Court.*

ous exploits. The participation of these men in Parliament reflects in part their legal training and in part their class origins.[94] It also suggests that this was a group of men who took their duty to their country seriously and did not wish to separate themselves from the active life in favor of the contemplative one. Evidence of participation in the Inns of Court and in Parliament, then, helps to confirm descriptive geography as a subdiscipline for members of those rising classes who came from the world of human affairs and remained engaged with that world.

The class origins of these students confirms this picture: of those whose class origins are known (forty-two of seventy-eight), 53 percent declared themselves to be members of the gentry class. A further 8 percent were sons of clergymen, and 6 percent were of merchant stock. Those newly empowered classes were thus more and more often the men making use of geographical information gleaned at university and applicable to their increasingly political and court-oriented lives.

The typical student of descriptive geography was a member of the gentry (or occasionally merchant) class, who attended university for a few years, studying within the Arts curriculum and perhaps achieving a B.A. He then went on to an active career, usually involving some service to crown and country and some time spent abroad. Descriptive geography was an attractive subject for such students because it was current and exciting, and it provided necessary skills and information that might prove invaluable for a rising career in the civil administration, in politics, or in mercantile and colonial endeavors.

❦

Descriptive geography, often combined with history, was a common study for sixteenth- and seventeenth-century undergraduates. They were most enthralled with sensational and titillating tales of far lands, "wonders" near and far, and horrible tortures, although they also read more mundane descriptions of Old World politics and economics and New World settlements. Many probably studied geography for the colorful illustrations it provided for political speeches or occasionally sermons, while a few engaged in the more scholarly work of amassing collections of descriptions, either from oral sources or by translating foreign works. The number of undergraduates who adopted descriptive geography as a vocation, or even as a serious avocation, was small, but the influence of this subdiscipline was nevertheless wide-ranging. Descriptive geography helped English students to identify themselves as separate from the Continental unrest they saw before them. It provided them with a model of inductive

94. Neale, *Elizabethan House of Commons.*

methodology, which would become more important as the human and collecting sciences developed. It encouraged them to regard the world as an endless source of wondrous tales and new goods, thereby creating a mentality that would condone and encourage the exploitation of foreign peoples and resources. In the short term, this study provided these rising men with court polish, vernacular language skills, economic information, and political comparisons. In the long run, it set in place a mentality of separateness and exploitation that would allow the growth of the English empire.

CHAPTER FIVE

Chorography: Geography Writ Small

We shall not cease from exploration
And the end of all our exploring
Will be to arrive where we started
And know the place for the first time.
(T. S. Eliot, *Little Gidding V, Four Quartets*)

When Sir Julius Caesar recorded "The Singularities of England" in his commonplace book, he was participating in chorography, the third branch of geography studied in the late sixteenth and early seventeenth centuries. He described London Bridge, one of the great wonders of the world, and the four naturally occurring hot baths of England, and he discussed the abundance of jewels and minerals to be found on the island.[1] These topics were both intriguing and inherently useful, indicating the strong practical bent of this third area of geographical study. Chorography was the most wide-ranging of the geographical arts, in that it provided the specific detail to make concrete the other general branches of geography.[2] This study of local history was particularly important in the development of an attitude that favored things English, adding to an ideology of separateness and superiority. English readers could identify themselves in its descriptions and create a solid definition of English people, places, and history against which to compare the wider world they had studied in descriptive geography. At the same time, it encouraged an even more focused identification, as its practitioners and readers concentrated on smaller local areas and developed loyalties to their county or locality, at odds with simple centralized nationalism.

The message of chorography was fundamentally practical, since it emphasized the surveying of estates close to home, the compilation of genealogies of local families, and the description of local attractions and commodities. Yet there was a theoretical thread running through it, especially in the emphasis on chronology, the mapping of time, which was related to

1. BL Add. MS 6038, f. 250a.
2. Carpenter draws this distinction in *Geography delineated forth in Two Books*, 3–4.

genealogy and local description but was founded on an attempt to develop mathematically a theoretical definition of absolute and relative time. The political motivations for and manifestations of the study of chorography were more overt than were those for mathematical and descriptive geography. By describing local places, chorography implicitly called to mind descriptions of the wider world, thereby encouraging the English to define themselves as self-contained and separate from their Continental counterparts. The emphasis on new surveying practices and theoretical suggestions for the improvement of techniques were motivated by attempts at political and economic aggrandizement on the part of middling gentry and merchant families. Descriptions of individual counties, such as Richard Carew's of Cornwall, stressed administrative practicalities through the prominence of musters and hundreds in their portrayals.[3] Even more important, local genealogies were drawn up in response to a search for ancient and respectable roots that would allow this group to enter the ranks of the armigerous gentry, as well as identifying them with their county or geographical area. Finally, the study of chorography, more perhaps than that of the other two subdisciplines of geography, encouraged the development of an inductive and public spirit in the human sciences. Chorographers classified time and place in an effort to reduce their surroundings to a manageable and controllable object. The methodology of incremental fact gathering was fundamentally important to chorographers in the development of their subdiscipline and, when combined with the political and economic motivations that drove the study, provided a science that accepted public scrutiny while placing a high value on thoroughness, tabulation, classification, and the development of a community of like-minded scholars. Chorography evolved from Continental sources, was encouraged by social and economic motives, and developed new techniques and methodologies for seventeenth-century natural science. It was a geographical study distinct from the two previously discussed, although, like them, it fostered an aspect of imperial thinking, in this case a self-definition that gave strength and confidence to the ability of the English to represent themselves to the larger world.

The Anatomy of Chorography

Chorography, like mathematical geography, owed its early definition as well as its subsequent reintroduction in the fifteenth century to Ptolemy's *Geographia.* Ptolemy had separated chorography from the more elevated study of geography proper, defining its purview as the description of ports,

3. Carew, *Survey of Cornwall,* ff. 90b–95b.

villages, peoples, rivers, and so forth.[4] Thus, the term *chorography*—or local history, as it might be called (although the term *local history* evokes a closer connection with history and natural history than is always warranted in this diverse and far-reaching study)—came to be applied in the sixteenth century to any study of local places or people. This study often, though not always, focused on one town or county and, through its detailed enumeration of local sights, marvels, and commodities, helped engender local pride and loyalty.

John Dee included chorography as one of the mathematical arts derived from geometry in his complex categorization of the mathematical sciences.[5] For Dee, the study of mathematics was essential to both the individual in search of spiritual excellence and the nation in search of European supremacy. Chorography, as Dee saw it, helped develop both observational skills and local pride: "Chorographie seemeth to be an underling and a twig of *Geographie:* and yet neverthelesse, is in practise manifolde, and in use very ample. This teacheth Analogically to describe a small portion or circuite of ground, with the contentes . . . in the territory or parcell of ground which it taketh in hand to make description of, it leaveth out . . . no notable or odde thing, above ground visible."[6]

Dee thus defined the function of chorography as the description of local detail rather than the "commensuration it hath to the whole," which was the job of geography proper. This subdiscipline included descriptions of all things above ground and also "thinges under ground, geveth some peculier marke: or warning: as of Mettall mines, Cole pittes, Stone quarries, etc."[7] Nothing was too small or insignificant to be worthy of a chorographer's notice.

The Renaissance study of chorography originated in Italy.[8] Flavio Biondi's *Italia Illustrata* (1482) influenced an entire generation of local historians.[9] A Renaissance humanist, Biondi applied the techniques he had mastered in dealing with classical texts to the evaluation of the Italian countryside, proceeding province by province, town by town. This investi-

4. Ptolemy, *Geographia,* sig. a1a.
5. Dee, *Mathematicall Praeface.*
6. Ibid., sig. a4a.
7. Ibid.
8. My description of chorography in this chapter owes much to Mendyk, "*Speculum Britanniae*" and "Early British Chorography." Mendyk describes well the work of sixteenth- and early-seventeenth-century British chorographers, although I take exception with his claim that the introduction of Baconian ideals into this subdiscipline, occurring with the foundation of the Royal Society in the 1660s, changed chorography into natural history. Rather, these Baconian tendencies existed long before the Civil War and themselves helped to shape the ideology of both the Lord Chancellor and the Royal Society.
9. Hay, *Annalists and Historians,* 109.

gation of the local setting as text spread throughout Europe in the fifteenth and sixteenth centuries, influencing Konrad Celtis, Georg Braun, and Francis Hogenberg in Germany, and even Jean Bodin in France.[10] Its influence came late to England, reaching British shores only in the later sixteenth century. There it mingled with a much older native chronicle tradition to form a chorographical study that was uniquely British and helped to develop pride in county and to breed familiarity with a rapidly developing inductive approach to the natural world.[11]

The native English tradition traced its origins to the medieval chronicle genre. While the romantic historiography of Geoffrey of Monmouth, who creatively told the myths of King Arthur, provided one strand in this development, more important were the local and historical chronicles of more fact-based scholars. Undoubtedly the most popular of these was the universal chronicle of Ranulf Higden. Higden's *Polychronicon* (1327–ca. 1360) provided a biblically based history of the world and of England that combined geography and history in an astonishing compilation.[12] *Polychronicon* was divided into seven books, corresponding to the seven days of creation, and the first book was taken up with the geography, natural history, peoples, and customs of the world. This provided the macrocosm in which the events of human history—the microcosm—were played out.[13] Since Higden's work was essentially a compilation, it became the means by which sixteenth-century antiquaries were acquainted with the earlier chorographies. Higden encompassed and superseded the earlier works of such Englishmen as Gildas, Bede, Henry of Huntingdon, William of Malmesbury, Matthew Paris, and Gerald of Wales.[14] Later chroniclers tended to supple-

10. Konrad Celtis, *Germana Illustra or De Origine situ moribus et institutis Norimbergae libellus incipit* (Nuremberg, 1502); Georg Braun & Francis Hogenberg, *Civitates Orbis Terrarum,* 6 vols. (Cologne, 1572–1617). Jean Bodin, *Methodus, ad facilem historiarum cognitionem* (Paris, 1566); *Les Six Livres de la Republique* (Paris, 1576).

11. Gransden sees this same mingling of medieval English and Renaissance Continental influence in late-sixteenth-century historiography. *Historical Writing in England,* 2:479.

12. Ranulf Higden, *Polychronicon,* written from 1327 until his death in the 1360s. First published in complete form as *Cronica Ranulphi Cistrensis Monachi (the book named P. Proloconycon [sic]) . . . compiled by Ranulph Monk of Chrestre . . .* [Westminster, W. Caxton] (1482, reprinted 1495). Selections of Higden's work were also printed in 1480 as *The Descrypcyon of Englonde,* and reprinted in 1497, 1502, 1510, 1515, and 1528. Reprinted in H. Savile, ed., *Rerum Anglicarum Scriptorum veterum,* 3 vols. (London, 1596). On Higden, see J. Taylor, *Universal Chronicle.*

13. Gransden, *Historical Writing in England,* 2:44–45.

14. Gildas (516?–570?), *Gildas Britannus Monachus . . . de calamitate, excidio & conquestu Britanniae . . .* (first published London, 1525); Venerable Bede (673–735), *Historiae ecclesiasticae gentis Anglorum* (first published [Strassburg, 1475?]); Henry of Huntingdon (d. 1155), *Historia Anglorum,* first published in H. Savile, ed., *Rerum Anglicarum Scriptores* (London, 1596); William of Malmesbury (d. ca. 1143), *De Gestis rerum Anglorum* and *De Antiquitate Glastoniensis Ecclesiae,* first published in *Rerum Anglicarum;* Matthew Paris, *Historia Maior* (first published, ed. Matthew Parker, London, 1571); Gerald of Wales (d.

ment Higden, rather than creating their own original works. All of these chroniclers combined a pride in English events and especially local places with a salvational or biblical view of the world.[15] The authors tended to be most concerned with the world closest to their monasteries or towns and often provided antiquarian information such as the original location of Roman roads and ruins, ancient place names, and linguistic history, as well as describing either historical or contemporary events.

This medieval tradition, identifiable by its continuing biblical thrust and local loyalties, combined with the work and methodology of Continental local historians such as Biondi, inspired by a reevaluation of classical literature and methods, to produce a new English chorography in the sixteenth century. This study was distinct from its medieval antecedents but was not simply a copy of its Continental sources of inspiration. Rather, sixteenth-century English chorography united a medieval belief in the ordered development of time and place, shown in local histories that were an adjunct to medieval chronicles, with a more modern interest in classifying and dividing that same time and place. This change in emphasis became possible because the location of absolute time allowed chorographers to concentrate on local segments, rather than feeling obligated to reiterate the whole. Just as economic gains could be won by a more accurate accounting of disputed property, as social advancement could be attained through careful genealogical analysis, as religious controversy necessitated the establishment of God's time, rather than simply local or relative time, so chorography developed to answer these concerns.[16]

The geographical subdiscipline of chorography developed in response to these problems in three separate yet clearly related directions: surveying and cartography supplied a geometrical and legal emplacement of local lands and county seats; antiquarianism, genealogy, and local history more generally provided new techniques drawn from classical sources and fueled both a curiosity concerning local and ancient sites, as well as natural history, and a social drive for antiquity or gentility; and chronology attempted, using a rigorously mathematical method, to set the world and its history, both natural and local, into a single schema of time measurement. The study and practice of surveying and cartography, mathematical in technique and in application, developed in the second half of the century as legal and economic arguments began to rely more on accurate descrip-

1223), *Topographia Hiberniae, sive Mirabilibus Hiberniae. Expugnatio Hiberniae Itinerarium Cambriae, seu . . . Baldvini Cantuar* (printed in Camden, *Anglica, Hibernia, Normannica . . .*, London, 1602).

15. Gransden, *Historical Writing in England,* 2:454–66.

16. Helgerson sees this transition from a chronicle tradition "to being a topographically ordered set of real-estate and family chronicles." *Forms of Nationhood,* 135.

tion and, increasingly, measurement rather than on prestige and power.[17] The study of chronology, a descendant of both medieval chronicles and the *computus* tradition of calendar reform, became imbued in the sixteenth century with a new sense of historical perspective and a passion for objective mathematical accuracy. The main branch of chorography, which included genealogy and antiquarianism, was perhaps motivated by notions of the desirability of objective description through the incremental collection of facts, leading to a categorization of individuals and of places that would allow the discovery of God's order. In addition, it gained much of its momentum from a desire on the part of its practitioners to establish their own categories of gentility and social value and from an increasing sense of local and national pride, shown by a tendency for local inhabitants to identify themselves with their country and increasingly with their county.[18] Chorography thus supplied an alternative face of imperial thinking, giving English people a sense of identity, placement, and purpose by encouraging them to distinguish themselves by their locale, whether that was England as a whole or a more local particularism.

The study of surveying in England was fundamentally practical, usually written in the vernacular and eschewing any theoretical discussion. Although most surveying texts began with a simplified precis of Euclid's geometrical axioms, the major thrust of these works was the description of surveying techniques and instruments, often including practice problems and an address where such instruments could be obtained. English surveying literature was aimed at an audience that was practical rather than abstract and so was guided by the marketplace rather than by the commonwealth of ideas.[19]

Surveying in sixteenth-century England was in transition between lin-

17. Mendyk claims that the English chorographical movement was a result of the break with Rome because the Church of England needed to establish its own historical precedents (*"Speculum Britanniae,"* 8). Rather, there may have been a relationship between chorographical interests and Henry's "Great Matter" that was personal and economic, an interest on the part of land-hungry gentry to take over more land with as little cost to themselves as possible. However, although written surveys of land were important by the 1530s, mapped surveys were not significant until Elizabeth's reign. Kain and Baigent, *Cadastral Map,* 5; Harvey, *Maps in Tudor England,* 83–93. See also P. Eden, "Three Elizabethan Estate Surveyors," 76.

18. Helgerson, in "The Land Speaks," contends that the proliferation of county maps in the seventeenth century and their changing iconography indicate a movement away from the crown toward county identification and loyalty. Mendyk also sees patriotism driving the investigation of such men as Leland, Camden, and Lambarde. *"Speculum Britanniae,"* 38–56.

19. Bendall suggests that the audience for the thirty-three surveying texts she examined, published from 1523 to 1827, varied from the practical would-be surveyor to the wealthy landowner. *Maps, Land and Society,* 123.

ear and geometrical methods on the one hand and angular or trigonometric methods on the other.[20] In the first half of the century, surveying was very primitive, often restricted to "viewing" or simply chain-measuring. This type of surveying continued well into the eighteenth century, but by the end of the sixteenth century, other more trigonometric methods had also been developed. The introduction of triangulation methods (see figure 24), the plane table, and the theodolite, as well as rules of acceptable surveying practice, began to transform surveying into an exact art.[21] Part of this transformation can be tied to the changing awareness on the part of landowners of the desirability of surveying and mapping their lands. In the first half of the sixteenth century, most surveys were written descriptions, containing measurements but no depiction. Only in the last quarter of the century did surveyors begin to use "platts" on a regular basis.[22] This occurred concurrently with the use of maps, increasingly measured (as opposed to sketched) maps, as evidence in court cases, and we must see both as symptomatic of the growing map culture of the last years of Elizabeth's reign, driven in part by the study of geography at university by these landowners and their sons. The discipline of geography helped these gentry conceptualize their own world in terms of maps rather than merely words or deeds.

Elizabethan surveying began with the work of Valentine Leigh and Leonard Digges.[23] Leigh attempted to combine a description of the duties of a land steward with those of a land surveyor and so described both the political laws of land division and geometrical rules for the laying out of land.[24] His descriptions of measurement were vague and demonstrated little acquaintance with trigonometric techniques, probably reflecting an audience ignorant of even basic concepts. Digges's work, on the other hand, described land measurement in a relatively detailed manner. He also included the first English description of the theodolite, a universal instrument to measure distances and angles that had long been in use on the Continent (see figure 25) and made use of trigonometric rather than geometric methods.

Edward Worsop, the next important English land surveyor, complained

20. Chilton, "Land Measurement," 19.

21. Bendall traces this development, seeing the introduction of the plane table and the theodolite, as well as more "professional" surveyors, as important to a slow transformation of surveying (*Maps, Land and Society*, ch. 5, esp. 129–38). This change did not occur overnight, of course, and many of the suggestions in surveying manuals were not used in practice for some time.

22. Harvey, *Maps in Tudor England*, 83–93.

23. Leigh, *Most Profitable Science of Surveying;* L. Digges, *Pantometria.*

24. Richeson, *English Land Measuring*, 74.

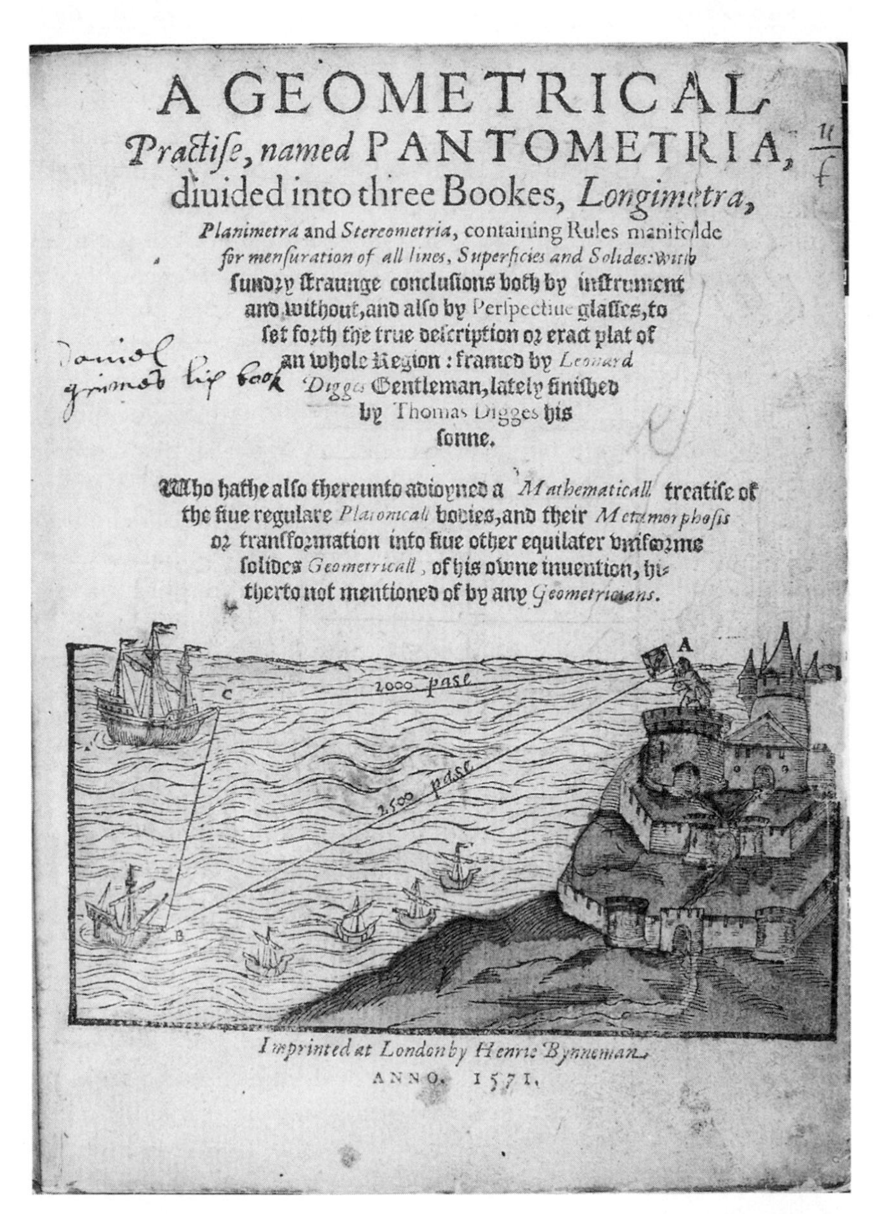

A GEOMETRICAL
Practise, named PANTOMETRIA,
diuided into three Bookes, *Longimetra,*
Planimetra and *Stereometria*, containing Rules manifolde
for mensuration of all lines, Superficies and Solides: with
sundry straunge conclusions both by instrument
and without, and also by Perspectiue glasses, to
set forth the true description or exact plat of
an whole Region: framed by *Leonard*
Digges Gentleman, lately finished
by Thomas Digges his
sonne.

Who hathe also thereunto adioyned a *Mathematicall* treatise of
the fiue regulare *Platonicall* bodies, and their *Metamorphosis*
or transformation into fiue other equilater uniforme
solides *Geometricall*, of his owne inuention, hitherto not mentioned of by any *Geometricians.*

Imprinted at London by Henrie Bynneman.
ANNO. 1571.

FIGURE 24. The method of triangulation, from Leonard Digges, *A Geometrical Practise, named Pantometria* (London, 1571), title page. (BL c.114.e.5. By permission of The British Library.)

of the prevalence of charlatans in the surveying business, men with little understanding and fewer scruples who had perpetrated foolish and gross errors in land surveying: "Every one that measureth Land by Laying head to head, or can take a plot by some Geometricall instrument, is not to be accounted therefore a sufficient Landmeater, except he can also proove his

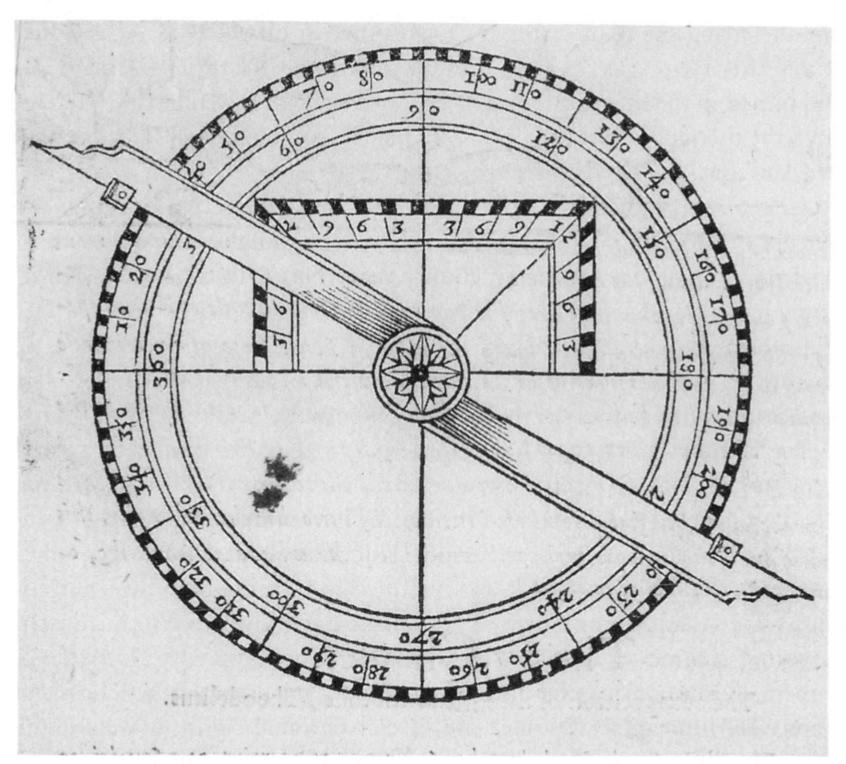

FIGURE 25. Theodolite, from Leonard Digges, *A Geometrical Practise, named Pantometria* (London, 1571), sig. H3b. (BL c.114.e.5. By permission of The British Library.)

instruments, and measurings, by true Geometricall Demonstrations."[25] Worsop, as Edward Wright had done with navigation, attempted to reform surveying through the distribution of written information while calling for proper organization and incorporation of qualified surveyors to protect their reputation and occupation.

The introduction of trigonometric techniques continued to lag behind Continental standards, as seen in the publication of *A Treatise Named Lucarsolace* (1590) by the New College graduate and friend of John Dee, Cyprian Lucar. Lucar's book, written for an audience of practical surveyors and landed gentlemen, explained the art of surveying but limited discussion to the use of geometry and the plane table.[26] He did not describe the theodolite, which had been known to Digges twenty years earlier.

Ralph Agas's *A Preparative to Plotting of Landes* (1596) provides the first real evidence of the use of the theodolite in English surveying. Agas, a practical surveyor, admitted that he had turned to the theodolite late in

25. Worsop, *Discoverie of Sundrie Errours,* sig. A2a.
26. Lucar, *Treatise Named Lucarsolace,* f. 2b. See Richeson, *English Land Measuring,* 81.

life, preferring the plane table for twenty-five of his thirty years as a surveyor. After 1591, Agas began to find the table unsatisfactory for large tracts and turned to the theodolite and to trigonometry for the solution.[27] Thereafter, the theodolite was used more frequently, although it did not become standard until the late eighteenth century.[28]

Concurrent with these more theoretical discussions of surveying, the practice of surveying for mapping was becoming much more common in England. The number of dated county maps now extant increase significantly between 1550 and 1600, suggesting both an improved survival rate and more tellingly an increased reliance on maps as symbols of county identity, as exhibits for litigation, and as proof of land ownership.[29] The most important products of this growing interest in local maps were Christopher Saxton's *Atlas* (1579), John Norden's *Speculum Britanniae* (1592), and John Speed's *Theatre of the Empire of Great Britain* (1611).[30] Saxton's *Atlas,* based on his own surveying and funded by government patronage,[31] provided the first clear image of the entire span of England, county by county (see figure 26). Norden's work was intended to be more grandiose and inclusive even than Saxton's, as well as to make use of the more sophisticated surveying techniques available by that time. Unfortunately, Norden was perpetually short of capital and never finished his great work.[32] Norden's maps were innovative for their use of conventional signs, as well as the alphabetical listing of place names, but he had great trouble finding a patron and had to finance the publication of the *Speculum* by writing best-selling works of piety (see figure 27).[33] Clearly, the audience for such large, expensive books of local maps would always be somewhat limited, so the

27. Richeson, *English Land Measuring,* 82.

28. Bendall, *Maps, Land and Society,* 134.

29. S. Tyacke, *English Map-Making,* 15. See Harley, "Meaning and Ambiguity"; P. Eden, "Three Elizabethan Estate Surveyors"; also Helgerson, "The Land Speaks." P. Barber, in "England II," argues that "map-consciousness" expanded from politicians and ministers to the country itself in the second half of the century.

30. See Tyacke and Huddy, *Christopher Saxton,* for a full discussion of Christopher Saxton's life, times, and work.

31. Ibid., 25; Evans and Lawrence, *Christopher Saxton,* 9, 66. Harvey also discusses Saxton and sees the publication of his *Atlas* as a cause of the map-consciousness of the later Elizabethan years (*Maps in Tudor England,* 84). Instead, I see both caused by the shared education in geography these gentlemen received as they attended Oxford and Cambridge in increasing numbers.

32. E. G. R. Taylor, *Late Tudor Geography,* 46.

33. Most popular, judging by the number of times they were reprinted, were *A Pensive Mans Practice* (1584), with at least sixteen editions by 1640, and *A Poore Mans Rest,* with its twentieth edition in 1684 (Lynam, "English Maps and Map-Makers," 16). Frank Kitchen, "Cosmo-choro-poly-grapher: An Analytical Account of the Life and Work of John Norden, 1547?–1625" (Ph.D. dissertation, University of Sussex, 1993), discusses the reasons for Norden's lack of success in obtaining crown or government patronage. (This thesis was not available for inclusion in this study.)

difference in the fortunes of these two English local cartographers can be directly related to government patronage. There were many reasons for the lack of government sponsorship for Norden, which had been so important for Saxton. Burghley had been extremely keen on Saxton's mapping of the entire country, obtaining each map as it was completed and creating his own atlas for administrative use.[34] Norden, on the other hand, had more trouble interesting Burghley or the queen, in part because the government was now concentrating its resources on the mapping of coastal defenses, much more important with the imminent threat of Spanish invasion.[35] In addition, Norden's involvement with the Earl of Essex did not improve his situation at court, especially after the earl's rebellion.[36] In any case, the early promise of government sponsorship of this nationalistic project was not borne out, and other means of financing chorography had to be found.

John Speed's work also helped spread an interest in county maps. More than Saxton or Norden, in fact, Speed was a true chorographer in the fuller sense of the term. His hugely popular book, *The Theatre of the Empire of Great Britain* (1611), was largely cartographic but combined with his *History of Great Britain* (1611) to provide a comprehensive work of chorography. Speed produced a concrete view of the counties of England, in both verbal and cartographic terms, and in so doing implicitly encouraged the English to conceive of their counties in the highly abstract format of geometrical maps, while identifying their own lives (as Speed described them) with those maps and counties.

This more general emphasis on antiquities, genealogy, and local history per se occupied the time and talents of the majority of English chorographers. This aspect of chorography proved instrumental in shaping English people's concept of their native land; it provided a focus for local pride and loyalty, as well as supplying a remarkable wealth of detail about one's city or shire. Developing from techniques of textual analysis, this study of chorography involved the comparison of sources for accuracy and authenticity, etymological studies of local phrases and place names, and the painstaking collection and cataloging of diverse facts and monuments. This incremental gathering of data, so fundamentally important to local history, suggests once again an early application of the inductive method, pointing to a methodology of science more strongly identified with the seventeenth than the sixteenth century. Induction, while most clearly articulated by Bacon,

34. Evans and Lawrence, *Christopher Saxton,* 6.

35. Colvin, ed., *History of the King's Works,* vol. 4. Helgerson discusses this change of fortunes for Norden (*Forms of Nationhood,* 125–27).

36. Kitchen discusses how Norden's attempt to separate himself from Essex's downfall has resulted in the creation of a biographical "double" that has plagued later interpreters of Norden's work. *Dissertations Abstracts International,* vol. 55, no. 1, 1994.

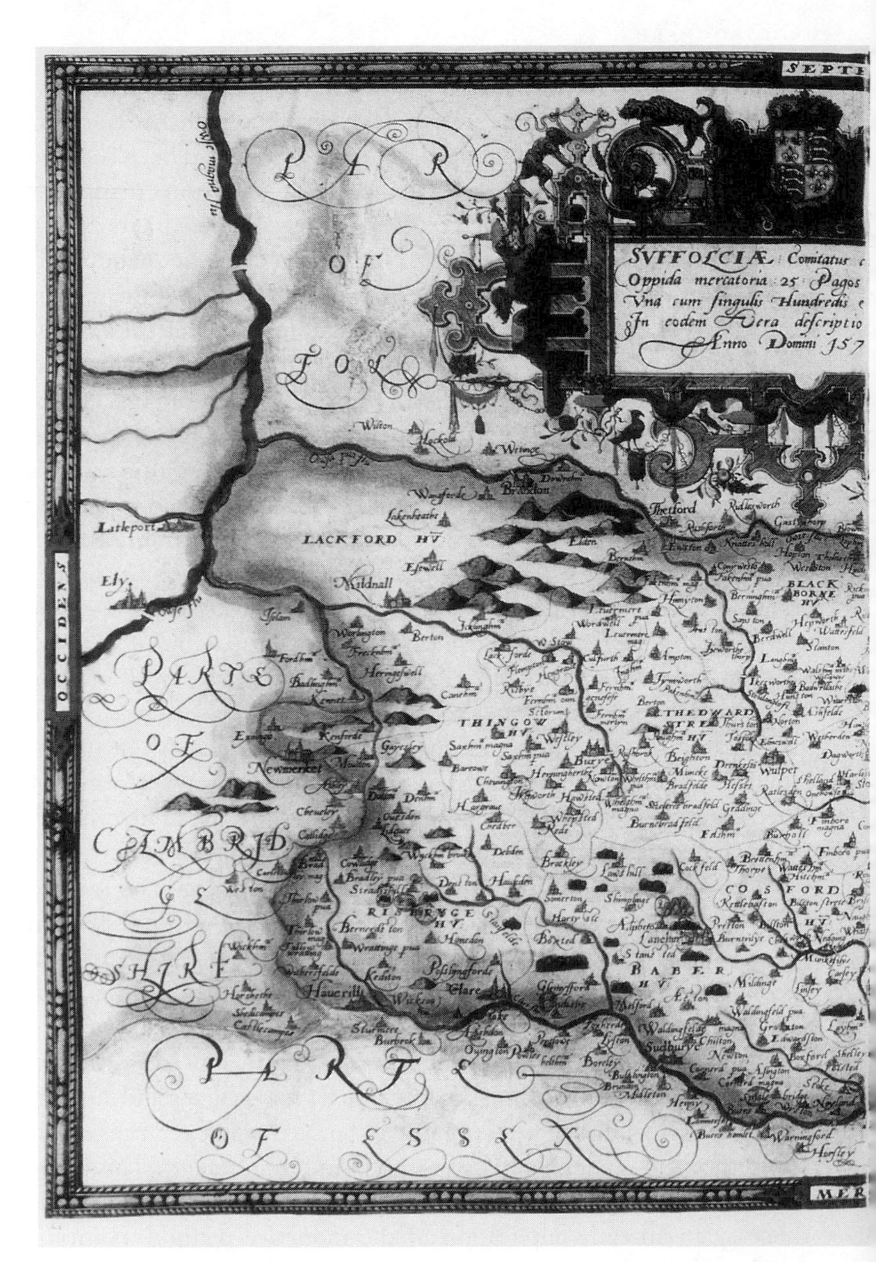

FIGURE 26. Map of Suffolk, Christopher Saxton, 1575, from *Christopher Saxon's* (Devonshire Collection, Chatsworth. Reproduced by permission of the

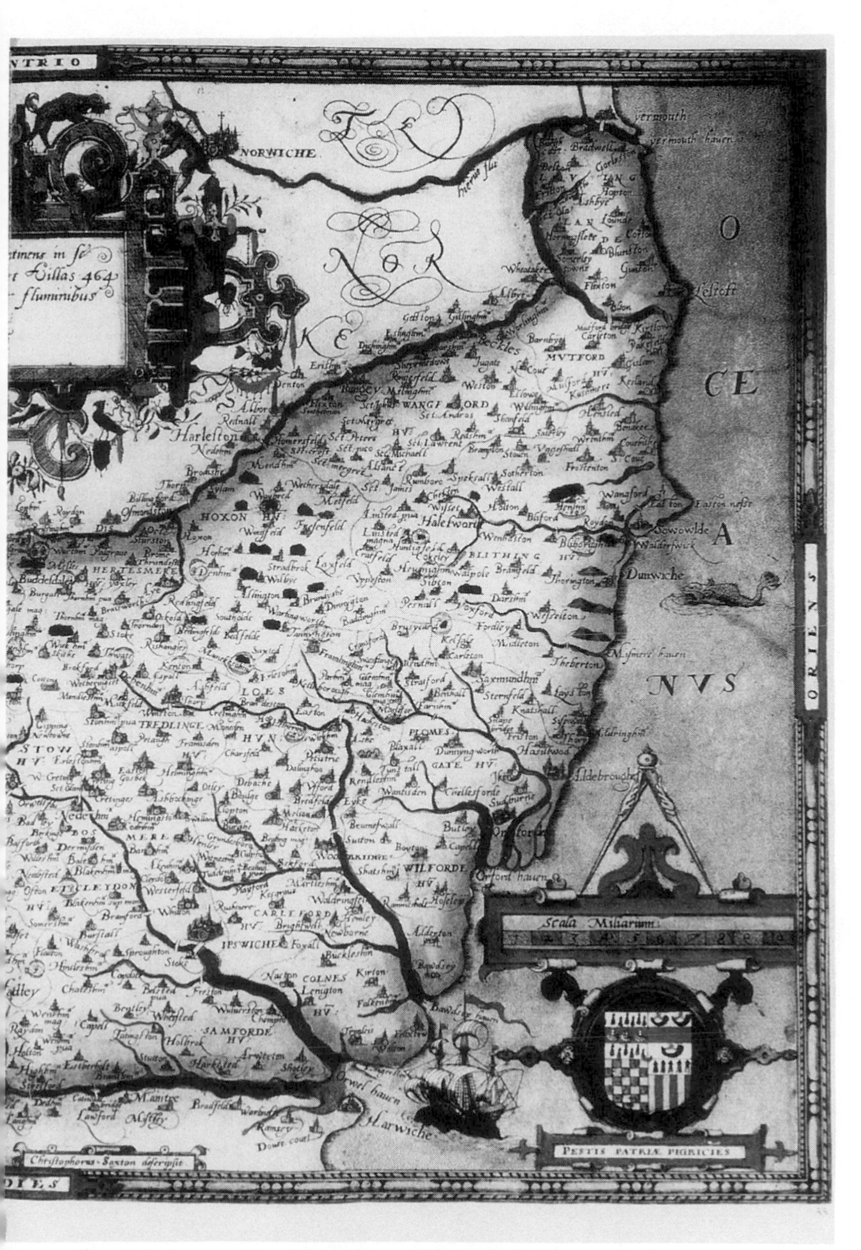

Sixteenth-Century Maps (Shrewsbury: Chatsworth Library, 1992).
Duke of Devonshire and the Chatsworth Settlement Trustees.)

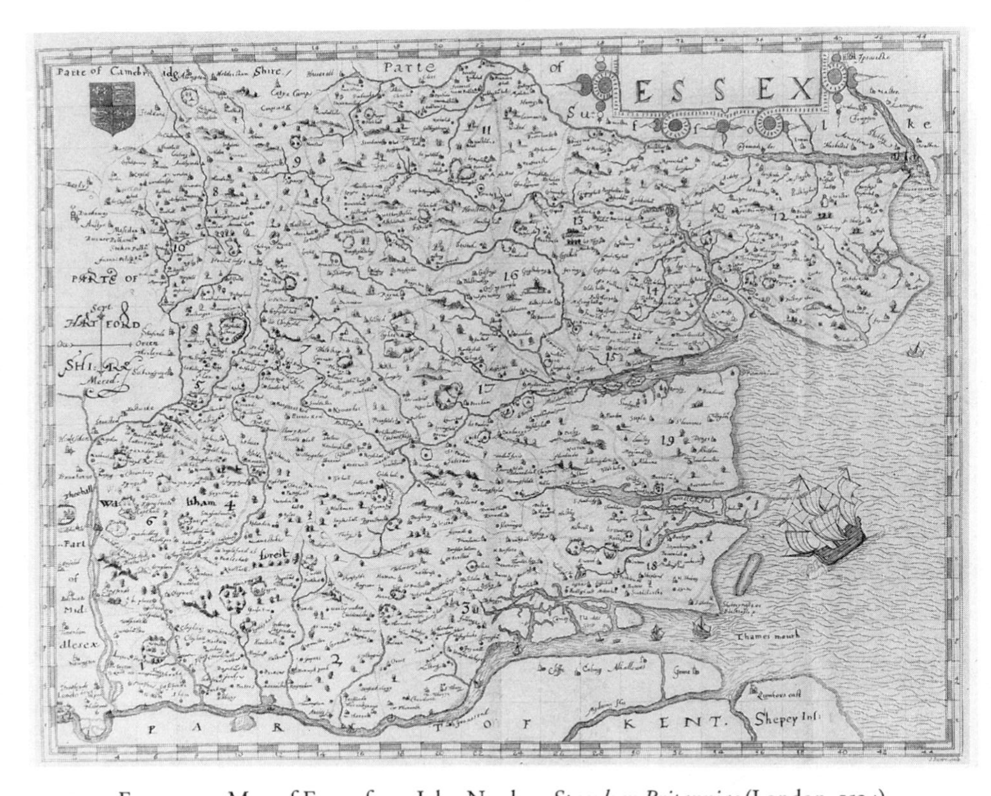

FIGURE 27. Map of Essex, from John Norden, *Speculum Britanniae* (London, 1594).

developed its methodology on an ad hoc basis from such wide-ranging studies as local history.

The most important figure of this "collecting" study in Elizabethan England was William Camden.[37] Camden's *Britannia,* first published in 1586 and enjoying a seventh Latin edition by 1607, amalgamated the volumes of chorographical information Camden had collected over the preceding fifteen years into one huge work. Just as Hakluyt's collection of voyages forever changed descriptive geography, so Camden defined and stabilized the genre of local history. In *Britannia,* Camden brought together the study of all aspects of human habitation: history, locale, linguistics, genealogy, and etymology. He greatly expanded his work in succeeding editions but never lost sight of the goal of local description.[38] He carefully defined the etymologies of each place name, in a manner reminiscent of medieval encyclopedists, but with a new sense of textual and critical rigor.[39] He quoted ancient and medieval authorities, but with a highly critical eye.[40] He described monuments, especially Roman inscriptions, churches, the countryside, and country seats. (See figure 28, Camden's rendering of Stonehenge.) In keeping with his heraldic position as Clarencieux King of Arms, Camden carefully included genealogies of prominent English families, although he was criticized for not including more genealogy.[41] Anything involving the British human condition, especially of the gentry or aristocratic classes, came under his purview. Following Biondi's structure in *Italia Illustrata,* Camden created an image of the antiquity of Britain and the inevitability of its development as an autonomous nation.[42] In *Britannia,* a new standard of critical treatment of the sources combined with a long-standing interest in the wealth and antiquity of Britain to create a book that set the pace for British local history in the centuries that followed. It helped the English define themselves by their setting and history in a way that aided burgeoning national consciousness and individual self-expression.[43]

37. For biographies of Camden, see Trevor-Roper, *Queen Elizabeth's First Historian;* Levy, "William Camden as a Historian"; also Levy, "The Making of Camden's *Britannia.*" Woolf discusses the distinction in early modern terms between historians and antiquaries, with Camden firmly in the antiquary camp (*Idea of History,* 13–23).

38. Camden, *Britannia* (London, 1586). The 1586 publication of Camden's *Britannia* was followed by seven more Latin editions: London, 1587; London, 1590; Frankfurt, 1590; London, 1594; 1600; 1607; Frankfurt, 2 vols., 1616. There were also two English editions: London, 2 vols., 1610; 1637.

39. Camden, *Britannia,* 2nd ed. (1587), for example, 649.

40. E.g., Bede on p. 381; Gerald of Wales on p. 608. Piggott claims Camden was among the first to question Geoffrey of Monmouth's myth of Brutus (*Ruins in a Landscape,* 34).

41. Piggott, *Ruins in a Landscape,* 36.

42. Rockett, "Historical Topography," 77–78. See also Rockett, "Structural Plan of Camden's *Britannia.*"

43. Camden, *Britannia* 1695, preface, sig. e1b. See also Piggott, *Ruins in a Landscape,* 9.

FIGURE 28. Stonehenge, from William Camden, *Britain, or a Chorographicall Description* (London, 1637), p. 252. (BL 1321.k.1. By permission of The British Library.)

Influenced by such geographers as Abraham Ortelius and Mercator, Camden relied on a huge network of compilers, including William Lambarde, Humphrey Lhuyd, Sampson Erdeswick, Richard Carew, Sir Robert Cotton, and Reginald Bainbrigg, to name but a few.[44] Most of these men contributed their own descriptions to the genre, as well as aiding Camden in his great work. William Lambarde, for example, the first Englishman to produce a chorographical study of a single county, described his own county of Kent in a work first published in 1570 that was copied by most chorographers up to the Civil War. This work grew out of Lambarde's interest in Old English law. It was a detailed description of notable sights and people of the county and exuded a sense of Lambarde's genuine pleasure in the description of native antiquities and genealogies. Lambarde was also interested in compiling a complete description of Britain, but he abandoned the task when he discovered his friend Camden to be working on an identical project. Instead, Lambarde supplied Camden with material on Kent and read Camden's manuscript for him.[45]

Camden also drew on material concerning Staffordshire. Sampson Erdeswick's *View of Staffordshire,* compiled from 1593 until his death in 1603, contained a wealth of local historical detail, some of which he had previously supplied to Camden and other sections of which were incorporated into the *Britannia* in its later editions. Erdeswick had an informal and conversational style, as if writing a letter to a friend. He described Staffordshire by following the courses of the major rivers and detailing each town, manor, or other habitation or marvel that lay along their paths. This was a clever organizing principle, allowing readers to set out vicariously on such a journey. It was this style that would soon lead to the geographical novel, similarly constructed to simulate a journey through space and time.[46]

44. See, e.g., Oxf. Bodl. MS Smith 71, for letters from Camden to Cotton concerning antiquities; Oxf. Bodl. MS Smith 86, f. 97, for collections from Dr. Dee; BL Lansdowne MS 121, ff. 160–64b, "Instructions for the pictes wall sent by Mr. Reginald Bainbrigg to Mr. William Camden"; and Oxf. Bodl. MS Smith 74, for letters from Camden to various foreign scholars, including Paul Merula and Abraham Ortelius. Powicke discusses Ortelius's visit with Camden in 1577 in "William Camden," 74.

45. Lambarde, *Perambulation of Kent.* See Oxf. Bodl. Rawl quarto 263, for Lambarde's copy with corrections and additions for the second edition. Many of Lambarde's editorial notes involved additions to genealogical trees or new marvels and details of various places. He also supplied a new index. On Lambarde's help to Camden, see Mendyk, *"Speculum Britanniae,"* 49. Also see Alsop and Stevens, "William Lambarde and the Elizabethan Polity."

46. I have discovered two manuscript exempla of Erdeswick's *A view of Staffordshire containing the Antiquities of the same County:* Oxf. Bodl. MS Gough Staffordshire 4; and BL Harley MS 1990. For evidence of Erdeswick's contribution to Camden, consider the following: "There is in the other Church-yard a Monument with Saxon caracters (as I take 'em) whereof I caused Worley to take a noat and send the same to Mr. Cambden to Westminster" (Oxf. Bodl. MS Gough Staffordshire 4, 48). See Adams, *Travel Literature.*

Richard Carew, a student with Camden at Christ Church, supplied another source of local material for the *Britannia* besides producing his own eloquent and comprehensive work. Carew's *Survey of Cornwall*, first circulated in manuscript form and only published in 1602 at Camden's behest, was a model of chorographical eloquence. In it, he described the fish, fowl, and animals of Cornwall, including "The Seale . . . not unlike a pigge, ugly faced, and footed like a moldwarp, he delighteth in musike, or any lowd noise, and thereby is trained to approach neare the shore, and to shew himselfe almost wholly above water [for which pains he is killed]." Likewise, a dog was described that had altruistically supplied its blind mastiff companion with daily meals. Probably more important to Carew and to Cornwall was his lengthy description of Cornish tin mining, corresponding to Dee's earlier definition of chorography.[47] He also discussed legal and political problems in the county, as well as devoting some time to the more usual etymologies of place names and genealogies of prominent families. Still, Carew's work begins to approach natural history, with its relative stress on flora and fauna, rather than remaining merely local history. It also demonstrates that an inductive methodology, combined with a relatively rigorous set of criteria for selection of data, had been introduced into English chorography well before the Civil War.

In a vein similar to Carew's and Erdeswick's work, William Burton's *The Description of Leicestershire, Containing Matters of Antiquity, Historye, Armorye, and Genealogy*[48] contributed much to this inductive fervor (see figure 29). Beginning while a student at Oxford, Burton spent much of his life gathering together all the descriptive and genealogical information he could locate concerning his native county. He listed all the place names and noble seats, all the agriculture, rivers, wondrous wells, and any extant antiquity that might be of interest. This tabulation of material indicates an identification with his county and a belief in the efficacy of such a method of data collection.

Another important English chorographer of this period was undoubtedly John Stow. Unlike the aforementioned men, Stow concerned himself with his city of London, rather than with a county. A merchant tailor by trade, he was inspired by Lambarde's work on Kent and "therefore . . . attempted the discovery of London, my native soyle and countrey."[49] In *A Survey of London* (1599), Stow described the great metropolis ward by ward,

47. Carew, *Survey of Cornwall*, sig. p¶ 4a, ff. 34b (the seal), 113a (the mastiff), 7b–18b (tin mining). The last indicates the close relationship between chorography and the emerging technology of mining and minerals, as exemplified in such works as Agricola's *De Re Metallica* (Basel, 1556). See also Mendyk, *"Speculum Britanniae,"* 77–79.

48. BL Add. MS 10,126.

49. Stow, *Survey of England*, 2nd ed. (1603), sig. A1a.

FIGURE 29. Title page, with Leicestershire and Antiquity personified, from William Burton, *The Description of Leicestershire* (London, 1622). (BL G.3495. By permission of The British Library.)

concentrating on the city itself rather than the burgeoning suburbs, and showed its antiquity and its vibrancy. Although Stow was concerned that the many changes taking place were deleterious to the well-being of the city, his main message in the *Survey* is his pride in his growing city and his view that only by remembering its history could London—and England—attain its greatest strength.[50]

Another genre of chorography was topographical poetry.[51] Most representative of this form was the work of Michael Drayton, whose *Poly-Olbion* (1613) provided a most entertaining circuit through England's counties, largely told as a romance of the rivers.[52] In this long poem, Drayton employed the work of chorographers such as Saxton, Camden, and various county authorities to tell in illustrated verse the history, local customs, and commodities of the different counties. This was entertaining material for its readers, rendered more scholarly by the learned explanations of John Selden.[53] Drayton's work shows that chorographic images of England were much more widely disseminated than just within the confines of scholarly texts and so affected the way a large portion of English people viewed their regions and their country.

Local descriptive works were also written by a number of other English chorographers. George Owen, Lord of Kemes, for example, produced a *Description of Pembrokeshire* in 1603 that took Richard Carew's book as its model. William Smith and Francis Tate wrote on cities and towns of England generally, while Edward Waterhouse compiled facts on Ireland, Sir Henry Spelman on Norfolk, and the famous Brian Twine on Oxford.[54] All were concerned with etymologies of place names, the antiquity of British towns and cities, and the basic divisions of the country, based on their own political and judicial bailiwicks. These local descriptions all seem to express a belief in the power of categorizing; by naming the parts of their world, chorographers perhaps felt that they could establish its natural order. Just

50. See Trevor-Roper, "John Stow," for a brief biography of Stow. Power, in "John Stow and His London," provides an interesting attempt to map the places described in Stow.

51. See J. G. Turner, "The Matter of Britain," for a comprehensive list of this genre.

52. Drayton, *Poly-Olbion.* Helgerson discusses both the poem and its illustrated title page in *Forms of Nationhood,* 117–22.

53. Drayton, *Poly-Olbion,* sig. A2a. "For most of what I use of *Chorographie* [claimed the author of the editorial glosses], joyne with me in thanks to that most Learned Nourice of Antiquitie my instructing friend Mr. *Camden Clarenceulx*" (sig. A3b).

54. William Smith, "The Particular Description of England. With the Portratures of Certaine of the Cheiffest Citties and Townes. 1588," BL Sloane MS 2596; Francis Tate, "The Antiquity, use and Privelidges of Cittyes, Borroughes, and Townes, Wrytten by Mr. Francis Tate of the Middle Temple London" (ca. 1598), CUL Ee. 2. 32, ff. 92–95b; Edward Waterhouse, *A Collection of the Description and Division of all the severall Shires and Townes in Ireland* (ca. 1600), CUL Dd. 3. 84, ff. 1–53; Sir Henry Spelman, *Icenia, or Description of Norfolk,* MS 1611. Part of this manuscript is copied in BL Stowe MS 163, ff. 50–63; Brian Twine, "Notes about Oxford," Oxf. Bodl. MS Wood D.32, ff. 163–73.

as the divine presence and purpose could be uncovered through the study of natural law in the political arena, so, too, could the proper classification of the countryside, that is, the establishment of a natural order based on names and measurement, reveal the divine hand in English local affairs. This revelation of order and therefore purpose was, in essence, the strength of this nascent inductive method. It gained currency in chorography because of its political and economic implications; chorographers, like lawyers in political philosophy, saw themselves empowered with certain privileges over nature by right of their taxonomic knowledge. Thus, the often tedious recitation of places, names, and dates, seen in every local history text, was essential in establishing objective criteria for the natural order of the world and man's control over it.[55] Likewise, this naming and ordering of the counties of England helped the English to establish control over their own bailiwick, an attitude that in turn was important in the wider imperial agenda.

Related to these studies of local space was that of chronology. Chronology, an often maligned branch of mathematical science, was vitally important in the sixteenth and seventeenth centuries, both as a link between human and cosmic affairs and as an attempt by scholars to render inert and mathematically controllable a larger part of their universe.[56] In the same way as local description was encouraged by a desire to order the world and therefore control it, chronology attempted to order time and gain control over history. The study of chronology stemmed from a desire to know, in absolute as well as relative terms, exactly when human events had occurred. It established the distinctions among ancient, medieval, and modern and, for Protestants, attempted to separate the chronology of the true church from that of the popish church in error.[57] For English chronologers concerned with questions of political power, chronology had to take into account the results of historical actions, such as the effect of the Norman conquest on immemorial English custom, as well as organizing events

55. Wilcox sees the sixteenth century as the beginning of this trend of looking, historically, for secondary causes rather than concentrating on divine ones (*Measure of Times Past,* 167). This is part of a distancing of people from their world and their God seen in mathematical terms in chorography and in causal terms in history. This might be seen as part of the so-called historical revolution, most clearly described by Fussner in *Historical Revolution, English Historical Thought;* and by Pocock in *Ancient Constitution.* For a discussion of this historiographical development, see Preston, "Was There an Historical Revolution?"

56. For an extremely interesting treatment of this question of time, see Grafton, "Joseph Scaliger and Historical Chronology"; "From *De Die Natali* to *De Emendatione Temporum*"; and *Joseph Scaliger.* For the larger picture, see Wilcox, *Measure of Times Past.* What follows relies in large part on these sources.

57. G. J. R. Parry makes the point that Harrison's chronology was intimately connected with his Puritan views of the problems with the Elizabethan Church. *A Protestant Vision,* 9.

sequentially.[58] This investigation of chronology in part resulted in the seventeenth-century development of the concept of absolute time, which existed independent of the world of human activity, thereby allowing chronologers to place human events within a time scale that was above petty human concerns and that revealed God's true time and purpose.

Chronology developed from two rigorous yet completely diverse studies: the humanist investigation of classical texts, and the mathematical study of astronomy and the calendar.[59] It attracted humanist historians, who were also interested in local history and description. In the fifteenth century, Leonardo Bruni and Flavio Biondi, both of whom were historians, humanists, and chorographers, began to impose a new chronology on Italian history.[60] Biondi, especially, by dating ancient events according to their relationship to the sack of Rome, introduced a new sense of change, of rise and fall, into history.[61] This new sense of time developed throughout the histories of Macchiavelli and Guicciardini, but the pinnacle of Renaissance chronology was reached by Joseph Scaliger and Domenicus Petavius. Both attempted to establish astronomical time as absolute time, and to use this time frame to pinpoint historical events (including Creation).[62]

The establishment of absolute time was a project that interested many scholars besides Scaliger and Petavius. In Britain, this investigation centered around James Ussher. Ussher is perhaps best known today for his estimate of the date of Creation, as follows: "In the beginning God created Heaven and earth which happened at the beginning of time (according to our chronology) in the first part of the night which preceded the 23rd of October in the year of the Julian Period 710."[63] This statement has in recent times been held up to ridicule and is now taken to exemplify the rigid, unscientific attitudes thought to be typical of religion in general.[64] This, of course, shows a complete misunderstanding of Ussher's work. Ussher based this figure on painstaking mathematical calculations using Scaliger's methods and in so doing raised British chronological research to its Renaissance apogee. Ussher attracted and corresponded with a group of scholars interested in correlating biblical chronologies with chronologies of histori-

58. Pocock, *Ancient Constitution,* 1–90.

59. Grafton, "From *De Die Natali* to *De Emendatione Temporum,*" 103.

60. Wilcox, *Measure of Times Past,* 163.

61. Ibid., 165.

62. Petavius developed a conventional point of reference, the birth of Christ, as the center of his chronology, although he also proved that Christ had not actually been born in that year. Wilcox, *Measure of Times Past,* 207.

63. Ussher, *Annales Veteris Testamenti,* quoted in Wilcox, *Measure of Times Past,* 187. This date corresponds to 4004 B.C.

64. This attitude can be traced to several influential nineteenth-century books that helped create the war between science and religion. See especially Draper, *History of the Conflict;* and A. D. White, *History of the Warfare of Science.*

cal events; with Irish, Greek, and Roman legends; and with semimythical events, such as the life and times of King Arthur. This chronological study thus united an interest in mathematics and astronomy with biblical exegesis, and with the collecting of old manuscripts and chronicles.

The most prolific of Ussher's coterie was Thomas Lydiat, already mentioned as a mathematical geographer. Lydiat wrote a series of chronological papers,[65] comparing Persian, Greek, and biblical chronologies, as well as corresponding with Ussher on several chronological topics.[66] In a similar manner, Robert Chambers (Roman Catholic, 1582 English College at Reims, 1592–3 at Rome) wrote to Ussher concerning the chronology of the Book of David.[67] Robert Vaughan (matriculated as a commoner, 1612, Oriel, Oxford), the Welsh antiquary, likewise composed "A Short English Annal of Ireland" in which he studied Irish chronology.[68] All these men shared an interest in mathematics, astronomy, and chronology, as well as an enforced isolation from English intellectual centers because of geography, poverty, or religion. Ussher was isolated in Ireland, Vaughan in Wales, Lydiat by his pecuniary embarrassments, and Chambers by his Catholicism. For these marginalized men, the English chronological research program was very powerful. It allowed them, lacking political or economic power as they did, to search for the classificatory power of chronology.

Local history, in the guise of surveying and cartography, of local description and genealogy, and of chronology, flourished in England in the years 1580 to 1620. Surveying, genealogy, and chronology all seemed to answer an English need for order and rigor in an increasingly confusing world and so developed into mature fields of study in the early seventeenth century. Surveying was needed to establish boundaries in a society that began to picture its holdings geometrically and to fight for them through the courts rather than with arms. Genealogy began to establish the placement of local families within a framework of antiquity and gentility. Chronology was used to place the puny events of the recent past within the mathematically rigorous schema of a common time frame. All three provided a practical setting for the development of a more secure sense of self-definition. At the same time, the more descriptive part of chorography did not advance nearly as rapidly as these three areas in the early seventeenth century, perhaps because of the lack of exactitude possible in this more amorphous subdiscipline. Local history needed the impetus of new observations and research programs in order to prosper. The saving spark of interest in local history thus appears to have been in descriptions

65. For example, Oxf. Bodl. MS Bodl. 666.
66. Oxf. Bodl. MS Add. C 297, ff. 9–10b; MS Bodl. 313, ff. 28–29, 68a–b.
67. Oxf. Bodl. MS Add. C 297, ff. 104a–b.
68. Oxf. Bodl. MS Add. A 380.

of flora and fauna, which helped to encourage the growth of local history, under the guise of natural history, only at the end of the seventeenth century.

Chorography at Oxford and Cambridge

Many students at Oxford and Cambridge were interested in the various aspects of chorography, reflected in their book ownership. Books of a chorographical nature appeared most often in the book lists of Corpus Christi College and Christ Church, Oxford, and of St. John's College and Peterhouse, Cambridge. These colleges were important for both mathematical and descriptive geography book ownership as well, indicating a cohesiveness to the study of geography, despite the diversity of the three subdisciplines.

In a college well endowed with geography books generally, chorography was an important branch of study at Corpus Christi College, Oxford.[69] Students here were interested in both the European tradition of local history and chorography developed closer to home. European chorographical works owned included three different works by Flavio Biondi, a description of the Holy Land by Christian Adrichem, two books on Roman antiquities, and two copies of J. J. Scaliger's chronological work, *De Emendatione Temporum.* As interested as Corpus men were in European chorography, they were even more intrigued with English local history. Several books were editions of medieval chorographical sources: for example, Ranulf Higden's influential *Polychronicon* and two copies of a more synthetic and consciously antiquarian work, *Rerum Britannicarum Scriptores Vetustiores* (1587). Modern monographs on the subject were also present on the book list: especially interesting were Camden's *Britannia* (1586), Carew's *Survey of Cornwall* (1602), Lambarde's *Perambulation of Kent* (1576), Stow's *Survey of London* (1599), Norden's *Speculum Britanniae* (1593), and Speed's *Theatre of Great Britain* (1611). Especially when combined with the more historical works of British chronology, such as Lively's *True Chronologie of the Persian Monarchie* (1597), and history per se, these turn-of-the-century local histories demonstrate a keen interest on the part of some Corpus Christi College students in specific descriptions of the British Isles. Corpus students owned the chronicle of the Venerable Bede (fl. 700), as well as modern histories by Polydor Vergil (1534), Raphael Holinshed (1577), and John Stow (1615) and the chronicles of Grafton (1563) and Cooper (1565).[70]

69. See appendix B for chorography books in the Corpus Christ College, Oxford, book list.

70. Venerable Bede, *Historia Ecclesiastica Gentis Anglorum* (Strassburg, 1475); Polydor Vergil, *Anglica Historia* (Basel, 1534); Raphael Holinshed, *The Chronicles of England, Scotlande, and Irelande* (London, 1577); John Stow, *The Chronicles of England, continued to 1614*

The ownership of these books concerned geographically or historically with small and discrete parts of England suggests that these students were beginning to identify England as a country and a culture separate from the rest of Europe and worthy of its own detailed analysis and resulting loyalty. As these book owners started to picture their country and eventually their county as the focus for their pride and fidelity, they became more and more enamored of this subdiscipline of local history, which instilled in them a newfound sense of nation and home. English local history thus proved to be the catalyst for a movement toward greater local loyalty and the separation of Englishmen conceptually from the Continent, while its study was in turn encouraged by such separation and loyalty.

The stress on local history at Christ Church was probably enhanced by its famous alumnus William Camden. Although no copy of Camden's definitive work on British chorography was present on the Christ Church list, there was a strong representation of the English study.[71] A copy of the so-called Matthew of Westminster's *Flores Historiarum*[72] provided a fine example of medieval English local history, while Henry Savile's *Rerum Anglicarum Scriptores Post Bedam Praecipui* (1596) and John Speed's *Theatre of the Empire of Great Britain* (1611) presented investigations of chorography that took into account more recent trends. The list also contained a respectable delegation of European chorographers, especially Flavio Biondi's *Roma Ristaurata* (1543), Guilio Ballino's *De' disegni delle piu illustri citta et fortezze del mondo* (1569), Girolamo Rossi's *Italicarum et Ravenatum Historia* (1603), Johannes Pontanus's *Rerum et Urbis Amstelodamensium Historia* (1611), and Georg Braun and Francis Hogenberg's *Civitates Orbis Terrarum* (1572). Braun's great atlas of European (and some Asian) cities, by 1617 in six volumes, provided an unparalleled collection of local scenes, costumes, and industry, which together supplied a spectacular view of European city life. Pontanus's book, as well, provided an important local history of Amsterdam, including laws, customs, fashions, and maps of the city. This book could also be seen as crossing the subdisciplines of chorography and descriptive geography, since the middle section describes the exploration by the Dutch of the Far North, India, and the Far East, com-

by Edmund Howes (London, 1615); Richard Grafton, *An Abridgement of the Chronicles of England* (London, 1563); Thomas Lanquet, *An Epitome of Cronicles . . . continued to the reign of Edwarde the sixt by T. Cooper* (London, 1565).

71. See appendix B for chorography books in the Christ Church, Oxford, book list.

72. Matthew Westminster, "name of an imaginary author, to whom is assigned, in a fifteenth-century manuscript, the chronicle 'Flores Historiarum,' compiled by various writers at the abbeys of St. Albans and Westminster, first printed, 1567" (*DNB*, epitome, 1389). Gransden intimates that this work was actually by Matthew Paris (*Historical Writing in England*, 2:436, 479). It was first printed by Matthew Parker, who began a strong effort to publish medieval chronicles.

plete with vivid illustrations.[73] These two areas of geography had clear connections with each other. There were also two genealogical works, both examining the German Nassau family, probably inspired by their leadership of the resisting Dutch: one by Hieronymus Henninges in four volumes (1588–93) and the other by Johannes Orler, *Genealogia Nassovica* (1616). There was thus a relatively high proportion of European chorographical works, especially Italian and Dutch ones, to the detriment of native *opera* in a college where more local English interest might have been generated through Camden's connection with its halls. Still, these European works were current and demonstrate an interest on the part of Christ Church students in the new and developing field of chorography.

St. John's Cambridge students showed a strong interest in chorography, of both local and European origin. The European books owned by these students and their college library included several descriptions of Rome, by Biondi, Fabricius, Raphael Maffei, and Marlianus, as well as descriptions of the Holy Land by Brocardus, Heyden, and Adrichem.[74] There was a particularly strong representation of descriptions of antiquities, especially by Apian, Carr, and Rosinus. English local history was not forgotten in this list, since St. John's students owned books by Camden, Stow, and Savile, as well as two genealogical works, Gerard Legh's *Accedens of Armory* (1562) and Augustin Vincent's *Catalogue of Nobility* (1622). This college had a strong representation of chorographical works, especially Continental sources, which indicates that this was a college where exposure to modern chorographical ideas and works was common.

Peterhouse also provided an atmosphere conducive to learning and reading about chorography. This book list also boasted a strong list of Continental authors, especially in antiquities and chronology.[75] The latter included J. J. Scaliger's important *De Emendatione Temporum,* as well as works by Bibliander, Bucholtzer, and Lucidus. The Peterhouse list of English local history contained more medieval exempla than was the case at St. John's. There were three copies of Geoffrey of Monmouth, two of Gerald of Wales, two of Ranulf Higden, and one of Matthew of Westminster. Modern works were slightly less popular, although there was a copy of Lambarde and perhaps one of Camden.[76] Clearly, students at Peterhouse were interested in chorographical topics, particularly antiquities, chronology, and descriptions of medieval Britain.

73. Pontanus, *Rerum et Urbis Amstelodamensium Historia,* 129–228.

74. See appendix B for chorography books on the St. John's College, Cambridge, book list.

75. See appendix B for chorography books in the Peterhouse, Cambridge, book list.

76. The identification of "Kamlin Historiae Britanniae," in [Richard?] Mote, Inventory 1592, Cambridge VCC, Bundle 5: 1589–1595, is uncertain.

Other colleges at both Oxford and Cambridge fostered an interest in local history. The Trinity College, Cambridge, book list contained a remarkable collection of major medieval British chorographical works: Geoffrey of Monmouth's *Historia Regis Britanniae* (ca. 1138), Gildas's *De Calamitate, Excidio et Conquestu Britanniae* (ca. 570), Gervase of Tilbury's *Otia Imperialia* (ca. 1211), Gerald of Wales's *Itinerarium Cambriae* (ca. 1180), William of Malmesbury's *De Gestis Regum Anglorum* (ca. 1140), and that most influential work, Ranulf Higden's *Polychronicon* (ca. 1350). When combined with the more modern local histories, such as Camden's *Britannia* (1586), Braun and Hogenberg's *Civitates Orbis Terrarum* (1572), Raphael Maffei's *Geographia* (1506), and Sir Lewis Lewkenor's translation of Gaspar Contarini's *Commonwealth and Government of Venice* (1599), Trinity emerges as the vital center for the study of local history. The book lists connected with St. John's and New College, Oxford, both contained significant chorographical collections. St. John's students owned books by Camden (four copies), Speed, Fougasses, Lively, J. J. Scaliger, and Reisner; and New College students read the works of Pietro Bembo, Speed, Stow, and Scaliger, as well as a sizable number of works on antiquities. Smaller collections of works of local history appeared on the book lists of All Souls, Oriel, and Queen's at Oxford and Corpus, Sidney Sussex, Gonville and Caius, and King's at Cambridge (see tables 2 and 4, above). One or two local history books also appeared on the lists of Christ's, Emmanuel, Queens', Clare Hall, and Pembroke Hall, Cambridge, and at Lincoln, Exeter, Balliol, Brasenose, Merton, Trinity, Magdalen, and University colleges, Oxford. There were thus a slightly smaller number of colleges whose students were known to have been interested in chorography than colleges with an interest in mathematical and descriptive geography, suggesting that local history was less frequently pursued, or at least that fewer students owned books on the topic. Even so, there was a significant interest in this geographical field, a field that engaged its practitioner with the present but also demanded a new perception of the historical remoteness of geographical settings and of antiquities, allowing them to be seen as relics of another age rather than merely as modern curiosities.[77] This removal of the student from his object of study, encouraged by an inductive method of fact gathering, helped to develop an ideology of objectivity that resulted in the growth of seemingly objective criteria for accepting information and in the belief in the power of classification to reveal natural order.

The relationship between chorography and chronology was a close one. No local history would be complete without a chronological genealogy of

77. See Burke, *Renaissance Sense of the Past.* This revolution in historical consciousness was first described for the English scene by Pocock and more recently articulated by Woolf in *Idea of History.* Pocock sees Henry Spelman as the first to articulate this view of the past.

all the important local families, and often local histories appear to have been little more than such genealogies in a geographical setting.[78] Thus, it comes as no surprise to find that many of the local history books on these Oxford and Cambridge book lists were either basically genealogical or contained numerous genealogies. In addition, many of the chronicles in the history section of the lists were indispensable for the study of chorography, as well as being of more general interest to descriptive geographers and to chronologers. This is borne out by the fact that all the college book lists noted for their collections of local history books also listed medieval chronicles. For example, Corpus Christi College and Christ Church, Oxford, and St. John's College and Peterhouse, Cambridge, book lists each included several chronicles by such authors as Walter de Hemingford (ca. 1346), Johannes Funck (1545), Helmoldus (d. 1177), Hartmann Schedel (1493), Johannes Nauclerus (ca. 1500), Konrad von Lichtenau (1472), and Thietmar, Bishop of Merseburg (ca. 1500).[79] These German "medieval" chronicles may have gained importance with the English Reformation, when theologians required a medium in which to embed the events of the restructured church in the framework of God's divine purpose. In other words, the motivation for this interest in the chronological ordering of events through time was similar to the more general appeal of local history in establishing God's natural classification of things, both topographically and chronologically.

The emphasis of chorography changed over time, from a treatment influenced by medieval chronicles in the early period to a discipline stressing family trees and estates in the later decades. By examining the most popular chorographical authors through time, this development becomes clear (see table 8).

From 1580 to 1610, there was a marked reliance on medieval chorographical authors, such as Ranulf Higden, Matthew Westminster, and Geoffrey of Monmouth, peaking in 1610. By 1620, however, this reliance had disappeared. Instead, new local history books appeared, with such au-

78. Mendyk sees a specific trend in chorography toward this genealogical bent, occurring at about the turn of the seventeenth century. This, he feels, showed the dissatisfaction with the older tradition of local history and allowed local historians by the late seventeenth century to move to descriptions of natural history (*"Speculum Britanniae,"* 82).

79. Walter de Hemingford, *Chronica 1066–1346* (first printed Paris, 1508); Johannes Funck, *Chronologia* (first printed Nuremberg, 1545); Helmoldus, *Chronica Slavorum* (ed. R. Reineck, Frankfurt, 1581); Hartmann Schedel, *Liber Chronicarum* (Nuremberg, 1493); Johannes Nauclerus, *Chronica usque ad annum 1500* (Tubingen, n.d.); Konrad von Lichtenau, *Chronicon Urspergensis* (Augsberg, 1472); Thietmar, Bishop of Merseburg, *Chronica* (printed Frankfurt, 1580).

Table 8 Chorography Books Owned, by Decade

	1580	*1590*	*1600*	*1610*	*1620*
Raphael Maffei	2	3	—	2	5
Ranulf Higden	2	3	1	2	1
Biondi (*Roma Instaurata*)	2	1	1	4	3
Camden	1*	4	4	5	4
Matthew Westminster	1	1	3	5	3
Geoffrey Monmouth	—	3	—	3	1
Adrichem	—	1	3	2	2
Rosinus	—	1	2	4	1
Speed (*Theatre*)	—	1†	—	3	2
Lily	—	—	3	1	—
J. J. Scaliger	—	—	2	3	2
Suidas	—	—	1	3	1
Stow (*Survey*)	—	—	1	1	3
Savile	—	—	—	4	1

*John Glover, M.A. Fellow St. John's Oxford (1578), from Inventory 1578, Oxf. V.C.C. Recorded as "Fragmentum Britannicum". Perhaps this refers to an early manuscript version of Camden's *Britannia* (published 1586).

†King's College, Cambridge, Chained Donors Book, f.9. Transcribed by F. L. Clarke in early twentieth century (manuscript at King's). Clarke dated donations and checked extant library books for publishing information. Richard Lancaster's donation (B.Th., chaplain to Prince Henry) is dated by Clarke as 1589. It reads as follows: "Speedi Chron: angliae et descript: comitat: authore eodem" and "Item his mapps". This cannot be Speed's published work (1611), but perhaps is a manuscript version. Alternately, these could be Saxton's maps and the scribe might have been confused.

thors as J. J. Scaliger, Georg Braun, Laurentius Schrader, John Speed, and John Stow.[80] Camden and Biondi in many respects provided a bridge between these two groups, since *Britannia* and of *Roma Instaurata* are owned in multiple copies throughout the period under discussion. Since Camden's work combined a healthy respect for the old chronologers and chorographers with a new interest in antiquities and history, he logically and topically provided the link between these two groups and helped to serve as a catalyst for the new movement in local history, developing in the 1620s and onward. Likewise, Biondi's work provided a model from the early Italian chorographical renaissance for the development of English local history at the beginning of the seventeenth century. Besides these two, however, there was relatively little overlap between the "medieval" English phase in local history and the modern European phase. One copy of Bembo and

80. J. J. Scaliger, *De Emendatione Temporum* (Paris, 1583); Georg Braun and Francis Hogenberg, *Civitates Orbis Terrarum* (Cologne, 1597); Laurentius Schrader, *Monumentorum Italiae libri 4* (Helmstadt, 1592); Speed, *Theatre of the Empire;* Stow, *Survey of London.* Mendyk's work agrees with this change in the seventeenth century, although he sees it as a real degeneration of the chorographical tradition, not halted until natural history was introduced into the field after the Civil War (*"Speculum Britanniae").*

one of Braun were owned in 1600, and one copy of Ranulf Higden and of Geoffrey of Monmouth appeared on the book list in 1620, but generally the two sections appear to have been discontinuous. Indeed, the proportion of local history books that could be classified as medieval chorography fell drastically in the seventeenth century, accounting for 44 percent of local history books in 1580 and only 14 percent by 1620. This decrease in medieval sources was compensated for by a dramatic increase in the proportion of local history books that dealt primarily with antiquities, rising from 9 percent in 1600 and 16 percent in 1610 to 27 percent by 1620. These were entirely Continental sources, often dealing with Roman antiquities. Their presence suggests that Oxford and Cambridge students were beginning to combine a burgeoning interest in such antiquities and history with a new sense of distance from that Roman past. Thus, the local history book lists demonstrate a marked replacement by the early seventeenth century of the medieval English chronicles so popular in the late sixteenth century, with the Continental sources containing both general descriptions of cities and specialized studies of antiquities.

Belatedly, then, English students after 1600 gained an awareness of the solid work of European and especially of Italian chorographers, allowing these students to replace their older British sources of local history with more current Continental ones. Until 1610, the majority of local history books (two-thirds in 1580, 49 percent in 1610) were by English authors (see figure 30). By 1620, English authors accounted for only one-third of all local history books, while Italian authors were responsible for nearly as many. Since German, French, and Dutch authorship of chorographical works on this book list remained relatively stable after 1600, the reduced reliance on native ideas in chorography can be seen to have been the result of a new Italianate influence, witnessed as well in the art and architecture of the first decades of the seventeenth century.[81]

The development of local history in terms of the number of books owned followed a radically different pattern from that of descriptive geography, which began, after the turn of the century, to rely more and more heavily on the native production of geographical descriptions. The study of local history thus should not be conflated with that of descriptive geography, since the two followed clearly divergent paths of topic choice and influence. Local history became more cosmopolitan and continental at the same time as descriptive geography became more insular. The result was a loss of English innovation in local history until the end of the seventeenth century, since its students were content to look abroad rather than to at-

81. Especially Constantino de'Servi, Salomon de Caus, and Inigo Jones. Strong, *Henry, Prince of Wales,* 88. See chapter 6 below for a more complete treatment of this phenomenon.

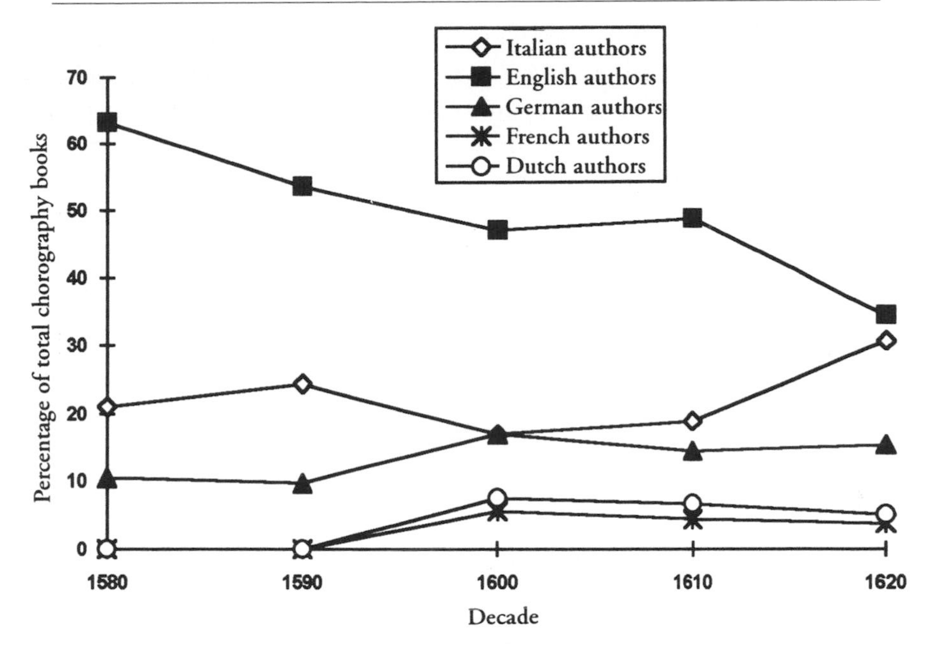

FIGURE 30. Countries of origin of chorography book authors.

tempt new tacks themselves, while descriptive geography developed as a vital English subdiscipline. The stellar names in English chorography—William Camden, Gervase Markham, William Lambarde—all did their most important work before the turn of the seventeenth century. The heyday of innovative chorography at Oxford and Cambridge thus occurred in the first thirty years of the period under investigation, rather than later.[82]

University Chorographers

The scholars interested in this study of chorography had much in common with those men who studied the other two branches of geography. Indeed, the same men sometimes read or wrote in two or three different areas. There was most often an overlap between mathematical geographers and chorographers such as Thomas Lydiat, though there were also descriptive geographers, such as Richard Verstegen, who moved at times into local history. Taken as a body, local historians had slightly different educational and career paths from those of adherents to the other two subdisciplines. Local historians appear to have attended university for longer periods of time than did descriptive geographers, though for shorter periods than

82. This accords well with Mendyk's contention in *"Speculum Britanniae"* that local history became less objective and more dilettantish in the early seventeenth century, only recovering in the 1660s.

mathematical geographers. More chorographers obtained degrees in theology than did the men in the other two subdisciplines, while, somewhat paradoxically, fewer became members of the clergy. The trait that most clearly set local historians apart from their geographical confreres was their relative isolation from the larger academic community. Once having attended university, these men often retreated from the mainstream, because of either distant geographical locations or relatively unpopular religious or political affiliations. These university chorographers, then, were a group of men, usually involved in the curricular Arts program, who were engaged in public affairs and yet were often isolated, by geography, finances, or religion, from the main community of scholars.

Of the 116 men known to have had some interest in chorography, more than two-thirds attended university. This was a much higher proportion than was the case with descriptive geographers, who tended to stress practical experience at the expense of university education, but was approximately equal to the proportion of students of mathematical geography who attended university. The majority of these university-educated men interested in local history (61 percent) attended Oxford. Oxford seems to have provided a more sympathetic environment for this subdiscipline than did Cambridge. In any case, university education seems to have been very important for students of both chorography and mathematical geography, the more mathematically based subdisciplines, more important than for descriptive geographers. This suggests that the universities provided the training and the environment in which these two subdisciplines could develop, while at the same time allowing for the vital contribution of political and social issues. Especially for chorography, appealing as it did to the new men of the universities, Oxford and Cambridge furnished significant support at a large number of colleges and halls.

The class breakdown of chorographers at the universities follows the pattern established within the two other geographical fields; twenty-five men came from gentle families, two were sons of clergy, six cited mercantile origins. In all, 67 percent of those whose class origins are known were from the "new classes" of the gentry and merchants. This suggests that chorography in particular, as well as geography in general, was an attractive study for these newly empowered classes. The close self-identification with place that chorography provided, as well as its practical utility, gave these increasingly important members of early modern English society a sense of their roles and location in the political and social spheres.

In a path reminiscent of mathematical geography students, a large number of the disciples of chorography attained a Bachelor's degree in Arts: fifty-two, or almost two-thirds of those who attended university. More than half of the university attendees remained for a Master's degree, while fully a quarter of these chorographically minded young men obtained the degree of Bachelor of Theology. These men were clearly not at the universities merely to pick up a little polish but were serious students with a prescribed set of courses and often with a clerical career in mind. Unlike the more desultory study of descriptive geography, local history was part of a serious intellectual and university-based enterprise.

Twenty-three of these men devoted their lives to the serious pursuit of chorography, either as antiquaries or as local historians. Some, like Richard Carew, were landed gentry and so made their living from their estates, but others, like John Norden, relied for their subsistence on the proceeds of their writing, either chorographical or devotional.

The most famous of these career chorographers was undoubtedly William Camden. Camden earned his living for a time as headmaster of Westminster School, but after his appointment as Clarencieux King of Arms in 1597 and his growing publishing fame, he was able to devote the rest of his life to antiquarian research, correspondence, and publishing.[83] The appointment of England's most important chorographer to this heraldic position of honor demonstrates more clearly than anything else the prominent position of genealogical research in the program of chorographical study. Although Camden himself used this position to investigate local history and British antiquities more generally, the public concern with genealogical investigation supplied the wherewithal he needed to continue his work. Camden pursued that work through his vast communication network; he knew every important English and European chorographer and conducted exhaustive correspondence with them all.[84] With his European connections, Camden was unusual in the English chorographical community. His audience was European, rather than strictly English, evident from his decision to publish in Latin. His subject matter, on the other hand, was exclusively British, and with this double-focused approach, Camden interpreted the English to themselves and to their Continental colleagues. This was, in part, an imperial act, since Camden was demonstrating the antiquity of the native British peoples, claiming their precedence over more recent Roman incursions.

The second great antiquarian of the age, and the man to whom we owe one of the finest early manuscript collections in England, was Sir Robert Bruce Cotton.[85] A close friend of Camden, as well as of John Selden, James

83. *DNB,* 8:279.

84. Oxf. Bodl. MS Smith 74, for correspondence between Camden and Ortelius, P. Merula, J. Gruterus, D. Grevile, inter alia.

85. For the best biography of Cotton, see Sharpe, *Sir Robert Cotton;* also Mirrlees, *A Fly in Amber.* Tite, in *Manuscript Library of Sir Robert Cotton,* has explored the development of Cotton's library, both during his lifetime and until it became part of the British Library collection.

Ussher, Francis Bacon, and Sir Walter Raleigh, to name but a few, Cotton (B.A. 1586, Jesus, Cambridge) devoted his life to the collection of manuscripts and antiquities. His property in Huntingdonshire maintained him in comfort, allowing him to purchase manuscripts and books and to spend his time aiding other scholars. While pursuing these scholarly activities, he did not neglect his more worldly duties. Cotton served as M.P. for each Parliament from 1601 to 1628 and was a staunch supporter of the rights of Parliament. Indeed, Cotton's careful establishment of his library, including monastic, Oriental, and political manuscripts, could also be seen as the performance of his duty to the commonweal, since Cotton saw himself creating a public lending library in lieu of the national library he had campaigned for unsuccessfully.[86] This close relationship between public service and private scholarship, representing England through his chorographical research, the scholarly community through his library, and his county through political activity, demonstrates once again the combination of theory and practice so much a part of the pursuit of geography. The understanding of the local land through geographical investigation was part and parcel of the governance ideology that encouraged the public responsibility of guarding local interests in Parliament. In addition, Cotton's interest in chorography was integrally related to his strongly held belief in the strength and antiquity of Britain and its institutions—and its superiority to other European nations.

John Selden (matriculated 1600, Hart Hall, Oxford; Inns of Court) was another of this group of antiquarians, consulting Cotton's manuscripts and himself responsible for the large collection of Oriental manuscripts donated to the Bodleian Library at his death in 1654.[87] Selden, like Cotton, supported the Parliamentary cause, serving as M.P. for Oxford during the Long Parliament and participating in the impeachment of Archbishop Laud.

Several other landed gentlemen seriously pursued a vocation in local history. These included William Lambarde, Sampson Erdeswick (admitted as gentleman-commoner 1553, Brasenose, Oxford), Richard Carew, and Brian Twine (B.A. 1599, M.A. 1603, Corpus Christi, Oxford). All of these men, as we have seen, wrote scholarly tracts on local description; Carew, Erdeswick, and Lambarde all devoted themselves to the counties in which their estates lay. Carew also represented the interests of two boroughs of Cornwall in Parliament. Local history became for these men, particularly Carew and Erdeswick, a pursuit compatible with their relative isolation in Cornwall and Staffordshire, making a virtue of the necessity of their exile

86. Tite, *Manuscript Library of Sir Robert Cotton,* ch. 3.

87. See "Catalogue of Mr. Seldens Library, given to Oxford," Oxf. Bodl. MS Add. C.40, 1–147.

from the intellectual centers of London and the universities, while encouraging them to identify with their counties and assume their governance responsibilities there.

Almost one-quarter of the university-trained chorographers held academic posts throughout their lives. None was as eminent as the professional antiquarians and chorographers already discussed, but this list does include Edward Brerewood (B.A. 1587, M.A. 1590, Brasenose, Oxford), Gresham Professor of Astronomy, author of a book on the origin of different languages,[88] and member of the Society of Antiquaries; William Lisle (B.A. 1589, M.A. 1592, King's College, Cambridge), an Anglo-Saxon scholar at Cambridge and correspondent with both Sir Henry Spelman (to whom he was related) and Sir Robert Cotton;[89] Thomas White (B.A. 1570, M.A. 1573, B.Th. 1581, D.Th. 1585, Christ Church, Oxford), the founder of the moral philosophy lectureship at Oxford, founder of Sion College in London, and donator to the Christ Church Library of several chorographical works;[90] Richard Kilbie (B.A. 1578, M.A. 1582, B.Th., D.Th. 1596, Lincoln College, Oxford), Regius Professor of Hebrew and another of the translators of the Authorized Version; and Alexander Neville, scholar of Anglo-Saxon law and secretary to Archbishop Parker. Others held less exalted posts as schoolteachers, which probably resulted in their relative isolation from the cutting edge of scholarship and suggests the recurring theme of chorographical study combined with some degree of separation from London and the universities.

Only 15 percent of the chorographers who attended university entered a career in the church. The most famous of these was Bishop Ussher.[91] James Ussher (M.A. 1601, D.Th. 1614, Trinity College, Dublin) became Archbishop of Armagh in Ireland, as well as the center of an extremely active group of chronologers in the early seventeenth century. He also corresponded with and visited Cotton and formed part of a circle of chorographers centered around the Inns of Court. Hugh Broughton was another theologian who was interested in chronology. In 1588, he published *A Concent of Scripture,* an attempt to settle biblical chronological disputes. He was a noted Hebraist and used his knowledge of this language to compare Jewish chronologies with the more usual biblical ones. William Harrison, a "hot Protestant" preacher, was deeply concerned with chronological is-

88. Edward Brerewood, *Enquiries touching the Diversity of Languages and religions through the chiefe partes of the World* (London, 1614).

89. For correspondence between Lisle and Cotton, see Oxf. Bodl. MS Smith 71, 113, 153.

90. See *DNB* and J. Foster, *Alumni Oxonienses,* for biographical information. Christ Church Donor's Register, 22.

91. For an early biography of Ussher, see Parr, *Life of James Ussher.* For more modern discussions of this influential man, see R. B. Knox, *James Ussher.*

sues, as both his "Great English Chronology" and "Chronological computations from the beginning of the world to his own time" attest.

These three examples indicate a close connection between chronological problems and theological ones.[92] Part of the biblical exegesis so important for sixteenth- and seventeenth-century theologians, wrestling with problems of chronology allowed these theologians to impose the newly valued methodology of measurement on biblical commentary.[93] Chronological studies attempted to distance the true church from the mistaken "popish" church by discovering God's time in a rational and mathematical manner. Just as the physical sciences were developing an ideology of quantification that encouraged measurement and categorization, so Ussher, Broughton, and Harrison were attempting to impose on theology a similar stress on rigorous quantification. It would prove somewhat less successful in theology than in physics, though biblical commentary continued to strive for objectivity and rigor in the following centuries.

As well as providing a new direction for biblical interpretation, chorography encouraged an interest in and identification with the local scene. Part of this interest was promoted by the relative isolation of some local historians, in geographical, financial, or religious terms. Political involvement with the local community also often furthered an interest in the chorography of that community. Several of the men interested in chorography espoused religious opinions other than those endorsed by the Church of England. Four are known to have been Romanists; five more were avowed Puritans.[94] These men, like those living on far-flung estates or with few financial resources, would have experienced some degree of marginalization from the scholarly community. The Catholics especially would have been limited in their public involvement, and local history probably provided an important outlet for their energies and political inclinations. Thus, these men on the religious margins could use the politically charged study of local history to develop their sense of identification with England and their counties. For others, this identification with the land resulted in an active political life. Fourteen chorographers, for example, served as

92. G. J. R. Parry, in *A Protestant Vision,* argues for this spiritual connection with chronology.

93. Christianson, in *Reformers and Babylon,* discusses one use of chronology, that of the end times. Most of the chronologers discussed in this study were less inclined to millenarianism.

94. The Roman Catholics included Richard Rowlands (alias Verstegen), Richard Stanyhurst (or Stonyhurst), Robert Chambers, and Lord William Howard (third son of Thomas Howard, fourth Duke of Norfolk). Those with hot Protestant sympathies included Abraham Hartwell the younger, William Harrison, John Norden, Robert Waldegrave (the Puritan printer), and John Whitgift, Archbishop of Canterbury.

M.P.'s at one time or another.[95] Thus service to the state combined well with interest in the local scene, and perhaps the two could be seen as complementary manifestations of a growing interest in local affairs and pride of place.

Many of those who studied chorography kept in contact even after their student days, most clearly demonstrated by the coteries of local historians that grew up in the late sixteenth century. As with mathematical geography, a loose but important network of scholars developed around the study of local history. The foci of this network were three men of great energy and resourcefulness—men with a research program into which their fellow chorographers could fit their own smaller efforts. These local historians, who made English chorography what it was, were William Camden, Robert Cotton, and James Ussher. These three men were friends and correspondents, sharing gossip and meals as well as manuscripts and geographical information.[96] Each collected around himself a large and somewhat amorphous group of men, interested in borrowing manuscripts or contributing and obtaining information. These three scholars served as a source of material for the student of local history, as well as acting as a clearing house for such information. This early manifestation of the republic of letters helped to develop a structure and a program for English chorographical and chronological research.

Ussher's circle was based on correspondence rather than on personal contact. Although Ussher did visit Sir Robert Cotton on occasion, his isolation in Ireland ensured that much of his chronological work was done by post. He corresponded with Cotton and Camden, borrowing and lending manuscripts to the former and exchanging information with the latter.[97] His most faithful correspondent was the redoubtable Thomas Lydiat, who

95. John Hooker (alias Vowell), M.P. Athenry (Irish Parliament), 1568; for Exeter, 1571, 1586–7. Richard Carew, M.P. Saltash, 1584; Michel 1597. Alexander Neville, M.P. Christchurch, 1584–86; Saltash 1601. Sir Thomas Chaloner, the younger, M.P. St. Mawes, 1586–87; Lostwithiel, 1604–11. Abraham Hartwell, the younger, perhaps M.P. 1586, 1593. Sir William Pole, M.P. Bossiney, 1586. John Lyly, M.P. Hindon, 1589; Aylesbury, 1593, 1601; Appleby, 1597. Sir Henry Spelman, M.P. Castle Rising, 1593, 1597–98; Worcester 1625. Sir John Dodridge, M.P. Barnstaple, 1588–89; Horsham, 1603–11. Sir Robert Cotton, M.P. Newtown, Isle of Wight, 1601; Huntington, 1604; Old Sarum, 1624; Thetford, 1625; Castle Rising, 1628–29. Francis Tate, M.P. Northampton, 1601; Shrewsbury, 1604–11. John Selden, M.P. Lanchaster, March 1621, 1624–25; Great Bedwin, 1626; Ludgershall, 1628; Oxford University, 1640–53 (Long Parliament). Sir John Stradling, M.P. St. Germans, 1624–25; Old Sarum, 1625; Glamorganshire, 1626. Sir James Ware, M.P. Dublin University, 1634–37, 1661 (Irish Parliament).

96. For correspondence between Ussher and Cotton, see Oxf. Bodl. MS Smith 71, 133; between Camden and Cotton, 35, 93, 117; between Camden and Ussher, 93.

97. Ussher exchanging books with Camden, Oxf. Bodl. MS Smith 71, 93; with Cotton, 133.

wrote concerning chronological calculations. Ussher also communicated with the chronologically minded men at the Inns of Court, especially those around William Crashaw.[98]

The largest group of local historians developed around Sir Robert Cotton, whose huge collection of manuscripts and magnificent library provided a remarkable research resource. Cotton began collecting manuscripts in about 1588, at the age of seventeen.[99] Soon he amassed a truly staggering collection of books and manuscripts concerning British antiquities.[100] Many local historians used this collection, either directly or through correspondence. Roger Dodsworth, the antiquary, for example, is known to have "studied in London in the library of Sir Robert Cotton."[101] Sir Thomas Chaloner, Sir Henry Bourchier, John Overall, Richard Montagu, Richard Carew, and Richard Verstegan, among others, all exchanged letters with Sir Robert concerning manuscripts and antiquities.[102] His closer friends and fellow antiquaries included such eminent men as William Burton (Robert's elder brother), William Lisle (the Anglo-Saxon scholar), John Selden, Sir Henry Spelman, and, of course, William Camden.[103]

Camden had been Headmaster of Westminster when Cotton attended the school in the 1570s, and he likely first interested the younger man in the study of British local history. Camden certainly attracted a large group of interested men who formed the nucleus of a cooperative and energetic stable of chorographers engaged for the next twenty years in a program of chorographical research. Camden's notebooks were full of correspondence from men spread throughout Britain who reported back on the exact wording of a newly found Roman inscription or the discovery of a new Roman or Saxon coin.[104] Camden corresponded with such local historians as his friend William Lambarde, John Meyrick (Bishop of Man and a Welsh antiquary), Reginald Bainbrigg (the Cambridge mathematician), and Lord William Howard (a Norfolk recusant and improving landlord).[105] In London, he was surrounded by men such as Cotton, John Speed, Sir John Stadling (poet and M.P.), and the mathematically and

98. Fisher, in "William Crashawe's Library," discusses this group. See chapter 3 above.

99. J. Evans, *History of the Society of Antiquaries,* 7.

100. Tite, *Manuscript Library of Sir Robert Cotton,* ch. 1.

101. *DNB;* Oxf. Bodl. MS Dodsw. 141, ff. 41–78b.

102. Oxf. Bodl. MS Smith 71. Letters to Cotton from Chaloner, 55; Bourchier, 143; Overall, 77; Montagu, 71, 107; Carew, 41–42; Verstegan, 57, 105.

103. Ibid., from Burton, 145; Lisle, 113, 153; Selden, 141. See also Fisher, "William Crashawe's Library," 120.

104. Camden, "Notes on English Antiquities," Oxf. Bodl. MS Smith 84.

105. Mendyk, *"Speculum Britanniae,"* 54; Meyrick acknowledged Camden in BL Cotton MS Julius F.10; Bainbrigg corresponded with Camden, BL Lansdowne 121, ff. 160–164b.

chronologically oriented group centered around William Crashaw at the Inns of Court.[106] In fact, this London group appears to have formed the nucleus for a somewhat mysterious group, the Society of Antiquaries.[107] According to Linda Van Norden's careful chronology, the Society of Antiquaries first met in about 1586, the date of the publication of Camden's *Britannia.*[108] The society held thirty-eight documented meetings at Derby House in London,[109] where the members debated questions concerning British antiquities. It had close connections with the Inns of Court, perhaps with Crashaw's group. In fact, the members of the Society of Antiquaries tended to be lawyers and heralds, university-educated, but not clerics.[110] The true aim of the society was "to establish a 'cultural longevity' for their country,"[111] to demonstrate the antiquity and nobility of their native land. These men had a new vision of Britain,[112] both as an island with ancient rights, traditions, and history and as a focus for the encouragement of local loyalties. There thus grew up from these beginnings a seemingly paradoxical loyalty to and identification with the local region on the one hand, seen in increased interest in town and county histories and maps, and on the other a stress on the antiquity of British institutions and traditions that would help prepare the ground for the revolt of county men and lawyers against the king and royal prerogative. The Society of Antiquaries did not breed regicides (with the exception of Sir Thomas Chaloner's son), but it did encourage an attitude toward the land and its people that might allow individuals to distance themselves from the crown, and perhaps begin to transfer their loyalties from king to country more generally and sometimes even to Parliament. The crown thus ceased to be the exclusive focus of ideals of national cohesion, in part because of the study of local history.

This enthusiasm for chorography slackened in the seventeenth century, and the Society of Antiquaries appears to have become moribund after

106. See Fisher, "William Crashawe's Library." Also BL Add. 36,294, f. 25, for correspondence with Stadling.

107. This society attracted much interest before 1960, though it seems to have become less popular since. See Gough, "Historical Account of the Origin"; van Norden, "Sir Henry Spelman on the Chronology"; Schoeck, "Elizabethan Society of Antiquaries"; Schuyler, "Antiquaries and Sir Henry Spelman." Probably the most complete, if commissioned, is J. Evans, *History of the Society of Antiquaries.*

108. Van Norden, "Henry Spelman on the Chronology," 158.

109. J. Evans, *History of the Society of Antiquaries,* 10.

110. Ibid., 11. This closely parallels my portrait of the typical chorography student.

111. Ibid.

112. These men usually spoke of Britain, which for them certainly included Wales and sometimes Scotland and Ireland. It clearly had a historical coherence, even when they thought politically only of England.

about 1607 or 1608.[113] As the European and especially Italian influences began to invade the consciousness of English scholars and chorographers, they turned from British antiquities and local history toward the glories of Greece and Rome.[114] But the seeds had been sown, and interest in British tradition and antiquities would reemerge politically in the 1640s and intellectually by the early eighteenth century.

THE STUDY OF local history followed a path largely independent of the other geographical subdisciplines. Founded on medieval sources and inspired by Italian Renaissance historians, local history quickly developed in late-sixteenth-century England as a result of the manifestation of a number of political and social needs that this study could address. The surveying of lands became more important in the latter half of the century as legal arguments began to rely on more accurate descriptions and increasingly measured platts. Genealogical investigation developed as more people became concerned with the antiquity of their families. Chronology grew as both the reformed church and local historians wished to establish an absolute time in which to place past events. Chorography generally burgeoned as naming and classifying began to take on the aura of power and self-definition. Local history depended on mathematics and the objective classification of names and places; it thus helped introduce inductive methodology into the study of the world and united this with the methodology of humanist textual analysis. The landscape became text in a way that tamed it and demystified it. Those who studied chorography identified themselves with the landscape and began to create a mindset that saw England as unique and superior. Often, this resulted in an emphasis on local loyalties; eventually, it produced an English population secure in its self-definition and willing to take on the world, either through Parliament or through trading ventures.

The interest and application of university students in this chorographical subdiscipline reached its apogee at the end of the sixteenth century. By that time, chorography had become a small but significant part of the Arts curriculum, studied by people with a teaching career in mind or who would return to their county estates. After 1600, other concerns than purely English chorography predominated, and, although the study of Continental local history continued with great enthusiasm, the native study diminished, only to be resurrected with new political events and the final merging of natural history with chorography at mid-century.

113. Van Norden, "Henry Spelman on the Chronology," 159.
114. J. Evans, *History of the Society of Antiquaries,* 15.

CHAPTER SIX

The Patronage of Patriotism: The "Third University" of London

The study of geography, in all three manifestations, encouraged a new way of envisaging England and the wider world in which it increasingly participated. This new imperial point of view was based on a belief in the power that came with information, measurement, and classification. Thus, geography, studied in the Arts curriculum by students increasingly destined for public positions of power, provided a series of images and attitudes toward governance and empire that gave these young men a posture of superiority toward other European and non-Western countries, as well as a self-identification as Englishmen. The men who absorbed these lessons from geography at Oxford and Cambridge set out into the wider political, social, and economic world better equipped with this newfound knowledge and understanding. As we have already seen, many took up professional careers as teachers, clerics, lawyers, and occasionally physicians. Others gravitated back to their country estates, where they took their governance duties seriously, serving as justices of the peace, members of Parliament, and lords of their manors. Still others gravitated toward London, that hub of economic and political power, and it is here that the lessons of geography were to prove most useful. For it was in London that the actual empire was to begin, with merchant ventures and changing court and government attitudes. In order to realize a tangible empire, the English needed an image of what that empire would be and an attitude that would facilitate its creation.

London in the late sixteenth and early seventeenth centuries was a booming metropolis.[1] Many of the newly formed trading companies had their headquarters there, and so it was in London that navigators, instrument makers, merchants, seamen, and investors congregated. This was also the printing center of England, encouraging the presence of skilled engravers and printers, especially important for the creation of the new geographical books and, in the 1590s, atlases. London was also the focus for much

1. For interesting analyses of early modern London, and especially why the potential crisis of the 1590s did not materialize, see Archer, *Pursuit of Stability;* and Rappaport, *Worlds within Worlds.*

of the country's political activity, both with the meeting of Parliament and the constant gathering of courtiers at the principal residences of the English monarchs. Thus, London attracted those men who had best absorbed the imperial message of geography, and, in turn, it was in London that we find the clearest evidence of a geographical research program aimed at imperial aggrandizement.

Geography at Gresham College

The most likely spot for such a geographical research center was Gresham College. Historians of science have long been inclined to see in Gresham College the embryonic foundation of both the Royal Society and the "new science" itself,[2] and since the early modern discipline of geography was so important for the changing face of seventeenth-century science, its presence at Gresham was significant. Unfortunately, Gresham was not quite the early modern community college historians like Christopher Hill wished it to be. Although it did supply a potentially important meeting place for scholars and craftsmen interested in geography and empire, it was not as significant for the application of the new ideologies as the royal court.[3]

Gresham College was founded by Thomas Gresham, a practical, wealthy, and strongly Protestant Elizabethan merchant. When Gresham died in 1579, he left his house and wealth to found a college for the edification of merchants and tradesmen in London. He wished for the college to teach useful things, in English, and stipulated in his will the subjects to be taught and the methods to be employed at his college. The law professor, for example, was to teach "subjects like wills, trusts, usury, contracts, sale and purchase, ships, seamen and navigation, monopolies, trade and merchants . . . which would be of direct interest to his audience."[4] The Professor of Astronomy was to teach astronomy, of course, but also geography and the art of navigation.[5] Since any Londoners could attend without charge, practical instruction in the geographical arts was available, at least in theory, to all those in London wishing to avail themselves of this opportunity.

2. Especially Johnson, "Gresham College: Precursor." Also Perver, *Royal Society;* Ornstein, *Role of Scientific Societies;* and Stimson, *Scientists and Amateurs,* on the Royal Society. Also C. Hill, *Intellectual Origins;* Charlton, *Education in Renaissance England;* Lawson and Silver, *Social History of Education,* on the new science and the relationship between science and Puritanism found at Gresham.

3. Feingold argues for the secondary nature of Gresham as compared with the universities (*Mathematicians' Apprenticeship,* ch. 5).

4. C. Hill, *Intellectual Origins,* 36.

5. Adamson, "Foundation and Early History of Gresham College," 46.

Unfortunately, the path to philanthropy is never straightforward, and Gresham's wishes could not be fulfilled until 1597. His widow remained in the house, and the college could not proceed until her death. Even then, wrangling over jurisdiction and over who actually controlled the college was to impede its work well into the seventeenth century. When the college was finally created, it was based on mid-sixteenth- rather than seventeenth-century ideas of practical education. As Ian Adamson reiterates, "It is a point sometimes overlooked by those whose eyes are on Gresham College but whose minds are on the Royal Society, that the College and its subject-range emanated from a man who was born 140 years before the Royal Society was founded."[6]

Adamson's exhaustive work on the early history of Gresham College has demonstrated that the college was neither intellectually innovative nor administratively sound from its very inception. There were constant struggles between the committee administering the college and its professors, with the result that no one really achieved control; regulations were seldom enforced, and appointments were often subject to patronage and corruption.[7] The high hopes of Thomas Gresham and the serious academic incumbents of the early professorships were seldom fulfilled.

Navigators and merchants may well have attended lectures at Gresham and learned something of the geographical discipline that held so much interest for the academics hired to fill the professorships. It is extremely difficult to assess just how great any such influence might have been, however, since there were no attendance records for classes at the college, no graduates per se, and no notebooks that might indicate keen attention to Gresham lectures.

Probably the most important role played by Gresham College was the employment it provided, in London, for several men keenly interested in the mathematical sciences in general and in geography specifically. Six Gresham professors, including three professors of astronomy, were interested in geographical topics. These included Edward Brerewood, a descriptive geographer who wrote on languages of the world; Edmund Gunter, Henry Gellibrand, and Henry Briggs, who were all mathematical geographers investigating the variation of the compass; Matthew Gwinne, a friend of John Florio and interested in descriptive geography; and Thomas

6. Ibid., 249. Feingold relies on Adamson's earlier work ("Foundation and Early History of Gresham College" and "Administration of Gresham College") in his evaluation (*Mathematicians' Apprenticeship,* 167, n. 5). Indeed, we might learn more about Gresham by comparing it with other sixteenth-century proposals than by looking at the Royal Society in the seventeenth century (e.g., H. Gilbert's *Queen Elizabethe's Achademy,* Elyot's *The governour,* and Ascham's *The Scholemaster).*

7. Adamson, "Foundation and Early History of Gresham College," 253.

Winston, a prominent member of the Virginia Company.[8] These men might well have raised the awareness of interested Londoners in the discipline of geography. In turn, they probably benefited from the contacts they established with craftsmen while in their professorships. Gresham thus was an important point of contact between scholars and craftsmen, encouraging practical men to see the deeper theoretical and imperial issues at stake and aiding the scholars in their appreciation of real-life problems and knowledge.

Still, Gresham College should not be seen as an institution completely separate from Oxford and Cambridge. Since all the Gresham professors with geographical interests obtained their degrees at one of the English universities and since many of them returned to one of these universities in their later careers (for example, Briggs returned to Oxford with his appointment as Savilian Chair of Astronomy), it is reasonable to see geographical interest at Gresham as an offshoot of that at the universities, rather than as a separately evolving extra-university tradition.[9] Indeed, according to Adamson, the Gresham professors themselves saw Gresham as a university college, though catering to a somewhat different clientele, and so saw the development of geographical ideas there as an extension of their university careers, rather than as a completely different venture.[10] Geographical investigation at Gresham College, then, which did exist at least at the professorship level, should be seen as complementary to, rather than in competition with, the work at the universities.

Gresham was not the location where geographical imperial ideology was aggressively applied—that required more power, money, and influence than that foundation possessed. To find such a radical program, we must turn to the royal courts. As we have already seen, Elizabeth's court provided a strong focus for imperial projects and proposals. Although it existed for a very short period, the court of Henry Fredrick, Prince of Wales, proved an even more striking example of geographical activity. Henry's court provides, in highly condensed form, as example of the type of political and imperial geographical activity that developed in a royal setting. It points to the strong connections between practical and theoretical, between schol-

8. Edward Brerewood, Professor of Astronomy 1596–1613; Edmund Gunter, Professor of Astronomy 1619–26; Henry Gellibrand, Professor of Astronomy 1627–36; Henry Briggs, Professor of Geometry 1596–1620; Matthew Gwinne, Professor of Physic 1596–1607; Thomas Winston, Professor of Physic 1615–42, 1652–55.

9. This argues against Christopher Hill's position for Gresham as the only bright spot in an otherwise gloomy pre–Civil War educational system. Feingold makes a similar point against Hill concerning the mathematical arts more generally (*Mathematicians' Apprenticeship,* ch. 5).

10. Adamson, "Foundation and Early History of Gresham College," 250.

arly geography and imperial aspirations, that were so important to the growth of geography in early modern England.[11]

Geography at Court

The royal courts had great potential for those interested in applying the imperial lessons of geography, since they could supply patronage, power, and a locale where like-minded geographers could meet. Especially at the court of young Henry, eldest son of James I, the conditions seemed ideal (although his lack of cash inhibited a full dispensation of patronage possibilities) and resulted in the creation of a community of geographers surrounding the prince. This group of court geographers supplied a link between the geographical discipline investigated within the universities and the political world where the shared geographical worldview could be put into practice. Henry's strong sense of imperial destiny suited the emerging ideology of superiority and otherness developing in the study of geography. This ideology attracted Henry to the geographers he gathered around him, while their presence at his court undoubtedly honed their imperial theories as well. The geographers who gravitated to Henry's court saw there an opportunity to use their geographical knowledge for the glory of England. This brief court, then, gives us a glimpse into the complementarity of university and political world as well as the application of imperial ideology to a larger political arena.

When Prince Henry died in 1612, a dynamic and innovative court died with him. Though only nineteen, he had already gathered around him a circle of the best and brightest individuals of English Renaissance art and science. In art and architecture, he had begun to cultivate a new Italianate style, while his scientific interests revolved around practical problems of war and overseas expansion.[12] Many of the most significant figures of early-seventeenth-century geographical investigation were associated with Henry and his court at Richmond Palace; had he survived to become king, English navigation, geography, and colonization might have been pursued much more aggressively and systematically in the pre–Civil War period.

11. Much more work needs to be done on the royal patronage of natural philosophy and the mathematical arts in early modern England. I am presently working on a study on Stuart patronage that should begin to fill this gap.

12. Most histories of Prince Henry have followed Cornwallis, *Life and Death of Henry Prince of Wales,* or the work of the antiquarian Birch, *Life of Henry, Prince of Wales.* Later work includes E. C. Wilson, *Prince Henry and English Literature;* and J. W. Williamson, *Myth of the Conqueror.* For the most complete and informative biography of this lost Renaissance figure, see Strong, *Henry, Prince of Wales.* Strong is rather too enamored of possible mystical elements in Henry's court but is really the first to understand the importance of Henry to art, theater, and architecture.

A large and coherent network of geographically minded individuals, larger and more coherent than that around any sixteenth-century English monarch, had its focal point at Henry's court in Richmond Palace. These men came from different backgrounds, though often sharing a university education, gentle birth, or the experience of Italian travel, but they all articulated a desire to improve the mapping of the world to England's benefit. Christopher Hill, one of the few historians to have recognized the importance of Henry's patronage for science, claims that this patronage was a result of Henry's violently anti-Catholic, extreme Protestant stance.[13] This represents only one side of the picture. Rather, the common denominator of this coterie of geographers was a growing sense of patriotism and imperialism, which Henry himself shared, and a "proto-Baconian" desire to render the more esoteric studies of mathematics and astronomy useful to England as it took its place on the international scene. Henry was a patron of those espousing notions of national aggrandizement and public utility of knowledge.[14] The result was a focus for concentrated practical and theoretical geographical research, aiding in the development of an experimental attitude soon to become popular in seventeenth-century science.[15]

The study of geography also provided a venue for Henry's court and English people more generally to fashion themselves as separate and superior to other people both within and outside the country. Henry's court provided an alternative to the corrupt and suspect court of his father, allowing English people to identify Henry with England's chivalrous past and encouraging them to think of him as a future imperial conqueror.[16] This separation of Henry's image from that of James was aided by the geographical vision and knowledge of Henry's court. By gathering around him

13. C. Hill, *Intellectual Origins,* 213–19, esp. 219.

14. Rosenberg, in *Leicester: Patron of Letters,* sees a similar impetus to Leicester's patronage activities fifty years earlier. Another political patron is examined in P. E. J. Hammer, "'The Bright Shininge Sparke,'" esp. ch. 6.

15. Studies of patronage in the early modern courts, especially the English courts, have largely centered on political and literary patronage. For a general view, see Trevor-Roper, *Princes and Artists;* Dickens, *Courts of Europe;* and Lytle and Orgel, eds., *Patronage in the Renaissance.* Good detailed examination of the political and cultural patronage system of the early Stuart courts includes Peck, *Northampton: Patronage and Policy* and *Court Patronage;* Orgel and Strong, *Inigo Jones;* and Smuts, *Court Culture.* None of these, however, deals with science at the courts; all mention it only to dismiss it. Little work has been done in English historiography to rival Moran, "German Prince-Practitioners," or the more recent Biagioli, "Galileo's System of Patronage" or *Galileo Courtier.* See Moran, ed., *Patronage and Institutions,* for several interesting expositions of scientific patronage.

16. Malcolm Smuts, in *Court Culture,* esp. ch. 2, sees Henry as the inheritor of the "cult of Elizabeth" after 1603. While James was forced to deal with Spain and by temperament favored a more diplomatic relationship with Europe, Henry still represented the hope for English hegemony. Smuts sees this movement by James toward Europe resulting in the rapid introduction of Baroque European culture and the distancing of the monarch from his people. The movement, according to Smuts, was partially retarded by Henry.

some of the leading geographers of his day, Henry fashioned himself into an imperial explorer and conqueror. The study of geography helped the English to identify themselves as separate from the rest of Europe, while it focused their aspirations on the authority of the prince in this conquering guise. Geography thus provided a powerful impetus for the creation of a seventeenth-century English psyche, both within and outside the court. Using images and descriptions of the world, these courtiers and those who watched them fashioned themselves as valiant English imperialists. In so doing, they engaged in the kind of cultural negotiations Stephen Greenblatt has described, using words, texts, and images to create a view of self in the face of powerful forces of conformity and control (as existed in James's court).[17]

Henry Fredrick passed through his most formative years under the watchful eye of the English nation. He embodied the hope for peaceful dynastic succession for a mere nine years, from his father James I's accession to the English throne in 1603, when Henry was just ten, until the prince's untimely death of typhoid in 1612 at the age of nineteen.[18] He occupied during that time a unique position in the minds of English patriots; he was the first heir in more than fifty years to grow up secure in his succession to the English throne. In fact, Henry was the first royal child of an English ruling family since Elizabeth's own childhood. The English took Henry, his younger brother Charles, and his sister Elizabeth to their hearts, and Henry soon embodied for them all that was necessary to make England the preeminent power of Protestant Europe and perhaps of the world.[19]

Henry was precocious from the very outset of his public exposure, though he inclined more to chivalric and sporting pursuits than to academic ones. This endeared him to a public seeking a chivalrous knight but caused his father, a scholar and a pacific man of compromise, much distress.[20] This fundamental conflict grew more evident with each passing year, and the whole history of Henry's court at Richmond Palace involved the increasing antagonism between James I and his son.[21]

The prince's court was initially set up by James, but soon Henry began appointing his own men. This circle of active Renaissance men is seen by

17. Greenblatt, *Renaissance Self-Fashioning.*

18. PRO SP 14/71, no. 32, Nov. 12, 1612. *The Funeralls of the High and Mightie Prince* (London 1613).

19. Strong, *Henry, Prince of Wales,* 7–11.

20. Ibid., 10.

21. See, for example, PRO SP 14/71, no. 42, f. 71, Nov. 26, 1612. Letter from John Chamberlain to Carleton, reporting that the king was keeping Charles under stricter supervision than his brother had been. The king "geves order that the younger Prince be kept within a stricter compasse then the former."

Sir Roy Strong, Henry's most recent biographer, as a focus for the ultra-Protestant war party in England, which favored engagement with Continental forces and increased imperial aspirations for the English crown.[22] Where James always attempted to govern by compromise, to the point of raising Henry as an extreme Protestant and his brother Charles as an Anglo-Catholic and of attempting to marry them to women of complementary religious faiths, Henry longed for decisive military and religious action. He saw himself as heir to the Protestant leadership of Elizabeth, who had been the focus of a powerful cult, based on native English virtues of chivalry and valor in warfare.[23] With the peace of 1604, those with military aspirations turned to Henry, who pictured himself in the ranks of such Protestant soldiers of Europe as Henri IV (whose pragmatic defection to Catholicism seems not to have moved Henry from his hero worship) and Prince Maurice of Nassau.[24] Hence, Henry provided a focus for deeply felt Protestant English patriotism, such as had been articulated in the works of Hakluyt during Elizabeth's reign.

Henry also used much of his influence to bring the Italian Renaissance to England.[25] Henry was fascinated by Italy and all things Italian. He demanded detailed descriptions of the country from his friends and courtiers who had the freedom to travel, which he lacked.[26] He learned a great deal about Italian Renaissance art and became a true connoisseur. He began the first royal art collection and sent courtiers to Medician Florence and Rudolfine Prague to collect pieces, often specifying works by well-known Italian artists such as Michelangelo and Raphael.[27] Strong argues that Hen-

22. Strong, *Henry, Prince of Wales,* 26, 66–70.

23. Smuts, *Court Culture,* 15–24.

24. Strong, *Henry, Prince of Wales,* 72, 73.

25. Ibid., 86.

26. These men included Sir Thomas Chaloner, his Lord Chamberlain, who had been in Florence from 1596 to 1598 as an agent for Essex; Sir William Alexander, Gentleman Extraordinary of the Privy Chamber, who also visited Europe and Italy before joining Henry's service; John Holles (1564–1637), Henry's Comptroller, who visited Italy at Henry's behest; Sir Robert Dallington (1561–1637); Sir John Danvers (1588?–1655), designer of Henry's Italian gardens; John Harington, second Baron Harington of Exton, who corresponded widely with Henry concerning Italy and Europe more generally and who was one of Henry's closer friends; and Edward Cecil (1572–1638), younger son of Robert.

27. Strong, *Henry, Prince of Wales,* 184. Although all the Tudors (especially Henry VIII) had owned royal portraits and patronized painters (see Millar, *Tudor-Stuart and Early Georgian Pictures,* 1:9–13), Henry was the first royal to collect in a conscious way, separate from portraits for power or marriage. Henry's connection with Rudolf's court is doubly interesting, given Rudolf's patronage of science and technology. See Moran, "German Prince-Practitioners"; for a more general portrayal of Rudolf's patronage, see R. J. W. Evans, *Rudolf II and his World.* It is tempting to see the development of Henry's research program in geography in part influenced by an importation of a similar attitude from Prague. In fact, there exist two candidates for the transmitter of these values: John Dee and Cornelius Drebbel. Dee had been at Rudolf's court from 1583 to 1586, and his influence on Henry's court

ry's greatest claim to fame was his importation to England of Italian styles and tastes in art, architecture, and performance, a movement that lost much headway after Henry's untimely death but changed the shape of England's culture irrevocably. Thus, focused on Henry's persona were two divergent geographical interests: the renewal and growth of "Merrie Olde Englande" and connections with and knowledge of Europe, especially Italy. Added to this was a third area of geographical investigation: the New World.

In his self-appointed role as the Protestant prince par excellence, preparing to vanquish Catholicism and to conquer the world, Henry was deeply concerned with English geographical expansion and colonization. Michael Drayton predicted, in his dedication to *Poly-Olbion,* that Henry would develop trade over three seas and control great kingdoms:

> Britaine, behold here portray'd, to thy sight
> Henry, the best hope, and the world's delight;
> Ordain'd to make thy eight Great Henries, nine:
> Who, by that vertue in the trebble Trine,
> To his own goodnesse (in his Being) brings
> These severall Glories of th'eight English Kings;
> Deep Knowledge, Greatnes, long Life, Policy,
> Courage, Zeale, Fortune, awfull Maiestie.
> He like great Neptune on three Seas shall rove
> And rule three Realms, with triple power, like Jove;
> Thus in soft Peace, thus in tempestuous Warres,
> Till from his foote, his Fame shall strike the starres.[28]

Henry was a strong supporter of the Second Virginia Company, which saw him as its political and spiritual leader.[29] Henry also had always defended Sir Walter Raleigh, that quintessential Elizabethan privateer, against James, stating that "no king but his father would keep such a bird in a cage."[30] Through these and other interests, Henry developed a research program at his court, albeit informal and unarticulated, in geographical investigation, especially concerned with colonial enterprises. Strong claims that "colonial expansion may be seen as an extension of the chivalrous tra-

---

was great. Likewise, Drebbel, a scientific inventor, served both Rudolf and Henry (though at different times). Further investigation into the lives of these two men and their work might well prove fruitful in linking these two patrons of science and technology. See Clulee, *John Dee's Natural Philosophy,* for some discussion of Dee's time in Prague; and Feingold, *Mathematicians' Apprenticeship,* 202, for mention of Drebbel.

28. Drayton, *Poly-Olbion,* sig. x3b.

29. Palumbo, "The Path to fame, the proofe of zeale." I owe this reference to Paul Christianson.

30. Birch, *Life of Henry, Prince of Wales,* 392.

dition of which the Prince's exercises in the tiltyard were also a part."[31] In other words, Henry's interest in geography and colonization might be seen as a natural conclusion to his attempt to become a sort of Castiglionian courtier.[32] Castiglione's courtier, however, was a somewhat emasculated individual, who placated his prince and made up for lack of real power and ambition by his smooth manner and prowess on the tiltyard. Henry's geographical interest, on the other hand, evolved more from a desire to emulate the great English heroes, especially Henry V, or the rival Spanish *conquistadores* than from the chivalric ideals of the Italian courtiers. Henry's enthusiasm for practical geographical problems thus sprang from a belief in his ability to act, rather than as a substitute for action. An interest in geography helped Henry to define himself as different from his father and separate from the Continental princes he emulated. In this way, a court of geographical protégés provided Henry with a plan for imperial expansion and an image of himself as a conquering prince. Henry's court thus developed a serious program of geographical investigation, thanks to the coterie of geographically engaged and active men with whom Henry surrounded himself, and became an important nexus of geographical activity outside the English universities.

At least thirty-seven men, connected in some way with Henry's court, are known to have shared an interest in some aspect of geography.[33] Ten of them were mathematical geographers, either occupied with the more practical concerns of instrument making, navigation, and ship design or exploring more theoretical questions concerning the mathematical construction of the globe and the relation of astronomy to geography. Almost two-thirds (twenty-three) were interested in describing the world, either through some firsthand experience of America, Europe, or the Orient or through a less engaged approach, involving textual analysis and gathering comparative tales of countries, peoples, and voyages. A further four could best be classed as local historians or chorographers, since they spent their time describing their own counties rather than traveling farther afield. Though this latter group seems less part of the aggressive outward orienta-

31. Strong, *Henry, Prince of Wales,* 62.

32. Baldesar Castiglione, *The Book of the Courtier,* first printed Venice, 1528; first printed in English, translated by Thomas Hoby, London, 1561. Henry's court would have known this work, providing some overlap between geography and Castiglionian chivalry.

33. See Cormack, "Twisting the Lion's Tail," 73 for a full list of these men. This compares with forty-one men interested in geography connected with Elizabeth's court, over a period of twenty-three years. Thus, Henry's court, though slightly smaller in raw numbers, was much more geographically concentrated than Elizabeth's had been. The biographical information for this section comes from five main sources: *DNB;* Venn, *Alumni Cantabrigienses;* Clark and Boase, *Register of the University of Oxford;* J. Foster, *Alumni Oxonienses;* and PRO State Papers, Domestic, James I. In addition, Strong, *Henry, Prince of Wales,* supplies many details from his more extensive primary research into Henry's court generally.

tion of Henry's policies, local histories reflected a pride of county and country that agreed well with Henry's attitudes. Still, the majority of geographically inclined courtiers surrounding Prince Henry were most interested in descriptive geography, which was partly a result of Henry's predilection for colonization and exploration but which in turn might have encouraged him to think more deeply about those topics.

As has already been demonstrated, university training was valuable for most people interested in pursuing the study of geography. University attendance also was becoming almost a prerequisite for court and government appointments, such as those held by many of the geographers around Henry. Thus, it is not surprising to find that a clear majority of these geographically inclined courtiers had university training. Two-thirds are known to have attended university, fifteen (41 percent) at Oxford and eight (22 percent) at Cambridge.[34] Of these twenty-five university-trained men, sixteen received the degree of Bachelor of Arts, and fifteen were granted a Master of Arts. Considerably fewer of them remained to study at the higher faculties. Four of these geographically minded men studied medicine, and a further six received higher degrees in theology.

These court geographers, in other words, match the picture of the new men attending university on their way to positions of power and prominence. The Arts faculty attendance indicates that they were serious scholars. The shared ideology they there developed, based in part on their geographical studies, equipped them well for the court connections they were now making. Their patronage paths also indicate that their view of empire was an important part of their intellectual world.

The men in Henry's geographical circle who did not attend university tended to pursue careers that involved the acquisition and use of geographical knowledge in a practical manner, either as travelers and explorers or in more down-to-earth metiers as gunners, naval commanders, instrument makers, or inventors. They often supplied the raw materials for geographical investigation in terms of descriptions or instruments, though they also on occasion took an interest in a broader understanding of the world map. Henry's court thus acted as an important site for the meeting of practical craftsmen and scholars, as well as an ideal locale for those who wished to combine these perspectives.

The religious affiliations of some of these men lend credence to the claims of Hill and Strong concerning the strong Protestant tendencies of the court. Seven of the twenty-five university-trained geographers held positions in the church. The religious predilections of most of these geo-

34. A few attended foreign universities; specifically, Sir William Alexander attended Stirling, Glasgow, and Leiden, while Theodore Mayerne de Turquet went to Montpelier (*DNB*).

graphically inclined courtiers are untraceable, but nine can be identified as having strong Protestant or even Puritan leanings.[35] Thus, Henry fostered at least some men with a combination of "hot Protestant" ideals and geographical interests in his court, suggesting some agreement with Christopher Hill's identification of science and geography at Henry's court with the rising Puritanism of the seventeenth century.

Those geographers at Henry's court who were neither mechanics nor theologians tended to be teachers, academics, and professional writers. Many of them taught geography as a topic in applied mathematics. Others were gentlemen of the Privy Chamber and dedicated descriptions of other countries to Henry. These were the Richmond geographers who were most deeply concerned with the discipline of geography as a complete entity and with increasing the theoretical and practical knowledge of the world. Eleven of these men instructed Henry in the geographical arts or presented books of geography, travel, or mathematics to him as courtiers. It is this more select group that forms the nucleus of this geographical court faction, both in terms of their influence on Henry and contact with one another and in terms of creating a program of research from the pressing geographical problems of the day. These men, as Henry's instructors, were more truly of the court than merely at the court, and, as geographers who combined theoretical and practical concerns, they were more seriously engaged with the geographical enterprise than the other career courtiers could be. To understand the place of geography at Richmond Palace and to appreciate its role in Henry's attitudes and policies, these men must be examined.

Perhaps the most noteworthy was Edward Wright, who combined a theoretical foundation gained from his university studies with practical firsthand experience of the problems of navigation. At about the turn of the century, Wright moved from Cambridge to London, where he established himself as a teacher of mathematics and geography. In the early seventeenth century, he is said to have become a tutor to Prince Henry, a

35. These included Sir Robert Dallington; John Harington, second Baron Harington of Exton; William Crashaw; Josuah Sylvester, a translator, merchant, and Groom of Chamber to Henry; John Prideaux (B.A. 1600, M.A. 1603, B.Th. 1611, D.Th. 1612, Exeter, Oxford), latterly Bishop of Worcester and Chaplain to Henry, James, and Charles; Daniel Price (B.A. 1601, M.A. 1604, B.Th. 1611, D.Th. 1613, Exeter, Oxford), also chaplain to Henry, James, and Charles; as was Joseph Hall (B.A. 1592, M.A. 1596, B.Th. 1603, D.Th. 1610, Emmanuel, Cambridge). I do not intend to become embroiled in the ongoing historiographical debate about the nature of Puritans. For my purposes, "hot Protestants" (i.e., those who lived their Protestantism with relish) should be regarded as Puritans. See especially Collinson, *Elizabethan Puritan Movement* and *Godly People* for the most cogent attempt to sort this out; for the seventeenth century, see McGee, *Godly Man in Stuart England;* and Hunt, *Puritan Moment.*

claim strengthened by Wright's dedication of his second edition of *Certaine Errours* to Henry in 1610.[36] Upon becoming tutor, Wright "caused a large sphere to be made for his Highness, by the help of some German workmen; which sphere by means of spring-work not only represented the motion of the whole celestial sphere, but shewed likewise the particular systems of the Sun and Moon, and their circular motions, together with their places, and possibilities of eclipsing each other. In it was a work by wheel and pinion, for a motion of 171000 years, if the sphere could be kept to long in motion."[37]

Henry had a decided interest in such devices and rewarded those who could create them.[38] Wright also designed and constructed a number of navigational instruments for the prince and prepared a plan to bring water down from Uxbridge for the use of the royal household.[39] In or around 1612, Wright was appointed librarian to Prince Henry, but Henry died before Wright could take up the post.[40] In 1614, Wright was appointed to lecture to the East India Company on mathematics and navigation. He was thus an academic who clearly respected practical experience. He was not content to investigate exclusively theoretical problems but felt compelled, probably for financial gain but also spurred by more intellectual motivations, to apply his learning to practical problems that helped further mercantile and expansionist causes of the East India Company and perhaps of Henry himself.

Another preeminent figure in mathematical geography connected with Prince Henry was Thomas Harriot.[41] Harriot attended Oxford at the same time as Wright was at Cambridge. He matriculated from St. Mary's Hall in 1577 and received his B.A. there in 1580. By 1582, he was in the employ of Sir Walter Raleigh, who sent him to Virginia in 1585. Harriot, like Wright, was an academic and theoretical geographer whose sojourn into the practical realm of travel and exploration helped form his conception of the vast globe and of what innovations were necessary to travel it. Harriot's

36. E. Wright, *Certaine Errors in Navigation,* 2nd ed. (1610), sigs. 3a–8b, X1–4; *DNB,* 21:1016; Birch, *Life of Henry, Prince of Wales,* 389.

37. "Mr. Sherburne's Appendix to his translation of Manilius, p. 86," in Birch, *Life of Henry, Prince of Wales,* 389.

38. Smuts, *Court Culture,* 149–50. Smuts especially mentions Salomon de Caus, whose *La perspective avec la raison des ombres et miroirs* (London, 1612) was dedicated "Au Serenissime Prince Henry" (*Court Culture,* 157).

39. Strong, *Henry, Prince of Wales,* 218. E. Wright, *Plat of part of the way whereby a newe River may be brought from Uxbridge to St. James, Whitehall, Westminster, the Strand, St. Giles, Holbourne and London* (MS, 1610). Identified by E. G. R. Taylor in *Late Tudor Geography,* 235.

40. Strong, *Henry, Prince of Wales,* 212.

41. J. W. Shirley, *Thomas Harriot: A Biography.*

FIGURE 31. A Weroans, or chieftain, of Virginia, from Thomas Harriot, *A Briefe and true Report of the New found land of Virginia,* in Theodore de Bry, *America Par I* (Frankfurt, 1590), sig. A1r. (BL G.6834. By permission of The British Library.)

description of Virginia, seen in his *Brief report of . . . Virginia* (1588),[42] was "the first broad assessment of the potential resources of North America as seen by an educated Englishman who had been there."[43] (See figure 31.) His view of Virginia was sympathetic and not colored by any desire to convert the natives. Harriot compiled the first word list of any North American Indian language (probably Algonquin),[44] a necessary first step of classifying in order to control, thus illustrating that inductive spirit never far from the heart of even the most mathematical geographer. He saw Virginia's great potential for English settlement, provided that the natives were treated with respect and that missionary zeal and English greed were kept to a minimum.[45] His advice concerning Virginian settlement was to prove important as the Virginia companies of the seventeenth century were established. Thus, Harriot's descriptive geography was less driven by Protestant enthusiasm than by the spark of inductive inquiry, united with enlightened, self-interested patriotism.

42. Reproduced verbatim in de Bry, *America. Pars I,* published concurrently in English (also Frankfurt, 1590), and in Hakluyt, *Prinicipal Navigations* (1598), 3:266–80.

43. Quinn, "Thomas Harriot and the New World," 45.

44. J. W. Shirley, *Thomas Harriot: A Biography,* 133.

45. The manuscript information concerning this expedition is gathered in Quinn, *Roanoke Voyages.* J. W. Shirley discusses Harriot's desire for noninterference in *Thomas Harriot: A Biography,* 152.

More important for Harriot were issues of the mathematical structure of the world. Indeed, his mathematics were bound up closely with his imperial attitude generally and the experience of his Virginian contacts in particular.[46] He was deeply concerned about astronomical and physical questions, including the imperfection of the moon and the refractive indexes of various materials.[47] Harriot, in fact, was inspired by Galileo's telescopic sightings of the moon and produced several fine sketches himself after *The Starry Messenger* appeared. He also investigated one of the most pressing problems of seventeenth-century mathematical geography: determining longitude at sea. Harriot worked long and hard on the longitude question and on other navigational problems, relating informally to many mathematical geographers his conviction that compass variation contained the key to unraveling the longitude knot.[48] He never published his results concerning the longitude, resulting in his relative obscurity in later years. E. G. R. Taylor claims that "Thomas Harriot's correspondence shows that he was generous in putting his learning at the disposal of others, and it can only be supposed that it was his patron who was responsible for keeping his contributions . . . private."[49] It is much more likely, however, that his reticence in publishing was at least in part a result of the easy access he maintained with the important English mathematicians of the day, such as Wright, which alleviated any need to publish farther afield. Henry's court helped to foster an informal scientific interchange, encouraging English science to remain more parochial than its Continental counterpart.

Harriot was most closely associated with the ninth Earl of Northumberland, but he was also connected with Henry, Prince of Wales, as a personal instructor in applied mathematics and geography, just as Wright was.[50] Both Harriot and Wright formed part of a mathematical circle that included Thomas Aylesbury, William Lower, and John Harington, one of Henry's courtiers.[51] It is likely that Wright and Harriot, along with others

46. Amir Alexander, "The Imperialist Space of Elizabethan Mathematics," paper read at History of Science Society Meeting, New Orleans, 1994, argued that Harriot's work on the continuum was influenced by his view of geographical boundaries and the "other": "The geographical space of the foreign coastline and the geometrical space of the continuum were both structured by the Elizabethan narrative of exploration and discovery."

47. J. W. Shirley, *Thomas Harriot: A Biography*, 381–416.

48. Harriot wrote a manuscript in 1596, entitled *Of the Manner to observe the Variation of the Compasse, or of the wires of the same, by the sonne's rising and setting* (BL Add. MS 6788). These concepts are discussed in J. W. Shirley, *Thomas Harriot: A Biography*, 228; E. G. R. Taylor, *Mathematical Practitioners*, 44.

49. E. G. R. Taylor, *Mathematical Practitioners*, 44.

50. J. W. Shirley, "Science and Navigation," 81.

51. For some of the correspondence of this circle, see Halliwell, *Collection of Letters*, esp. 41–45.

of this group, met at Henry's court. As two university-trained contemporaries, with very similar interests and experiences, they would have gained much from their association. Given their mutual interests, it would have made sense for them to discuss matters of mutual geographical and mathematical interest while at court together.

A third member of this close circle of mathematical geographers connected by Henry's court and tutelage was the mathematician William Barlow. Barlow was slightly older than Harriot and Wright, graduating with a B.A. from Balliol College, Oxford, in 1565. He entered holy orders in 1573 and spent the rest of his life in ecclesiastical and teaching positions. There is speculation that he may have spent some years at sea between 1565 and 1573, accounting for his later interest in navigation, but this seems to be at the level of conjecture.[52] Barlow himself does not corroborate this experience at sea: "Touching experience in these matters, of my selfe I have none: For . . . by naturall constitution of body, even when I was yong and strongest, I altogether abhorred the sea: howbeit, that antipathie of my body against (as the Italian termed it) so barbarous an Element, could never hinder the sympathie of my minde, and heartie affection towards so worthie an Arte as Navigation is."[53]

Barlow is best known in geography for his two books on navigation, *The Navigator's Supply* (1597) and *Magneticall Advertisements* (written in 1609, published 1616).[54] The former, which gave a graphical method for creating a Mercator map projection, was probably owned by Henry himself;[55] the latter was dedicated to Sir Thomas Chaloner, one of the most politically important of the geographical courtiers surrounding Prince Henry.[56] Chaloner, Henry's Governor and later his Chamberlain, wrote a learned treatise on niter,[57] as well as cornering the alum market during Henry's lifetime (from alum mines on his northern properties), and the wresting away of this monopoly by Charles was probably part of the motivation for his son, also Thomas, to turn regicide. Barlow's book, thus dedi-

52. "Most of his biographers assume that he spent the greater part of these years at sea, but on no better ground, it would appear, than the interest he showed in navigation" (*DNB*, 1:1153).

53. Barlow, *Navigator's Supply*, sig. a4a. See also Strong, *Henry, Prince of Wales*, 218–19.

54. Barlow, *Navigator's Supply*, dedicated to the Earl of Essex. *Magneticall Advertisements*, manuscript version dedicated to Sir Thomas Chaloner, printed version to Sir Dudley Digges (who was also connected with Henry's court). Chaloner persuaded Barlow to publish the latter book (*Magneticall Advertisements*, sig. A3a).

55. Bill for binding for James and Henry "6 of the Bishope of Lincolns [Barlow's] bookes" (PRO SP 14/51, no. 11, n.d. [1609?]).

56. Sir Thomas Chaloner was appointed Henry's Governor in 1603 (PRO SP 14/3, no. 15, Aug. 9, 1603) and later his Chamberlain.

57. Chaloner, *Virtue of Nitre*.

cated, dealt in a popular way with the innovative magnetical theories discussed by Barlow's friend William Gilbert in *De Magnete* (1600).[58] Barlow first met Henry in 1605, when the prince visited Oxford. On that occasion, Barlow disputed before Henry and James on the question, "given out by the king himself, . . . that it is greater to defend than to enlarge the bounds of an empire,"[59] a geographical topic more to James's liking than to either Henry's or Barlow's. Barlow soon joined the court at Richmond and served for seven years as chaplain and tutor to Prince Henry, even employing his own workmen to make navigational instruments for the prince.[60] It was probably at court that he met Chaloner, as well as encountering Edward Wright, with whom he had a huge pamphlet battle over the primacy of navigational discoveries.[61] Even for a conservative divine such as Barlow (confirmed anti-Copernican as he was),[62] Henry's court provided a point of contact for the mathematically and geographically inclined.

There were, of course, other less mathematically minded geographers at Prince Henry's court. Sir Robert Dallington, of Corpus Christi, Cambridge, a Gentleman of the Privy Chamber in ordinary to Prince Henry, traveled through France and Italy and described some of the tantalizing details of his trip to Henry.[63] He also translated part of Guicciardini's *History of Italy* for the prince, as well as writing his own *Survey of France* (1604).[64] Thomas Lorkin, latterly of Peterhouse, Cambridge, Regius Professor of Physic at Cambridge from 1564 to 1591, and father-in-law of Edward Lively, was connected with the Richmond Court through Adam Newton, Henry's principal tutor. He wrote to Henry concerning the state of affairs in France, in reward for which he received a pension of eighty pounds a year, probably from the prince.[65] John Florio, son of an Italian Protestant refugee, taught Henry Italian and, commissioned by Richard Hakluyt, translated Cartier's accounts of his discovery of New France into

58. William Gilbert, the younger half-brother of the author of *De Magnete,* dedicated and presented to Prince Henry a manuscript version of *De Mundo* in 1610. Feingold, *Mathematicians' Apprenticeship,* 201.

59. Birch, *Life of Henry, Prince of Wales,* 52.

60. Barlow, *Magneticall Advertisements,* 86.

61. Barlow, in ibid., took issue with Mark Ridley over the unauthorized use Ridley had made of William Gilbert's work (E. G. R. Taylor, *Mathematical Practitioners,* no. 129). The result was a pamphlet war, with Barlow and Gilbert siding against Ridley and Wright. In later editions of Barlow's work, Ridley's reply was often appended.

62. Barlow, *Navigator's Supply,* sigs. G3b–4b.

63. This trip produced Dallington's *A Survey of the Great Duke's State of Tuscany in the year of our Lord 1596,* written in 1596, printed in London, 1605.

64. Sir Robert Dallington, *Aphorismis Civile, and Militare . . . out of Guicciardine,* manuscript presented to Henry in 1609, printed in London, 1613, dedicated to Prince Charles. *A method for Travell, shewed by taking the view of France as it stood in . . . 1598* (London, 1604).

65. Strong, *Henry, Prince of Wales,* 74; Birch, *Life of Henry, Prince of Wales,* 251.

English from Ramusio's Italian collection.[66] Likewise, Sir Arthur Gorges, an extensive translator as well as ship's commander under Raleigh (his account of the "Islands Voyage" was published by Purchas),[67] was connected with Henry's court as Gentleman of the Privy Chamber. His *Discourse tendinge to the wealth and strength of . . . Great Brittayne* (1610), dedicated to Henry, embodies a growing English chauvinism and suggests that sense of the potential of English supremacy, so much part of Henry's agenda.[68] Even Pierre Erondelle, a displaced Huguenot, dedicated the English translation of Lescarbot's *Nova Francia,* which Hakluyt had encouraged him to undertake, to Prince Henry.[69]

Finally, any discussion of the court of Prince Henry is incomplete without mention of Thomas Lydiat. A Master of Arts from New College, Oxford, Lydiat became chronographer and cosmographer to Prince Henry after dedicating his book, *Emendatio Temporum Compendio Facta ab Initio Mundi* (1609) to the prince.[70] Lydiat received a pension from Henry, unfortunately not outlasting the prince's life, indicating Henry's concern with mathematical geography in all its forms.[71] Lydiat was most interested in chronology, as his dedicated book shows, but was eager to dispense advice in all three geographical areas. He probably met Sir Thomas Chaloner and Adam Newton, with both of whom he corresponded, in connection with Henry's court.[72]

By gathering all these men around him, Henry acted as patron to geographers who combined theory with practice. Many of them were interested in the structure of the world and new methods of mapping or navigating it, but all were equally involved in using personal experience to aid in their understanding of the globe. As scientific clients, they delivered the promise

66. Curtis, *Oxford and Cambridge,* 140. This was printed in Hakluyt, *Prinicipal Navigations* 1598–1600, 3:201–40. For a somewhat dated biography of Florio, see Yates, *John Florio.*

67. Oxf. Bodl. Ballard MS 52, ff. 64–122b. Purchas 1625, part 2, 1938–79.

68. Strong, *Henry, Prince of Wales,* 41.

69. "To the Bright Starre of the North, Henry Prince of Great Britaine" (Erondelle 1609, sig. ¶ 1a). He acknowledges Hakluyt's encouragement (sig. ¶ 2a).

70. Birch, *Life of Henry, Prince of Wales,* 148.

71. Ibid., 389. "Dignitatis tuae studiosus Thomas Lydyat; . . . utinam appellari dignus, Serenissimi Principis Henrici. Chronographus et Cosmographus." Thomas Lydiat to Sir Adam Newton, 26 July–5 August 1611, Oxf. Bodl. MS Bodl. 313, f. 84b. I owe the reference to this treatise to Apt, who has transcribed it in "English Reception of Kepler's Astronomy."

72. Lydiat's works included *Praelectio Astronomica de natum coeli, et condicionibus elementorum tam autem de causis praecipuorum motuum coeli et stellarum* (London, 1605); *Tabulae ab ortu mundi ad Vespasianum* (manuscript for Ussher), Oxf. Bodl. MS Add. C 297; "Observations to be made by the Masters of Shipps, etc. in their East Indies, and other long voyages beyond the Equinoctiall line southward att Sea & Land respectively," Oxf. Bodl. MS Bodl. 313, ff. 13–14. He corresponded with Chaloner (Oxf. Bodl. MS Bodl. 313, ff. 28–29).

of practical results, as well as increased intellectual status for their patron and a clearer articulation of England's imperial future. These court geographers presented the wide world to Henry for his understanding and his control.

Many of these geographers shared with Prince Henry a substantial interest in the settlement of the New World. Proposals for expeditions to find the northwest passage came from Edward Wright, Thomas Harriot, and Thomas Lydiat.[73] Proposals to settle America issued from Harriot, Wright, Michael Drayton, William Crashaw, and George Chapman, among others.[74] The establishment of colonies in Africa and East India was proposed by Lydiat.[75] Part of the motivation for these proposals was probably a shrewd reading of Henry's mood and desire to achieve some sort of Protestant hegemony that would have Henry at its head. But all these men were sincerely motivated by a burgeoning patriotism—a heartfelt desire to use their esoteric mathematical and descriptive knowledge to further the interests of the country they were beginning to identify as separate from Europe, able to stand alone and above the shattering problems of the Continent.[76] This desire to make knowledge serve the needs of the English mercantile community and especially of the English state can be seen as "proto-Baconian," foreshadowing the Lord Chancellor's absolute insistence on the duty of the natural philosopher to apply his knowledge to useful purposes and to use knowledge as a tool of power over nature. This concept was articulated rather than invented by Bacon, and its roots are to be sought in the Continental influences beginning to invade England in the seventeenth century, rather than merely in some native insight.

73. Edward Wright, *Sea-Chart of the N.W. Passage,* MS, 1610; Thomas Harriot, *Three Reasons to Prove a N.W. Passage,* MS, 1611; Thomas Lydiat, Oxf. Bodl. MS Bodl. 313, f. 32.

74. Harriot, *Briefe and true report;* E. Wright, *Voyage of the Earl of Cumberland to the Azores* (MS 1589, published 1599); Michael Drayton, *Ode to the Virginia Voyage (Poems Lyrical and Pastoral)* (London, 1606); Crashaw, *Sermon preached on 21 February 1609* (a sermon Strong claims "Casts both King and Prince in an imperial messianic role as 'new Constantines or Charles the Great'"; *Henry, Prince of Wales,* 62) and publishing Whitaker's *Good Newes from Virginia* (1613); and George Chapman, *The Memorable Maske . . . of the Inns of Court* (London, 1613).

75. Oxf. Bodl. MS Bodl. 313, ff. 13–14, 32.

76. E.g., Harriot, *Briefe and true Report,* sigs. F3a–4b; E. Wright, *Certaine Errors in Navigation* (1599), sigs. A2a–3a, 2¶ 1a–3¶ 2b; Crashaw, *Sermon preached on 21 February 1609,* sig. C3aff., esp. "And tho Virginia . . . Thou hast thy name from the worthiest Queene that ever the world had: thou hast thy matter from the greatest King on earth and thou shalt now have thy forme from one of the most glorious Nations under the Sunne" (sig. L2a); Erondelle, *Nova Francia,* in his dedication to Prince Henry, "For who may better support, and manage magnanimous actions, such as be the peopling of lands, planting of Colonies, erecting civill Governementes, and propagating of the Gospell of Christ . . . then those whom the King of Kings, hath established as *Atlasses* of kingdoms and Christian commonweales?" (sig. ¶ 1b).

This insistence on the application of academic study to practical life probably owed more to the humanistic traditions that had gradually become more important in sixteenth- and seventeenth-century political theory than to any new value of scientific engagement.[77] Humanist political philosophy had long insisted that it was the duty of every citizen to eschew the contemplative life and to use contemplation actively to engage in the affairs of state. This could lead to monarchical or more democratic interpretations, but always stressed the necessity for the humanist training of the nation's ruling class and the need for true engagement in government and political affairs.[78] This attitude on the part of Henry's geographers, which might otherwise be seen as heralding a new relationship between knowledge and power, resonated to a note first sounded by the older Italian concept of the duty of the philosopher. Ideas of civic humanism had been developing in England throughout the sixteenth century, transforming to suit the very different realities of Tudor life. In the last two decades of Elizabeth's reign, however, Continental influences receded with the war against Spain, and native English ideas took their place.[79] With the peace of 1604, the overtures to the Hapsburg courts, and the new flexibility on the part of the Vatican, this trend was reversed.[80] In the consequently rapid adoption of the Italian Renaissance by Henry's court, humanist ideas increased in popularity, while combining with newer and sometimes contradictory ones of personal and state power and self-promotion. Italian ideas of the philosopher-king combined with fledgling capitalism and native notions of practicality and service to the monarch to provide a rationale and motivation for the study of such an outward-looking science as geography. Thus, the stress on the applicability of knowledge in geographical investigation (almost a sine qua non of geographical research) reveals an English society at once Continental and insular: fascinated with Italian ideas, even if slightly out of step with the Continent, yet moving inexorably toward an imperial attitude that would promote things English at all costs.

A lively coterie of geographical scholars, especially mathematical ones, thus sprang up around Richmond Palace, drawn by the rewards of royal patronage and the opportunity to engage in an expansion of geographical knowledge, both in theory and in reality. The interest and tutelage of Prince Henry provided a focus for some of the most innovative geographers of the day, many of whom shared a university education and thus a

77. Dowling, *Humanism in the Age;* Ferguson, *Articulate Citizen;* and Caspari, *Humanism and Social Order.* More recently, Shapiro, in "Early Modern Intellectual Life," has shown how important humanism was to seventeenth-century England.

78. Caspari, *Humanism and Social Order,* 207–8.

79. Smuts, *Court Culture,* 20.

80. Ibid., 118. This move is also described in Smuts, "Political Failure of Stuart Cultural Patronage."

set of assumptions about the geographical world and England's place therein. This active community of court geographers was able to establish a geographical network using personal contacts and the patronage system already in place. This resulted in a research program that was rigorous and innovative, drawing for its methodology on notions of practicality and the importance of personal experience that were to suffuse much of late-seventeenth-century scientific thought. Henry's encouragement of geographical research provided a focus for the energy of these mathematical and descriptive scholars, while their sense of duty to the state drove them to use their geographical knowledge to promote the ends of Protestant hegemony as envisaged by Henry.

❦

THERE WAS A close relationship between the study of geography and the development of ideas of English power and imperial growth. The intimate connection between the scholarly world of Oxford and Cambridge and the complex world of London and the court was possible in part because geography provided the justification for the growing sense of separateness and superiority increasingly experienced by English courtiers and governors. Geography, learned at least initially at university, gave these men a set of tools with which to interpret the world as it related to England. The lessons of geography made it an attractive discipline for a king-in-waiting such as Prince Henry because it stressed that the world was knowable and controllable—and that England and its prince could be instrumental in controlling that world. The men joining the princely court as geographers were demonstrating that they had internalized the imperial message of geography. They were eager to use geographical knowledge for the good of the state just as they used the state, in the guise of the patronage system, for the good of geographical knowledge. These two were intrinsically linked; just as the engaged life increased these geographers' knowledge of the terraqueous globe, geography also provided a key to the development of an English imperial identity and ideology.

The study of geography was situated both within a university milieu and in the political world outside the university towns. While the extra-university world must be taken into account when explaining the growth of this early modern scientific discipline, it would be a mistake to assume that the "new science" grew up only outside the halls of academe. Neither Gresham nor the court alone supplied the site for the alteration taking place in scientific topics or procedures. Rather, we must see the complementarity of the various environments. The universities were of intrinsic importance to the development of geography, but the outlet of London and the court was necessary as well. Geography, as a model of an interactive science involving both theory and experiential research, must by its very nature be both in the world and separate from the world. As a mathematical science, in contrast to the more traditional natural philosophy, geography necessitated a new way of interpreting the world, melding nature and mathematics in an integral and innovative way. Procedural and methodological values of mathematical geography developed especially rapidly in the atmosphere of the court, but they originated in the Oxford and Cambridge coteries and colleges. Similarly, the descriptive part of geography lent itself to fact-gathering missions, which involved much engagement with the world at large, even if the inductive methodology that resulted could be developed without such activity. Geography both inside and outside the universities helped prepare the way for a new vision of science that was to develop in the second half of the seventeenth century. It provided a model of mathematical application, which would be used fruitfully by mechanical philosophers; it helped develop a tolerance for and even reliance on inductive tabulation of facts, which would prove a necessary method for the natural and life sciences; and it promoted an ideology of the utility of knowledge, which helped transform English science into Baconianism. Because of its essentially public nature, geography provided the perfect avant-garde for a new science that would depend on public acceptance, public demonstration, and the consensual growth of knowledge. Geography helped prepare the groundwork for a shift in scientific ideology as far-reaching as it was fundamental.

CONCLUSION

Geography and the Idea of Empire

From the late seventeenth century on, England began to realize the potential of imperial thinking. The first step in such empire building was the subjugation of the Irish and union with Scotland, resulting in the creation of Britain by 1704. This was followed by a rapid expansion in influence and control over large segments of the world; the British Empire was a reality by the end of the eighteenth century. But in order to conquer the world in this way, the English first needed a vision of themselves as an imperial nation. This self-image as an independent and omnicompetent country had to precede the acquisition of an empire, and so the English needed an imperial ideology before they could begin to construct an empire in deed. The creation of this ideology of empire was facilitated by the study of geography at the early modern universities.

As a university education became more important for those men aspiring to positions of power in government, what they learned at university contributed more and more to a far-reaching vision of what England could become. As Lawrence Stone has shown, gentle and mercantile families began to send their sons to Oxford and Cambridge in greater numbers in this period. The increase in student numbers changed the structure of the universities, making them more reliant on the collegiate structure, and influenced the curriculum as well to reflect the interests of the new scholars and supply the needs of these potential leaders in governance and commerce.

Geography was part of that revitalized curriculum in the early modern English universities. Now often obscured by its failure to develop into a "hard science," geography was an extremely important pursuit for many men in sixteenth- and early-seventeenth-century England. It developed from a general interest in the world as part of the cosmos, a study more properly called cosmography, into the discipline of geography, which was centered on political society and driven by concerns that were both intellectual and pragmatic. The topic of geography was the earth and its inhabitants, an area of increasing interest to those students soon to be engaged in that world.

Geography became more specialized through time, as seen by the growth of the three subdisciplines of mathematical geography, descriptive geography, and chorography. Mathematical geography was the most common type of geography in the early period, but by 1620, descriptive geography had become most popular at the universities and beyond. In part this was because of the exciting discoveries taking place, and in part it reflects the changing prosopography of the university community. Descriptive geography provided these new students with a vision of England and the rest of the world that advanced their career aspirations and influenced their attitude toward the wider world. Each of the three subdisciplines developed relatively independently as the forty-year period progressed, although there was much crossover in terms of participants and interest. All three were driven by practical and theoretical concerns, and all developed a methodology and ideology that would be applicable to many more disciplines in the following century.

Geography was an important part of the formal Arts curriculum in this early modern period. The study of geography was included in some of the university and college statutes and encouraged by individual instructors. The significant role of geography is also demonstrated by the majority of geography students who achieved a B.A. or an M.A. Most of the students who studied geography did so in the process of following the statutory requirements for the Arts degrees, as the group biography makes clear. Many students, indeed, fulfilled the full seven-year program needed to achieve the M.A. The academic career patterns of these geographically inclined students demonstrate two important conclusions about the early modern universities. First, geography was part of the formal education received by the majority of young men at Oxford and Cambridge, and this suggests that the men I have identified represent only a fraction of the total interest in geographical topics within the scholarly ranks. Geography was thus encouraged and studied by serious students following the curriculum, whether they planned a career in the church, in academe, or elsewhere. Second, men who were destined for a more active political life, the new men who flooded the colleges as commoners and fellow-commoners, still tended to pursue the formal Arts curriculum, even if they did not intend to sit for the degrees. Thus, we should not see this change in demographics resulting in the creation of a completely different, casual educational stream. Rather, the introduction of these young men helped to modify the existing curriculum, adding topics of more immediate relevance to their lives, but continuing to insist on the more rigorous four- to seven-year Arts program.

Geography was studied by a large number of men at many different colleges at Oxford and Cambridge. This study of the earth and its inhabitants was not simply the work of a few remarkable individuals but engaged the interest of a broad range of men, at university and in the world outside. This large group of people studying geography was essential to the development of the discipline, providing both its audience and its paymaster. While not all students of geography went on to write geography treatises, many more would have been interested in purchasing such books because of their introduction to geography at university. This larger group of students studied geography at Oxford and Cambridge in part because it was an interesting topic, but also because it had become a method of participating in the imperial mood of early modern English society.

The analysis of books owned by students and masters at Oxford and Cambridge, as well as our knowledge of the lives of men interested in geographical topics, shows that some colleges and foundations provided special encouragement for the pursuit of geographical studies. Corpus Christi College, Christ Church, and St. John's College, Oxford, and Peterhouse, St. John's, and Corpus Christi colleges, Cambridge, stand out as foci of geographical emphasis. These colleges represent the best-known examples of loci of geographical interest, rather than its exclusive domain. Many more students and colleges, whose records are less complete, were undoubtedly involved in the teaching and study of geography in this period, and so these six colleges open a window on the widespread reality of geography teaching and interest at both Oxford and Cambridge.

The study of geography encouraged the emergence of a number of coteries of like-minded individuals, both at the universities and outside. Meeting first at university, geographically inclined men established connections and communities that would last well beyond the years of their Arts education. These networks helped link academics and practical men and helped transform the discipline of geography into an interactive science, requiring the integration of theory and practice in order to explain the world.

This discipline of geography, present in the universities in both formal and informal venues, attracted many new students of early modern Oxford and Cambridge because it supplied a vision of themselves, of England, and of the wider world that supported a new imperial identity. While such an identity was not universally acknowledged or supported, the very existence of a discourse of superiority and exploitation, combined with new ways of analyzing and interpreting the events and structures of the world beyond their island, helped prepare these men for the creation and expansion of an empire. All three geographical subdisciplines helped develop this complex and increasingly imperial worldview through three underlying assumptions: a belief that the world could be measured, named, and therefore controlled; a sense of the superiority of the English over other peoples and nations and thus the right of the English nation to exploit other areas of the globe; and a self-definition that gave these English students a sense of themselves and their nation. Mathematical geography, by mapping the world and measuring its parts, claimed to define and thus control within a rigid frame the diverse peoples and countries of that world. Both descriptive geography and chorography categorized and named regions, flora, fauna, and people. Just as Adam had achieved dominion over the plants and animals of the world by being granted the privilege of naming them, geographers likewise owned what they classified. It is small wonder that the first successful English settlement of the New World would be named Jamestown, or that that area of the world would be called Virginia. The study of geography also encouraged the separation of the English from their Continental counterparts. As English people began to see themselves as a nation separate from the Continental theater, they could study that theater with more interest and yet remain disinterested observers. Thus, geography became part of the self-fashioning of the English people, allowing them to see themselves as entirely separate from the Continent while at the same time maintaining a connection through "objective" observation. This self-fashioning took place through the study of England as well, where local history helped county men to separate themselves from the country as a whole and yet to see it as an interesting and important focus of inquiry. The ideology of supremacy intrinsic in geographical investigation led to a distancing from newly discovered tribes, from European wars, and from court debacles. Thus, geography supplied an attitude toward England and the rest of the world that encouraged the growth of the English empire.

The people who learned these important lessons from geography went on to careers that allowed them to put their schooling to work. Many students of geography lived engaged lives, adopting political careers either through a civic obligation to serve the state or because of a belief in personal and state aggrandizement. Geographers became lawyers, politicians at various levels of government, and participants at court, thereby assisting in the governance of England. Others became teachers, clerics, or physicians, all serving the commonweal and promoting a vision of England as strong, independent, and often superior.

This engagement with the world resulted in a close relationship between scholars and craftsmen, as well as between this community of geographers and the royal courts. This coalescence of men who were actively engaged in extending the boundaries of English influence, both through navigational and surveying practice and through political machinations and diplomacy, with those who articulated and helped define the ideology of empire and governance through the academic discipline of geography produced a fruitful exchange that helped create the idea of empire. This can be seen in the close connections between geographers such as Dee and Wright and new mercantile ventures such as the East India Company. This union of theory and practice, often through the medium of royal patronage, also changed the basis on which the natural world was studied. The patronage of geography, especially at the court of Henry, Prince of Wales, but also at the other royal courts, allowed the imperial message of geography to enter the discourse of governance and national expansion. Geography was attractive to these princes specifically because of its message of England's independence and superiority, and in turn geographers promoted such an image as they sought continuing support for their work from the leaders of the court. Thus, the courts supplied an important venue for those who had studied geography at the universities; the ideology they had gained at Oxford and Cambridge was precisely that which was best suited to the royal agenda, and they in turn were able to use that agenda to promote the geographical enterprise.

The discipline of geography was thus important in two facets of early modern English life. Not only did it help create a shared ideology of the nascent English empire, but geography also provided a meeting place for mechanics and philosophers, helping to change the protocols and values of the study of the natural world. Because of geography's important position as an interactive science, it is not surprising that it developed a methodology and a set of values that would become important in the seventeenth century for many other scientific disciplines. Geography combined a mathematization of the world, intrinsic in the development of a geometric grid in which to contain the hemispheres, with an inductive methodology and an ideology of utility and power through knowing and classifying. Geography was essentially a mathematical science, both in its methods, borrowed from astronomy, and in its mapped results, but it offered a synthesis of "objective" and "subjective" knowledge, empirical data and personal experience, thereby personalizing the world while legitimating personal experience with scientific understanding. Whether mapping space or time, geography attempted to categorize by quantification. At the same time, its inductive method developed in part because of the nature of its investigation. Understanding compass variation required observations at different points of the globe, which could only be compared fruitfully in tabular form. Descriptions of new lands could only be amassed by incremental gathering of information. Genealogical research required laborious charting and graphing of individual facts and figures. Thus, geography developed as a "collecting" science, as well as a mathematical one. Since geography was so closely engaged with political and economic concerns, its practitioners developed an ideology that counted the knowledge gained to be useful and in fact claimed that the goal of such intellectual activity must always be utility. Geography was a study that imbued its practitioners with power, the power that came through knowing. Taxonomic and Adamic power over the world was available to the geographer, evident especially in the hidden ideology of the descriptive geographers, which proclaimed the supremacy of England over other Continental powers, of Protestantism over Catholicism, and of the Old World over the New, while claiming such knowledge to be objective and disinterested.

In examining the discipline of geography in this period, the most striking and most important aspect of the study was the intrinsic role played by practical economic, social, and political motivations and information in its development. Christopher Hill, more than thirty years ago, was right when he claimed that sixteenth-century science could not be separated from practice. But where Hill and the Mertonians wanted to see Puritanism and the rising middle class as responsible for every scientific and curricular innovation, we can now see that any interpretation must be much more finely tuned and more complex. Science and technology are not competing knowledge systems, and so to see this early phase of the scientific revolution as a battle between those who do and those who know is to misunderstand the relationship between theory and practice. The study of geography makes clear that there was a delicate and symbiotic relationship between the theoretical problems of a geometrical construction of the globe and the need to navigate its waters and lands. The driving force behind the problems that were posed was essentially practical, concerned with trade, communication, and control of a global empire, but knowledge was important as well. Geographers felt the need for elegant proofs as well as pragmatic applications. Part of the theoretical knowledge had to be obtained, in the case of geography especially, by going out and doing, seeing, experiencing. This is why geography could not remain only within the universities, although it was given much of its impetus and all of its theoretical superstructure there.

Geography, then, provides a new model for understanding the development of early modern science. Part of the older university tradition and an integral part of that intellectual and theoretical world, it provided an important link to the wider world and to the new ideologies and methodologies of science and politics alike that were becoming so important. This interactive discipline provided a fruitful point of contact between the college and the court, between the world of ideas and the world of affairs. In the process, geography provided a new vision of England for those governing the country, and a new set of values for the investigation of the natural world.

APPENDIX A
Sources for Book List

Oxford

Bodleian Library

MSS Eng. misc. d.47. "Literary Commonplace Book of the Revd Lyonell Day, BD 1611–28."
MSS Rawl Qe 31. Thomas James. "Catalogus in Librorum in Bibliothecae Bodl. 1602–3."
MSS Savile 95. List of some books owned by Henry Savile.
———. 101. Books owned by Henry Savile.
MS Wood D. 10. Dr. John Reinolds' books willed to friends (1607).

Oxford University Archives

"Box of Transcripts of all inventories mentioning books."
Vice-Chancellor's Court inventories.

All Souls' College Library

"Catalogus librorum ad Bibliothecam Collegii Animarum omnium Fidelium defunct: de Oxon: Spectantium et donatorum eorundem." Benefactor's Register 1604–5.

Balliol College Library

"Catalogus Librorum et Benefactorum Bibliotheca Collegii de Baliolo Oxon."

Christ Church College

Donor's Register 1614–1841.

Corpus Christi College

MS D/3/1. Library catalog 1589.
MS D/3/2. "Catalogus Librorum Collegii Corporis Christi."
MS D/3/3. Donors Book.
B11/1/2. "Catalogus Librorum Collegii Corpus Christi" in book of Wills and Donations.

Magdalen College

MS 777. "Catalogus Benefactorum Bibliotheca."

Merton College Library

P.3.35. Manuscript Catalogs of books ca. 1586–1600.
Register 1.3.

New College Library

3582. Library Benefaction Book 1617–1909.

Oriel College Library

Benefactors Book.

Queen's College Library

MS 556. Benefactor's Register from 1621; 1659 transcription.

St. John's College

Library Benefactors Book.
Archives—Muniments. X 19. Inventory of Henry Price, Fellow in 1601.
———. X 26. Mr. John Lees Will, October 1609.

Trinity College

Benefactors Book, mid-sixteenth century to mid-eighteenth century.

University College

Benefactors Book (abortive), prepared 1674.
Roll HH.5.6. Inventory of Master's Lodging and Fellow's Chambers, May 12, 1587, 29 Eliz.

CAMBRIDGE

Cambridge University Archives

Registry, MS 31.1, item 10.
Vice-Chancellor's Court probate inventories (2–9).
Cambridge University Library. Oo.7.52. "Catalogus librorum quos habet Bibliotheca Publica Academ. Cantabrig."

Christ's College

Donation Book, 1623.

Corpus Christi College

MS 575. Library list by Matthew Parker, ca. 1593.
MS 576. Library list, ca. 1607.

Emmanuel College

Library catalogs 1597, 1621, 1622, 1626, 1628, 1632.

Gonville and Caius College

MS 645/352. Perhaps compiled by William Braithwaite, 1618.

King's College

Chained Donors Book.
IB catalog, transcribed by F. L. Clarke.

Pembroke College

Benefactors Book, 1617.

Peterhouse

MS 400. "Nomina librorum qui erant in Bibliotheca Collii ante Doctorum Perne mortuum [1588]."
MS 405. Library list, ca. 1610.
Unnumbered MS (probably close copy of 405).

St. John's College

Cambridge University Library. MS Dd. 5.54. Library catalog ca. 1620.
MS U.3. Library List and Donors Book.

Sidney Sussex College

MS 91. Donors Book.

Trinity College Library

R.17.8. "Memoriale Collegio Stae. et Individuae Trinitatis in Academia Cantabrigiensi dicatum 1614."

APPENDIX B

Geography Books Owned by Students, Fellows, and Libraries of Selected Colleges

Christ Church, Oxford

The list of books owned by Christ Church members was taken from the following sources: John Badger, Inventory 1577, Oxford Vice-Chancellor's Court (VCC); Gieles Dewhurst, Inventory 1577, Oxford VCC; Thomas Morrey, Inventory 1584, Oxford VCC; Sir [Matthew] Parkin, Inventory 1589, Oxford VCC; Caleb Willis, will proved 1599, "Booklists from Wills and Inventories in Christ Church Registers," GG, f. 178a; Christ Church Donor Register 1614–1841 (some identified by Ker 1986).

Mathematical Geography

Apianus, Petrus. *Cosmographia.* Ed. Gemma Frisius. [First edition Antwerp, 1529.]
Bertius, Petrus. *Tabula Georgraphica.* Amsterdam, 1618. Or *Theatrum Geographiae veteris.* Leiden, 1618, folio.
Globos, Caelestem, et Terrestr.
Grynaeus, Simon. *Novus Orbis Regionum ac Insularum veteribus incognitarum, una cum tabula Cosmographia.* Basel, 1537.
Hues, Robert. *Tractatus Globis caelestis et terrestris eorumque Usu.* Amsterdam: Hondius, 1617, quarto.
Magini, Giovanni Antonio. *Geographia tum veteris, tum novae.* Cologne, 1601, quarto.
———. *Primum Mobile, duodecim libris contentum, in quibus habentur trigonometria sphaericorum et astronomica, gnomonica, geographicaque problemata.* Bonn, 1609, folio.
Mercator, Gerard. *Atlas sive Cosmographicae meditationes de Fabrica Mundi et Fabricati Mundi.* Amsterdam, 1611, folio.
Merula, Paulus. *Cosmographiae generalis libri 3: item geographiae particularis libri 4; . . . cum Tabulis Geographicis aeneis.* 2 vols. Leiden, 1605, folio.
Ortelius, Abraham. *Theatrum orbis terrarum.* Antwerp, 1575.
———. *Thessaurus Geographicus.* Antwerp, 1587, folio.
Ptolemy, Claudius. *Geographia.* Strasbourg, 1522.
———. *Tabulae Geographia per Gerardum Mercatorem.* Frankfurt, 1600, folio.

Descriptive Geography

Aemylius, Paulus (Veronensis). *De rebus gestis Francorum. Libri x.* Basel, 1569, folio.
Alberti, Leandro. *Descriptio totius Italiae.* Cologne, 1567, folio.
Avity, Pierre de. *The Estates, Empires and Principalities of the World.* Trans. Edward Grimstone. London, 1611, folio, two copies.
Belgicarum Rerum Annales a diversis auctoribus. Frankfurt, 1580, folio.
Bertius, Petrus. *Commentariorum rerum Germanicarum libr tres, etc.* [with illustrations and maps]. Amsterdam, 1616, quarto.
Bizzari, Pietro. *Rerum Persicarum historia.* Frankfurt, 1601, folio.
Bongers, J., ed. *Rerum Hungaricarum Scriptores Varii.* Frankfurt, 1600, folio.
Broniowski, Martin. *Tartariae Descriptio . . . cum tabula geographica eiusdem chersonesus Tauricae.* Cologne, 1595, folio.

Bry, Johann Theodore de. *India Orientalis. Parts I–VII.* Frankfurt, 1598–1606, folio.

Bry, Theodore de. *America. Pars VI: "Sive, Historiae ab H. Benzeno scriptae . . . Peruäni regni . . . de Fortunatis insulis". Cum tabula Peru.* Frankfurt, 1596, folio.

Buchanan, George. *Rerum Scoticarum Historia.* [First edition Edinburgh, 1582, folio.]

Bunting, Henri. *Itinerarium et chronicon ecclesiasticum talius sacrae scripturae.* Madaburg, 1598, folio.

Camden, William. *Anglica, Normanica, Hibernica, Cambrica a veteribus scripta.* Frankfurt, 1603, folio.

———. *Annales rerum Anglicarum et Hibernicarum regnante Elizabetha.* Antwerp, 1596, folio.

Cluvier, Philip. *Germania Antiqua.* Leiden, 1616, folio.

Cober, Tobias. *Observationum medicarum castrensium Hungaricarum decades tres.* Frankfurt, 1606, octavo.

Crusius, Martinus. *Annales Sueuicae.* Frankfurt, 1595 (1596), folio, two copies.

Du Bellay, Martin. *Commentariarum de rebus Gallicis libri x.* Frankfurt, 1575, folio.

Fazellus, Thomas, ed. *Rerum Sicularum scriptores in unum corpus congesti.* Frankfurt, 1579, folio.

Freher, Marquard. *Corpus Francica Historiae veteris et sincerae.* Hanover, 1613, folio, two copies.

———. *Germanicarum rerum scriptores hactenus incogniti.* Frankfurt, 1600, folio.

Gaguin, Robert. *Rerum Gallicarum Annales.* Preface by J. Wolf. Frankfurt, 1577, folio.

Gender, Jacob. *Scriptores varii de emperio Ottomanico evertendo, et bello contra Turcas gerendo consilia tria in Latinum conversa.* Frankfurt, 1601, octavo.

Guicciardini, Lodovico. *Description de tous les Pais Bas, autrement appellés la Germanie Inférieure.* Amsterdam, 1613.

Hentzner, Paulus. *Itinerarium Germaniae, Galliae: Angliae; Italiae.* Breslae, 1617, quarto.

Heuter, Pontius. *Rerum Belgicarum Libri quindecim.* Antwerp, 1598, quarto.

Italiae illustratae, seu rerum, urbiumque Italicarum scriptores varii notae melioris; nunc primum collecti simulque editi. Frankfurt, 1600, folio.

Karius, Petrus. *Germania inferior coloribus.* Amsterdam, 1617, folio.

Knolles, Richard. *The Generall Historie of the Turkes.* Second edition London, 1610, folio.

Krantz, Albertus. *Rerum Germanicarum historici clarissimi, ecclesiastica historia, sive Metropolis.* Frankfurt, 1583, folio.

Kromer, Martin. *De Origine et rebus gestis Polonorum libri xxx.* Cologne, 1589, folio.

Lazius, Wolfgang. *De gentium aliquot migrationibus, sedibus fixis, reliquis, linguarumque initiie.* Frankfurt, 1600, folio.

Lescarbot, Marc. *Nova Francia; or the description of that part of New France which is one continent with Virginia.* Trans. Peter E[rondelle]. London, 1609, quarto.

Leunclavius, Joannes. *Historiae Musulmanae Turcorum, de monumentis ipsorum exscriptae, libri xviii.* Frankfurt, 1591, folio.

Lindenbrog, Erpoldus. *Scriptores Rerum Germanicarum Septentrionalium, vicinorumque popularum, veteres diversi.* Frankfurt, 1609, folio.

Lonicerus, Philip. *Chronicorum Turcicorum tomus primus.* Frankfurt, 1578, folio.

Maffei, Giovanni Petri. *Historia Indica.* Cologne, 1593, folio.

Magnus, Olaus. *Historia de Gentibus Septentrionalibus.* Basel, 1567, folio.

Martyr, Petrus (Anglerius). *De novo orbe, or the History of the West Indyes.* London, 1615, quarto.

Meurs, Johannes von. *Rerum Belgicarum libri quatuor.* Leyden, 1614, quarto.

Molina, Ludovicus. *De Hispanorum primogeniorum origine ac natura.* Cologne, 1601, folio.

Moryson, Fynes. *At Itinerary (containing his ten yeeres travell through the 12 dominions of Germany, Bohmerland . . . and Ireland).* London, 1617, folio.

Münster, Sebastian. *Cosmographia Universalis.* Basel, [first Latin edition, 1550], folio.

Pantaleon, Heinrich. *Prosopographiae heroum atque illustrium virorum Germaniae.* Basel, 1565, folio.

Pistorius, Johann. *Illustrium veterum scriptorum, qui rerum a Germanis gestarum historias reliquerunt.* Frankfurt, 1583; vol. 2 1584, folio.
———. *Polonicae Historiae corpus: hoc est, Polonicarum rerum latini scriptores, quotquot extant.* 3 vols. Basel, 1582, folio.
Pius II (Aeneas Sylvius). *Opera omnia.* [First edition Basel, 1551, folio.]
Pontanus, Johannes. *Originum Francicarum libri vi in quibus praeter Germaniae ac Rheni chronographiam, Francorum origines ac primae sedes . . . ordine deducuntur.* Harderwyk, 1616, quarto.
Possevino, Antonio. *Moscovia, et alia opera.* 2 parts. [Cologne], 1595, folio.
Postel, Guillaume. *De orbis terrae concordia.* [Basel], n.d.
Purchas, Samuel. *His Pilgrimage.* London, 1613, folio, three copies (one 1617 ed.).
Reineck, Reinerus. *Historia Orientalis: hoc est rerum in oriente a Christianis, Saracenis, Turcis et Tartaris gestarum diversorum Auctorum.* [First edition Helmstadt, 1602 quarto.]
Reusner, Nicolaus. *Rerum memorabilium in Pannonia sub Turcarum Imperatoribus a capta Constantinopoli.* Frankfurt, 1603, quarto.
Schottus, Andreas. *Hispaniae Illustratae, seu Rerum Urbiumque Hispaniae, Lusitaniae, AEthiopiae, et Indiae, Scriptores varii.* Frankfurt, 1603–6, folio.
Schrijver, Pieter. *Inferioris Germaniae Provinciarum Unitarum Antiquitates . . . item picturae operum ac monumentorum veterum, nec non Comitum Hollandiae, Zelandiae & Frisiae, Eicones, eorumdemque historia.* Leiden, 1611, quarto.
Stanihurst, Richard. *De rebus in Hibernia gestis.* Antwerp, 1584, quarto.
Strabo. *Geographia a G. Xylander. Graeco-latin.* Basel, 1571, folio.
———. *Geographia cum commentarum I. Casaubon.* Paris, 1620, folio.

Chorography

Adrichem, Christian. *Theatrum terrae sanctae.* [First edition Cologne, 1593, folio], two copies.
Angelocrator, Daniel. *Chronologiae Danielis Angelocratoris . . . prodromus . . . Refutatur hic etiam.* Hamburg, 1597, quarto.
Bale, John. *Illustrium mauoris Britanniae scriptorum, summarium.* Basel, 1559, folio.
Biondi, Flavio. *De Roma Triumphante.* Basel, 1559, folio.
Boissard, Jean Jacques. *Romanae urbis Topographia et Antiquitates.* Frankfurt, 1597, folio.
Braun, Georg, and Francis Hogenberg. *Civitates Orbis Terrarum.* [First edition Antwerp, 1572.]
Henninges, Hieronymus. *Theatrum Genealogicum.* Vols. 1–4. Magdeburg, 1598, folio.
Krentzheim, Leonhard. *Observationum Chronologicarum . . . libri iiii in quibus . . . sacrarum et profanarum literarum nodi chronologici . . . dissolvuntur.* Lignicii, 1605, folio.
Matthew of Westminster. *Flores Historiarum, per M. Westmonasteriensem collecti: praecipae de rebus Britannicis.* Frankfurt, 1601, folio.
Orlers, Jan. *Genealogia . . . comitum Nassoviae in quo origo, incrementa et res gestae ab iis ab anno 682 ad . . . 1616.* Leyden, 1616, folio.
Pontanus, Johannes. *Rerum et urbis Amstelodamensium historia.* Amsterdam, 1611, folio.
Ranulph Higden. *Polychronicon s.xiv.* [First printed edition London, 1482.]
Rosinus, Joannes. *Antiquitatum Romanarum corpus absolutissimum, in quo praeter ea quae Ioannes Rosinus delineauerat.* Paris, 1613, folio.
Rossi, Girolamo. *Italicarum & Ravenatum Historiarum libri xi.* Venice, 1603, folio.
Savile, Henry, ed. *Rerum Anglicarum scriptores post Bedam praecipui, etc.* London, 1596, folio.
Scaliger, Joseph Justus. *Eusebii . . . chronicorum canonum omnimodae historiae libri duo.* Paris, 1606, folio.
Schrader, Laurentius. *Monumentorum Italiae, quae hoc nostro saeculo & a Christianis posita sunt, libri quartuor.* Helmstadt, 1592, folio.
Speed, John. *The Theater of ye Empire of Greate Britaine.* London, 1611, folio.
Valerianus, J. Pierius. *Hieroglyphica.* Basel, 1575.

CORPUS CHRISTI COLLEGE, OXFORD

From the following sources: Roger Charnoke, Inventory 1576, 77 books named, Oxford Vice-Chancellor's Court (VCC); Oxford Bodleian MS Wood D.10; Corpus Christi College, MS D/3/1, library catalog (1589); "Catalogus Librorum Collegii Corporis Christi," MS D/3/2; MS D.3.3; Oxford VCC, box of transcripts, GG, f. 291b; Miles Windsor, will proved 1624, Oxford VCC, from "Extracts from filed wills concerning books," 86.

Mathematical Geography

Agricola, Georgius. *De mensuris & ponderibus.* [First edition Basel, 1533, quarto.]
———. *De Re Metallica libri xii.* Basel: Froben, 1556, folio.
Ailly, Petrus de. *Imago Mundi.* [First printed edition Louvain, 1480.]
Alunno, Francesco. *La fabrica del mondo.* [First edition Vinegia, 1548], folio.
Apianus, Petrus. *Cosmographia.* [First edition Ingolstadt, 1529], two copies.
Bartholomaeus, Anglicus. *De proprietatibus rerum.* [First printed edition Basel, 1470.]
Garcaeus, Johann. *Meteorologia.* Wittenberg. [First edition, 1568.]
Gilbert, William. *De Magnete.* London, 1600, folio.
Giraldus, Lilius Gregorius. *De re nautica libellus.* Basel, 1540, octavo.
Grynaeus, Simon. *Novus Orbis Regionum ac Insularum veteribus incognitarum, una cum tabula Cosmographia.* Basel, 1537, folio.
Honterus, Johannes. *Rudimenta Cosmographica.* [First edition Zurich, 1546], octavo.
"Mappes of England."
"Mappes of Oxford and Cambridge."
Ortelius, Abraham. *Theatrum Orbis Terrarum.* [First edition Antwerp, 1570], folio, two copies (one 1584).
———. *Thesaurus Geographicus.* Antwerp, 1587, folio.
Ptolemy, Claudius. *Commentarius super quadripartitum Ptolemaei, cum commentarie Bethanii et Albasino de Nativitate; Duae chartae cosmographicae, una totius mundi alteraque Europae.* (Not traced.)
———. *Geographia.* [First printed edition Vincenza, 1475], two copies.
———. *Geographiae Universae opus.* Cologne, 1597, quarto.
———. *Tabulae Geographia per Gerardum Mercatorem.* Brussels, 1578, folio.

Descriptive Geography

Aelianus, Claudius. *De historia animalium.* [First printed edition Lyons, 1562], octavo.
———. *De Varia Historia.* Trans. J. Lauro. [First printed edition Venice, 1550, octavo], three copies.
Aemylius, Paulus (Veronensis). *De rebus gestis Francorum.* [First edition Paris, ca. 1520], folio.
Antoninus, Augustus. *Itinerarium provinciarum omnium.* [First printed octavo edition Venice, 1550?], octavo.
Ascham, Roger. *A report and discourse written by R. Ascham, of the affaires of Germany.* London, [1570], quarto.
Bizzari, Pietro. *Rerum Persicarum historia.* [First edition Antwerp, 1583, folio.]
Boethius, Hector. *Scotorum Historia.* [First edition Paris, 1526], folio.
Bonfinius, Antonius. *Rerum Hungaricarum.* Basel, 1543, folio.
Bry, Theodore de. *America. Pars I: "Admiranda narratio, fida tamen . . . Virginiae . . . Anglico scripta sermone a Thoma Hariot."* Frankfurt, 1590, two copies.
Buchanan, George. *Rerum Scoticarum Historia.* [First edition Edinburgh, 1582, folio], three copies (one Frankfurt, 1584, octavo).
Camden, William. *Annales rerum Anglicarum et Hibernicarum regnante Elizabetha.* London, 1615, folio.
Cervarius Tubero, Ludovicus. *Ludovici Tuberonis . . . Commentariorum de rebus quae tem-*

poribus eius in illa Europae parte, quam Pannonii et Turcae eorumque . . . Frankfurt, 1603, quarto.
Chalcocondylas, Laonicus. *De Origine et rebus gestis Turcorum, libri decem.* Basel, 1556.
Chamier, Daniel. *Epistolae Iesuiticae.* Geneva, 1599, octavo.
Conestaggio, Girolamo Franchi. *Dell'unione del regno di Portogallo alla corona di Castiglia.* Venice, 1592, octavo.
Coryate, Thomas. *Coryats Crudities: hastily gobled up in five moneths travels.* London, 1611, quarto.
Crusius, Martinus. *Aethiopicae Heliodori historiae epitome.* Frankfurt, 1584, octavo.
Curio, Caelius Augustinus. *Caelii Augustini Curionis Sarracenicae historiae libri iii. in quibus Sarracenorum, Turcarum, Aegypti Sultanorum, Mamalucorum, . . . qui in Perside regnant, origines & incrementa, septingentorumque.* Basel, 1567, folio.
Danaeus, Lambertus. *Geographiae poeticae libri quatuor.* [Geneva], 1580, octavo.
Dicaearchus (Messenius). *Geographica quaedam, sive de vita Graeciae, eiusdem Descriptio Graeciae, versibus iambis, ad Theophrastum.* [Paris], 1589, decimo.
Discursus de rebus Gallicis. Halcyonia, 1589, octavo.
Does, Johan van der. *Bataviae Hollandiaeque Annales.* Leiden, 1601, quarto.
Du Bellay, Martin. *Commentariarum de rebus Gallicis libri x.* [First edition Frankfurt, 1575, folio.]
Dubravius, Johannes. *Historia Boiemica.* Basel, 1575, folio.
Du Tillet, Jean (Sieur de la Bussière). *Commentariorum de rebus Gallicis libri duo.* Frankfurt, 1579, folio, two copies.
Eytzinger, Michael. *De leone Belgico eiusque, topographica atque historica descriptione liber.* Cologne, 1583, folio.
Fazellus, Thomas, ed. *Rerum Sicularum scriptores in unum corpus congesti.* Frankfurt, 1579, folio.
Fragoso, Juan. *Aromatum, fructuum et simplicium aliquot medicamentorum ex India utraque . . . in Europam delatorum.* Strassburg, 1600 octavo.
Frangi, M. *Histoire de la Mappe-Monde Papistique.* Delphe, quarto (not traced).
Fumée, Martin. *Histoire generalle des troubles de Hongrie et Transilvanie.* Paris, 1608, quarto.
Gaguin, Robert. *De francorum regum gestis scripsit annales.* [First edition Paris, 1521, octavo.]
———. *De origine et gestis francorum Compendium.* [First edition Paris, 1499 (1495), folio.]
Gretser, Jacob. *De Sacris et religiosis Peregrinationibus.* [First edition Ingolstadt, 1606, quarto], two copies.
Guicciardini, Lodovico. *Description de tout le Pais-Bas.* Trans. F. de Belleforest. Anvers, 1567, folio.
Hakluyt, Richard. *Principal Navigations, Voyages, Traffiques and Discoveries of the English Nation.* 3 vols. London, 1598–1600, folio.
Hall, Joseph (alias Mercurius Britannicus). *Mundus alter et idem sive Terra Australis antehac semper incognita longis itineribus peregrini Academici nuperrime lustrata. [with 5 maps].* [First edition London, 1605, octavo.]
Heliodorus (Emesenus). *L'Histoire Aethiopique de Heliodorus.* Trans. [J. Amyot]. Paris, 1547, folio.
Herberstein, Sigmund. *Rerum Moscoviticarum commentarii.* [First edition Vienna, 1549, folio.]
Heuter, Pontius. *Rerum Burgundicarum libri sex.* [First edition Antwerp, 1584, folio], three copies (one 1598, quarto).
Historicum opus, in quatuor tomas divisum: quorum, tomus I, Germaniae antiquae illustrationem continet. Basel, 1574, folio.
Hoeniger, Nicolaus, ed. *Aulae Turcicae, Othomannicique imperii, descriptio . . . Author Antoine Geuffroy.* Basel, 1573, octavo.
Junius, Adrianus. *Batavia.* Leiden, 1588, quarto.

Knolles, Richard. *The Generall Historie of the Turkes.* London, 1603, folio.
Koran. Machumetis Sarracenorum Principis vita ac Alcoranum dicitur. [First Latin edition Basel, 1543, folio.]
Kromer, Martin. *De Origine et rebus gestis Polonorum libri xxx.* [First edition Basel, 1558, folio.]
Leo, Johannes. *De totius Africae Descriptione libri ix.* Antwerp, 1556, octavo.
Lery, Jean de. *Histoire d'un voyage fait en la tarre du Bresil.* La Rochelle, 1578, octavo.
Leunclavius, Joannes. *Annales Sultanorum Othmanidarum, a Turcis sua lingua scripti.* Frankfurt, 1588, quarto.
Linschoten, Jan Huygen van. *His discourse of voyages into ye East & West Indies. Translated from Dutch by W. P[hilip].* London, 1598, folio, two copies.
———. *Navigatio ac itinarium J. H. Linscotani in Orientalem sive Lusitanoram Indiam.* The Hague, 1599, folio.
Lopez de Gomara, Francisco. *Histoire Générale des Indes Occidentales & Terres neuues.* Trans. M. Fumée. Paris. [First edition, 1568, octavo.]
Maffei, Giovanni Petri. *Historiarum Indicarum libri xvi.* [First edition Florence, 1588], folio, two copies (one Venice, 1589).
Magnus, Olaus. *Historia de Gentibus Septentrionalibus.* [First edition Rome, 1555, folio.]
Manasses, Constantinus. *Annales.* Trans. I. Leuenclaius. [First edition Basel, 1573, octavo.]
Martyr, Petrus (Anglerius). *De Rebus oceanicis et novo orbe, decades tres.* [First octavo edition Cologne, 1574.]
Meteran, Emmanuel. *Historia Belgica nostri potissimum temporis.* [Antwerp? 1598], folio.
Monardes, Nicolas. *De Simplicibus Medicamentis ex occidentali India delatis.* Antwerp, 1574, octavo.
Münster, Sebastian. *Cosmographia.* Basel, 1559.
———. *La Cosmographie universelle, contenant la situation de toutes les parties du monde avec leurs proprietez et appartenances. Augmentée, ornée et enrichée, par F. de Belle-forest.* 2 vols. Paris, 1575, folio.
Notitia utraque dignitatum cum orientis, tum occidentis ultra Arcadii, Honoriique tempors. Et in eam G. Panciroli commentarium. Venice, 1593, folio.
Orta, Garcia de. *Aromatum et simplicium aliquot medicamentorum apud Indos nascentium historia. In epitomen contracta a C. Clusio.* [First edition Antwerp, 1567, octavo.]
Pausanias. *Description of Ancient Greece [in Greek].* Venice, 1516, folio.
Philippson, John (Sleidanus). *De statu religionis & reipublicae Carolo quinto Caesare commentarii.* Strasbourg, 1555, two copies.
Pistorius, Johann. *Illustrium veterum scriptorum, qui rerum a Germanis gestarum historias reliquerunt. (Germanicorum scriptorum tomus alter.)* Frankfurt, 1583, folio, two copies.
Pius II (Aeneas Sylvius). *Boemicae Historia.* [First edition Rome, 1475. (*Cosmographia.* Vol. 3. Venice, 1503, quarto.)]
———. *Opera omnia.* [First edition Basel, 1551], folio.
Plutarch. *Peri potamon kai oron. De Fluminibus et Montibus.* Ed. Gelenius Sigis. [First printed edition Basel, 1533.]
Possevino, Antonio. *Moscovia.* [First edition Cologne, 1587], folio, two copies.
Procospius (Caesariensis). *De rebus Gothorum, Persarum ac Vandalorum libri vii.* [First Latin edition Basel, 1531, folio.]
Reineck, Reinerus. *Chronicon Hierosolymitanum. Quae operis subiecti est pars prima.* Helmstadt, 1584, quarto.
Rerum Hispanicarum Scriptores aliquot. 3 vols. Frankfurt, 1579, folio.
Reusner, Nicolaus. *Hodoeporicorum, sive, Itinerium totius fere orbis, lib. vii.* Basel, 1580, octavo.
Schottus, Andreas. *Hispaniae Illustratae, seu Rerum Urbiumque Hispaniae, Lusitaniae, AEthiopiae, et Indiae, Scriptores varii.* Frankfurt, 1603–6, folio.
Simler, Josias. *De Helvetiorum republica.* Paris, 1577.

Solinus, Caius Julius. *Polyhistor [De situ orbis terrarum et de singulis mirabilibus] . . . Hic Pomponii Melae de situ orbis libros . . . adiunximus. Accesserunt nova scholia, quae loca autoris utriusque obscuriora . . . illustrant, tabulae geographicae, etc. (Solini vita per Camertem aedita).* Basel, 1538, folio.

Stanihurst, Richard. *De rebus in Hibernia gestis.* Antwerp, 1584, quarto.

Strabo. *De Situ Orbis [also named Geographia].* [First printed edition [Venice], 1494], four copies (one Basel, 1571, one Geneva, 1587).

Thevet, André. *Les Singularitez de la France Antarctique, autrement nommée Amerique.* [First edition Paris, 1558, octavo.]

Thomas, William. *The Historie of Italie.* [First edition London, 1549], quarto, two copies.

Vadianus, Joachim. *Epitome trium terrae partium, Asiae, Africae et Europae.* Zurich, 1534, octavo.

Vitriaco, Jacobus de. *Libri duo. Quorum prior Orientalis, sive Hierosolymitanae: alter Occidentalis historiae nomine inscribitur.* Douay, 1597, octavo.

Chorography

Adrichem, Christian. *Theatrum Terrae Sanctae.* Cologne, 1593, folio.

Arius, Jo. *De antiquitatibus variis.* (Not traced.)

Ballino, Giulio. *De' disegni delle piu illustri citta & fortezze del mondo.* Vinegia, 1569, quarto.

Biondi, Flavio. *De Roma Triumphante.* [First edition Mantua, 1472? folio], two copies (one Basel, 1559).

———. *Roma ristaurata, et Italia illustrata.* Venice, 1543, octavo.

Boissard, Jean Jacques. *Romanae urbis Topographia et Antiquitates.* Frankfurt, 1597, folio.

Brulus, Jo. Michaelis. *Florentina historia.* Quarto. (Not traced.)

Camden, William. *Britannia.* [First edition London, 1586, quarto.]

Carew, Richard. *Survey of Cornwalle.* London, 1602, quarto.

Gildas. *Opus novum: De Calamitate, excidio et conquestu Britanniae.* Ed. P. Vergilius and R. Ridley. [First published edition London, 1525, octavo.]

Gruterus, Janus. *Inscriptiones Antique totius orbis Romani.* 2 parts. [Heidelberg], 1602–3, folio.

Justinianus, Petrus. *Rerum Venetarum historia.* Venice, 1560, folio.

Lambard, William. *A perambulation of Kent: conteining the description, . . . of that shyre.* London, 1576, quarto.

Lively, Edward. *A Chronologie of the times of the Persian Monarchie.* London, 1597, octavo.

Marlianus, Raimundus. *C. Iulii Caesaris rerum ab se gestarum commentarii, etc. [cum] Veterum Galliae locorum, populorum, urbium, montium ac fluviorum brevis descriptio.* Paris, 1544, octavo.

Matthew of Westminster. *Flores Historiarum.* London, 1570, folio, two copies.

Norden, John. *Speculum Britanniae. The first parte an historicall discription of Middlesex. [with others following].* London, 1593, quarto.

Price, Sir John. *Historiae Brytannicae defensio.* London, 1573, quarto.

Ranulf Higden. *Polychronicon.* [First printed edition London, 1482], MS.

Rerum Britannicarum id est Angliae, Scotiae, vincinarumque insularum ac regionum Scriptores Vetustiores ac praecipui. Heidelberg, 1587, folio, two copies.

Rosinus, Joannes. *Romanarum Antiquitatum libri decem ex variis scriptoribus . . . collecti.* [First edition Basel, 1583], folio.

Savile, Henry. *Rerum Anglicarum scriptores post Bedam praecipui, etc.* London, 1596, folio.

Scaliger, Joseph Justus. *De emendatione temporum.* [First edition Paris, 1583], folio, two copies.

Schrader, Laurentius. *Monumentorum Italiae, quae hoc nostro saeculo & a Christianis posita sunt, libri quatuor.* Helmstadt, 1592, folio.

Speed, John. *The Theatre of the Empire of Great Britaine: presenting an exact geography of . . . England, Scotland, Ireland.* London, 1616, folio.

Stow, John. *A Survey of London.* London, 1599, quarto.

Suidas. *Historica caeteraque omnia que ad cognitionem rerum spectant.* Basel, 1581, folio, three copies (one in Greek).
Verstegan, Richard (Rowlands). *A restitution of decayed intelligence: in antiquities. Concerning the English Nations.* Antwerp, 1605, quarto.
William (of Newburgh). *Rerum Anglicarum libri quinque.* Antwerp, 1567, octavo.

Peterhouse, Cambridge

These books are from the following sources: Thomas Lorkin, Inventory 1591, Cambridge Vice-Chancellors Court (VCC) probate inventories (5); Robert Soame alias Some, Inventory 1608/9, Cambridge VCC (7); "Nomina librorum qui erant in Bibliotheca Collii ante Doctorum Perne mortuum [1588]," Peterhouse MS 400; MS 405; and unnumbered manuscript (library list, undated).

Mathematical Geography

Agricola, Georgius. *De Mensuris & ponderibus.* [First edition Basel: Froben, 1533, quarto], two copies.
———. *De re metallica libri xii. Eiusdem de animantibus subterraneis liber, ab autore recognitus.* [First edition Basel, 1556, folio.]
Apianus, Petrus. *Cosmographia.* Ed. Gemma Frisius. [First edition Antwerp, 1529], three copies (one 1533).
———. *Introductio geographica.* Ingolstadt, 1533, folio.
Digges, Leonard. *A Prognostication of right good effect.* London, 1555, folio.
Finé, Oronce. *De Mundi Sphera sive Cosmographia.* [First edition Paris, 1542, octavo], two copies.
Gemma Frisius. *De Principiis Astronomiae et Cosmographiae.* Antwerp, 1530.
Grynaeus, Simon. *Novus Orbis Regionum ac Insularum veteribus incognitarum, una cum tabula Cosmographia.* Basel, 1537.
Honterus, Johannes. *Rudimenta cosmographica.* [First edition Zurich, 1546, octavo.]
Loritus, Henricis (Glareanus). *De Geographia.* [First edition Venice, 1534, octavo], two copies.
"Map, on generall, & a map of France & an other litel."
"Map of the world, one of England, one of Grece and a picture of veritie and one old map."
"Maps, divers & picture & tables."
Mercator, Gerard. *Atlas sive Cosmographicae meditationes de fabrica mundi et fabricati mundi.* [First edition Dusseldorf, 1595], folio.
———. *Galliae (Belgii Inferioris-Germaniae) tabulae geographicae.* [First edition Duysburg, 1585, folio.]
Ortelius, Abraham. *Theatrum Orbis Terrarum.* [First edition Antwerp, 1570], folio, three copies (one in English, London, 1603).
Ptolemy, Claudius. *Geographia.* [First printed edition Vincenza, 1475], three copies.
Vicomercatus, Franciscus. *In quatuor libros Aristotelis meteorologicorum commentarii. [With text]* Greek & Latin. [First edition Paris, 1556, folio.]

Descriptive Geography

Aelianus, Claudius. *De historia animalium.* [First edition Lyons, 1562, octavo.]
———. *Variae historiae libri xiiii.* [First edition Venetia, 1550, octavo.]
Alberti, Leonardo. *Descrittione di tutta l'Italia.* [First edition Bologna, 1550, folio.]
Bellutius, Philippus. *Itinerarium.* Rome, [ca. 1525], quarto.
Billerbeg, Franciscus de. *Most rare and straunge discourses, of Amurathe the Turkish emperor that nowe is: with the warres betweene him and the Persians . . .* London, [1584], quarto.
Bizzari, Pietro. *Rerum Persicarum historia.* [First edition Antwerp, 1583, folio], two copies.
Boethius, Hector. *Scotorum Historia.* [First edition Paris, 1526], folio.
Buchanan, George. *Rerum Scoticarum historia.* Edinburgh, 1582, folio.

Curio, Caelius Augustinus. *Caelii Augustini Curionis Sarracenicae historiae libri III.* Basel, 1567, folio, two copies.

Drechsler, Wolfgang. *De Saracenis et Turcis chronicon, &c.* Strasbourg, 1550, octavo.

Du Tillet, Jean (Sieur de la Bussière). *Commentariorum de rebus Gallicis libri duo.* Frankfurt, 1579, folio.

Fazellus, Thomas, ed. *Rerum Sicularum scriptores in unum corpus congesti.* Frankfurt, 1579, folio.

Fontanus, Jacobus. *De bello Rhodio, libri tres.* [First edition Rome, 1524, folio.]

Gerbelius, Nicolaus. *In descriptionem Graeciae Sophiani, praefatio.* Basel, 1545, folio.

———. *Pro declaratione picturae Graeciae Sophiani.* Basel, 1549 [1550], folio.

Germanicarum rerum quatuor celebriores vetustioresque chronographi: I. Turpinus, Rhegino Abbas Prumiensis, Sigebertus Gemblacensis Lambertus Schaffnaburgensis. Frankfurt, 1566, folio.

Giovio, Paolo. *Descriptio Britanniae, Scotiae, Hyberniae, et Orchadum.* Venice, 1548, quarto.

Goropius, Johannes (Becanus). *Opera, hactenus in lucem non edita. Hermathena. Hieroglyphica. Vertumnus. Gallica. Francica. Hispanica.* Antwerp, 1580, folio.

———. *Origines Antwerpianae.* Antwerp, 1569, quarto.

Guagninus, Alexander. *Sarmatiae Europeae descriptio. (Lithuania, &c.).* [Cracow, 1578], folio.

Heliodorus (Emesenus). *Aethiopicae historiae libri decem, nunc primum a Greco sermone in Latinum translati, S. Warschewiczki interprete.* Basel, [1552], folio.

Herberstein, Sigmund. *Rerum Moscoviticarum commentarii.* [First edition Vienna, 1549, folio], two copies (one Basel, 1571).

Heuter, Pontius. *Rerum Burgundicarum libri sex.* [First edition Antwerp, 1584, folio], two copies.

Joannis Cantacuzeni. *[Emperor of the East] . . . contra Mahometicam fidem Christiana et Orthodoxa Assertio . . . Latinitato donata. R. Gualthero . . . interprete.* Basel, 1543, folio.

Junius, Adrianus. *Batavia.* Leiden, 1588, quarto.

Koran. Machumetis Sarracenorum Principis vita ac Alcoranum dicitur. [First Latin edition Basel, 1543, folio.]

Krantz, Albertus. *Rerum Germanicarum historici clariss. Regnorum Aquilonarium, Daniae, Sueciae, Norvagiae, Chronica.* Frankfurt, 1575, folio.

Kromer, Martin. *De Origine et rebus gestis Polonorum.* [First edition Basel, 1555, folio.]

———. *Oratio in funere Sigismundi, eius monis primi, Polonorum . . . Regis . . . habita, etc. In Maciejowski, Samuel. De Sigismundo primo rege Poloniae etc. duo panegyrici funevres* . . . Mainz, 1550, octavo.

Lazius, Wolfgang. *De gentium aliquot migrationibus, sedibus fixis, reliquis, linguarumque initiis et immutationibus ac dialectis, libri xii.* [First edition Basel, 1557, folio.]

Lily, George. "Virorum aliquot in Britannia, qui nostro seculo eruditione et doctrina clari memorabilesque fuerunt elogia. Omnium in quos, variante fortuna, Britanniae imperium translatum brevis enumeratio." In Paolo Giovio, *Descriptio Britanniae.* Venice, 1548, quarto.

Linthprandus de reb: gest: ab Europae imperatorib: & regib. (Not traced.)

Lonicerus, Philip. *Chronicorum Turcicorum tomus primus (-tertius).* Frankfurt, 1578, folio, two copies.

Lupanus, Vincentius. *Commentarii de magistratibus Francorum.* [First edition Paris, 1551, octavo.]

Magnus, Joannes. *Gothorum Sueonumque historia.* [First edition Rome, 1554, folio.]

Magnus, Olaus. *Historia de Gentibus Septentrionibus.* [First edition Rome, 1555, folio], three copies.

Mela, Pomponius. *De Situ Orbis, adjectis Joachimi Vadiani . . . in eosdem scholiis: Addita quoque in Geographia Catechesi: et Epistola Vadiani ad Agricola.* Vienna, 1518, folio, two copies.

Mela, Pomponius, and Caius Julius Solinus. *Pomponius Mela [De Situ Orbis]. Julius Solinus [De situ orbis terrarum et de singulis mirabilibus]. Itinerarium Antonini Aug. Vibius*

Sequestor. P. Victor de regionibus urbis Romae. Dionysius Afer de situ orbis. Prisciano interprete. [First edition Venice, 1518, octavo.]

Monardes, Nicolas. *De Simplicibus Medicamentis ex occidentali India delatis.* Antwerp, 1574, octavo.

———. *Ioyfull Newes out of the New Founde World.* Trans. J. Frampton. London, 1577, quarto.

Moryson, Fynes. *At Itinerary (containing his ten yeeres travell through the 12 dominions of Germany, Bohmerland . . . and Ireland).* London, 1617, folio.

Münster, Sebastian. *Cosmographia.* [First edition Basel, 1544.]

Mutius, Huldericus. *De Germanorum prima origine, moribus, institutis.* Basel, 1539, folio.

Nebrissensis, Aelius Antonius. *Rerum a Ferdinando & Elizabe Hispaniarum regibus gesta R. decades duas, &c.* [First edition n.p., 1545, folio.]

Orta, Garcia de. *Aromatum et simplicium aliquot medicamentorum apud Indos nascentium historia. In epitomen contracta a C. Clusio.* [First edition Antwerp, 1567, octavo.]

Osorio, Geronymo. *De Rebus; Emmanuelis regis Lusitaniae virtute gestis libri xii.* Cologne, 1574.

Pausanias. *Description of Ancient Greece [in Greek].* [First edition Venice, 1516, folio.]

Perez de Moya, Juan. *Varia Historia de sactas e illustres mugeres en todo genero de virtudes.* Madrid, 1583, folio.

Pistorius, Johann. *Polonicae Historiae corpus: hoc est, Polonicarum rerum latini scriptores, quotquot extant.* 3 vols. Basel, 1582, folio.

Pius II (Aeneas Sylvius). *Historia Bohemica.* [First edition Rome, 1475, quarto], two copies.

———. *Opera omnia.* [First edition Basel, 1551, folio.]

Possevino, Antonio. *Moscovia, et alia opera, de statu hujus seculi, adversus catholicae Ecclesiae hostes.* 2 parts. [First edition Cologne, 1587], folio, two copies.

Postel, Guillaume. *De Etruriae regionis originibus, commentatio.* Florence, 1551, quarto.

Ramusio, Giovanni Battista. *Delle navigationi et viaggi.* [First edition Venice, 1550], folio.

Reineck, Reinerus. *Chronicon Hierosolymitanum. Quae operis subiecti est pars prima.* Helmstadt, 1584, quarto.

———. *Pars secunda. continens duorum priorum familiae Luceburg. Imperatorum historiam, &c.* Helmstadt, 1585, quarto.

Reipublicae Romanae in exteris provinciis, bello acquisitis, consitutae commentariorum libri duodecim . . . Frankfurt, 1598, folio.

Rhenanus (Beatus). *Rerum Germanicarum libri tres.* [First edition Basel, 1531, folio.]

Richardi, *Confutatio legis Mahomet. Mundus Novus.* (Not traced.)

Solinus, Caius Julius. "Commentaria" [by J. R. Vellinus, Camers] in C. I. Solini, *Polyhistora, &c. [With the text].* Basel, 1557, folio.

Stanihurst, Richard. *De rebus in Hibernia gestis.* Antwerp, 1584, quarto.

Strabo. *De Situ Orbis [also named Geographia].* Many editions; first printed edition Venice, 1494.

Theologia Muscovitarum, Russorum, & Tartarorum. (Not traced.)

Thevet, André. *The New Found Worlde, or Antarchke, wherin is contained beastes, fishes . . .* Trans. from the French. London, 1568, quarto.

Vecerius, Conrad. *De Seditionibus Siciliae historia. Item eiusdem de rebus gestis Imperatoris Henrici VII libellus.* [First edition the Hague, 1531, quarto.]

Ziegler, Jacobus. *Quae intus continentur. Syria, Palaestina, Arabia Petraea, Aegyptus, Schondia, Holmiae excidii historia.* Strasbourg, 1532, folio, two copies.

Zosimus, the Historian. *Historiae Novae libri vi. numquam hactenus editi: quibus additae sunt historiae Procopii Caesariensis, Agathiae Myrinaei, Jornandis Alani . . .* Basel, [1576?], folio, three copies.

Chorography

Apianus, Petrus, ed. *Inscriptiones sacrosanctae vetustatis.* Ingolstadt, 1534, folio.

Camden, William. *Britannia.* [First edition London, 1586, quarto.]

Carr, Nicholas. *De scriptorum Britannicorum paucitate oratio.* London, 1576, octavo.

Fabricius, Georgius. *Rerum Misnicarum libri VII. Electorum Saxoniae lib. I. Marchionum Misnensium lib. I. Annalium urbis Misnae lib. III. Siffidi Misnensis presbyteri epitomes lib. II.* Leipzig, [1569], quarto.
Fitzherbert, John. *Boke of Surveying.* [First edition London, 1523, quarto.]
Geoffrey of Monmouth. *Historia Regis Britanniae.* [First printed edition Paris, 1508], three copies.
Gerald of Wales. *Itinerarium Cambriae: auctore Giraldo Cambrense.* London, 1585, octavo.
———. *Topographia Hiberniae.* MS (not published in sixteenth century).
Lambard, William. *A perambulation of Kent: conteining the description, . . . of that shyre.* London, 1576, quarto.
Lazius, Wolfgang. *Vienna Austriae. Rerum Viennensium Commentarii.* Basel, 1546, folio.
Lucidus, Johannes. *Opusculum de emendationibus temporum ab orbe condito ad usque hanc aetatem.* [First edition Venice, 1537], quarto.
Maffei, Raphael (Volaterranus). *Commentariorum Urbanorum.* [First edition Rome, 1506], folio.
Marlianus, Bartholomeus. *Urbis Romae Topographia.* [First edition Rome, 1544.]
Matthew of Westminster. *Flores Historiarum.* London, 1570, folio.
Price, Sir John. *Historiae Brytannicae defensio.* London, 1573, quarto.
Ranulf Higden. *Polychronicon.* [First printed edition London, 1482], two copies.
Reisner, Adam. *Ierusalem, ex sacris literis et approbatis historicis ad unguem descripta.* Trans. Ioh. Heyden. Frankfurt, 1563, folio.
Rosinus, Joannes. *Romanarum Antiquitatum libri decem.* [First edition Basel, 1583, folio.]
Scaliger, Joseph Justus. *De emendatione temporum.* [First edition Paris, 1583], folio.
Segar, Sir William. *The Booke of honor and armes.* London, 1590, quarto.
Smetius, Martinus. *Inscriptionum Antiquarum quae passim per Europam, liber. Accessit auctarium a J. Lipsio.* 2 parts. [Leiden], 1588, folio.

St. John's College, Cambridge

Sources include: John Hatcher Inventory 1587, Cambridge Vice-Chancellor's Court (VCC), Box 4, in Plan Press II; Philip Kettle, Inventory 1605/6, 13 books named, Cambridge VCC (7); Edmund Roberts, Inventory 1580, Cambridge VCC (3); Philip Stringer, Inventory 1605, Cambridge VCC (7); St. John's College Cambridge Library Catalogue, Cambridge University Library, MS Dd. 5.54; library list and individual book donations, St. John's College Cambridge, MS U.3.

Mathematical Geography

Agricola, Georgius. *De re metallica libri xii.* [First edition Basel, 1556], folio.
Apianus, Petrus. *Cosmographia.* [First edition Ingolstadt, 1529.]
Blundeville, Thomas. *His Exercises.* [First edition London, 1594], quarto.
Digges, Leonard. *A prognostication everlasting.* London, 1555, folio.
Gilbert, William. *De Magnete.* London, 1600, folio.
Honterus, Johannes. *Rudimenta Cosmographica.* [First edition Zurich, 1546], octavo.
"Mappe of brillen [Britain?]."
"Mappe of England."
"Mapps, five."
Ortelius, Abraham. *Theatrum Orbis Terrarum.* [First edition Antwerp, 1570], folio, two copies.
Ptolemy, Claudius. *Geographia.* [First printed edition Vincenza, 1475.]
Recorde, Robert. *The castle of knowledge.* [First edition London, 1556, folio.]
Warner, William. *Albions England: Or Historicall map of the same island.* [First edition London, 1586], quarto.

Descriptive Geography

Aemylius, Paulus (Veronensis). *De rebus gestis Francorum. Libri iiii.* [First edition Paris, ca. 1520], folio.

Ascham, Roger. *A report and discourse written by R. Ascham, of the affaires of Germany.* London: J. Daye, [1570], quarto.

Belgicarum Rerum Annales a diversis auctoribus. Frankfurt, 1580, folio.

Benzoni, Girolamo. *Historia Indiae Occidentalis, tomis duobus comprehensa . . . Hieronymo Benzone . . . et Ioanne Lerio . . . testibus oculatis, autoribus. Ex eorum autem idiomate in latinum sermonem Urbani Calvetonis et G. B. studio conversi.* [Geneva], 1586, octavo.

Boaistuau, Petris. *Le Theatre du monde, ou il est faict un ample discours des miseres hum[a]ines* . . . Paris, 1561, sextodecimo.

Borlandi historia et iconde comitum Hollandiae. Folio. (Not traced.)

Broniowski, Marcin. *Tartariae Descriptio . . . cum tabula geographica eiusdem chersonesus Tauricae.* Cologne, 1595, folio.

Buchanan, George. *Rerum Scoticarum historia.* Edinburgh, 1582, folio.

Callimachus, Philippus (Experiens). *Historia de rege Vladislao.* Augsburg, 1519, quarto.

Camden, William. *Anglica, Normanica, Hibernica, Cambricaa a veteribus scripta.* Frankfurt, 1603, folio.

Contarinis, Giovanni P. *Historia delle cose successe dal principio della guerra da selim ottomano a Venetiani fina al di della gran giornata vittoriosa contra Turchi.* Venice, 1572, quarto.

Does, Johan van der. *Bataviae Hollandiaeque Annales.* Leiden, 1601, quarto.

Ens, Gaspar. *Rerum Danicarum Friderico II gestarum historia. Accesserunt epigrammata Ioannis Lauterbachii.* Frankfurt, 1593, folio.

———. Perhaps *Rerum Hungaricarum historia.* Cologne, 1604, octavo.

Gaguin, Robert. *De origine et gestis francorum Compendium.* [First edition Paris, 1499 (1495), folio.]

Geuffroy, Antoine. *Aulae Turcicae, Othomannicique imperii, descriptio.* 2 vols. Basel, 1573, octavo.

Groot, Hugo de. *Liber de antiquitate Reipublicae Batavicae.* Leiden, 1610, quarto.

Harriot, Thomas. *A brief and true report of the new found land of Virginia.* London, 1588, quarto.

Heliodorus (Emesenus). *Historiae Aethiopicae libri decem.* Basel, 1552, folio.

Heroldt, Johannes. *Originum ac Germanicarum antiquitatum libri.* Basel, 1557, folio.

Heuter, Pontius. *Rerum Burgundicarum libri sex.* [First edition Antwerp, 1584, folio.]

Hortensius, Lambertus. *De bello Germanico libri septem.* Basel, 1560, quarto.

———. *Lamberti Hortensii . . . secessionum civilium Ultrajectinarum et bellorum ab ann. XXIIII. supra M.CCCC., usque ad translationem Episcopatus ad Burgundos, Libri septem.* Basel, [1546], folio.

Hotoman, François. *De furoribus Gallicis.* Edinburgh[?], 1573, quarto.

[Hurault, Michel]. *A Discourse upon the present state of France.* London, 1588, quarto.

Leo, Johannes. *De totius Africae Descriptione libri ix.* Antwerp, 1556, octavo.

Leunclavius, Joannes. *Annales Sultanorum Othmanidarum, a Turcis sua lingua scripti.* Frankfurt, 1588, quarto.

Lonicerus, Philip. *Chronicorum Turcicorum tomus primus.* [First edition Frankfurt, 1578], folio.

Lorchensis de rebus danicis quaere Huttorum. Folio. (Not traced.)

Ludolphus, de Suchen. *De terra sancta et itinere jherosolomitano* . . . [First edition Strasbourg, 1474? folio.]

Maffei, Giovanni Petri. *Historiarum Indicarum libri xvi.* [First quarto edition Florence, 1589.]

Mariana, Juan de. *Historia de Rebus Hispaniae.* [First edition Toledo, 1592, quarto.]

Martyr, Petrus (Anglerius). *De Rebus oceanicis et novo orbe, decades tres.* [First Latin edition Cologne, 1574], two copies.

Mercliantius, Iac. *Descriptio Flandriae.* Octavo. (Not traced.)

Meteran, Emmanuel. *Historia Belgica nostri potissimum temporis.* [Antwerp? 1598], folio.

Meyer, Jacob. *Compendium chronicorum Flandriae.* Antwerp, 1561, folio.

Münster, Sebastian. *Cosmographia.* [First edition Basel, 1544], folio.

Newes of Strange countries. (Not traced.)

Osorio, Geronymo. *De Rebus; Emmanuelis regis Lusitaniae virtute gestis libri xii.* Cologne, 1574, octavo.

Patten, William. *The expedicion into Scotland of Edward, duke of Somerset.* London, 1548, octavo.

Philippson, John (Sleidanus). *Tabule in xxvj libros historiarum Ioannis Sleidani de stat religionis et reipublicae Christianae.* Strasbourg, 1558, octavo.

Pius II (Aeneas Sylvius). *Cosmographia in Asiae & Europae eleganti descriptione.* [First quarto edition Venice, 1503], two copies.

———. *Opera omnia.* [First edition Basel, 1551, folio.]

Possevino, Antonio. *Moscovia, et alia opera, de statu hujus seculi, adversus catholicae Ecclesiae hostes.* 2 parts. [First edition Cologne, 1587], folio.

Postel, Guillaume. *De Orbis Terrae Concordia.* [Paris, 1544].

Purchas, Samuel. *His Pilgrimage. Or relations of the world and the religions observed in all ages.* [First edition London, 1613], folio.

Ramusio, Giovanni Battista. *Delle navigationi et viaggi.* [First edition Venice, 1550], folio.

Reineck, Reinerus. *Chronicon Hierosolymitanum. Quae operis subiecti est pars prima.* Helmstadt, 1584, quarto.

Rerum Hispanicarum Scriptores aliquot. 3 vols. Frankfurt, 1579, folio.

Rhenanus (Beatus). *Rerum Germanicarum libri tres.* [First edition Basel, 1531, folio.]

Saligniaco, Bartholomaeus de. *Itinerarii Terre Sancte: inibique sacrorum locorum: ac rerum clarissima descriptio.* [First edition Lyon, 1525, octavo.]

Schrijver, Pieter. *Inferioris Germaniae Provinciarum Unitarum Antiquitates . . . item picturae operum ac monumentorum veterum, nec non Comitum Hollandiae, Zelandiae & Frisiae, Eicones, eorumdemque historia.* Leiden, 1611, quarto.

Stanihurst, Richard. *De rebus in Hibernia gestis.* Antwerp, 1584, quarto.

Vadianus, Joachim. *Epitome trium terrae partium, Asiae, Africae et Europae.* Zurich, 1534, octavo.

Werner, Georg. *De admirandis Hungariae aquis. Included in Possevino, Antonio. Moscovia, et alia opera.* [First edition Cologne, 1587], folio.

Chorography

Adrichem, Christian. *A Briefe description of Hierusalem . . . with a beautiful and lively map of Hierusalem.* Trans. Thomas Tymme. London, 1595, quarto.

Apianus, Petrus and Bartholomaeus Amantius. *Inscriptiones sacrosanctae vetustatis non illae quidem Romanae, sed totius fere orbis summo studio ac amximis impensis terra marique conquisitae feliciter incipiunt.* Ingolstadt, 1534, folio.

Biondi, Flavio. *De Roma instaurata. De Italia illustrata.* [First edition Venice, 1503, folio.]

———. *De Roma Triumphante.* [First edition Mantua, 1472? folio.]

Brocardus (de Monte Sion). *Descriptio Terrae Sanctae. Item Itinerarium Hierosolymitanum B. de Saligniaco.* Magdeburg, 1587, quarto.

Busaeus, Johannes. *Pro calendario gregoriano disputatio apologetica.* Mainz, 1585, quarto.

Camden, William. *Britannia.* [First edition London, 1586, quarto], three copies.

Carr, Nicolas. *De scriptorum Britannicorum paucitate oratio.* London, 1576, octavo.

Digges, Leonard. *A boke named Tectonicon briefelye shewyne the exacte measurynge, and speady reckenynge all maner lande, squared tymber . . .* London, 1556, quarto.

Fabricus, Georgius. *Roma. Itinerum liber unus . . .* Basel, 1551, octavo.

Fulvio, Andrea. *De urbis [Romae] antiquitatibus libri quinque.* Rome, 1545, octavo.

Geoffrey of Monmouth. *Historia Regis Britanniae.* [Paris, 1508.]

Gildas. *Opus novum: De Calamitate, excidio et conquestu Britanniae.* Ed. P. Vergilius and R. Ridley. [First edition London, 1525, octavo], two copies (one 1555).

Heyden. *Descriptio Hyeroselem: et Palestiniae.* Folio. (Not traced.)

Iugalphi historia quaere Malmesb. Folio. (Not traced. Perhaps in Henry Savile, *Rerum Anglicarum scriptores post Bedam praecipui, etc.* London, 1596, folio.)

Legh, Gerard. *The Accedens of Armory.* London, 1562, octavo.

Loritus, Henricis (Glareanus). *Liber de asse, & partibus eius.* [First edition Basel, 1550], folio.
Maffei, Raphael (Volaterranus). *Commentariorum Urbanorum.* [First edition Rome, 1506], folio, two copies.
Marlianus, Bartholomeus. *Urbis Romae Topographia.* [First edition Rome, 1544.]
Pomponius Laetus, Julius. *De Romanae Urbis vetustate.* [First edition Rome, 1510], quarto.
Price, Sir John. *Historiae Brytannicae defensio.* London, 1573, quarto.
Scaliger, Josephus Justus. *De emendatione temporum.* [First edition Paris, 1583], folio.
Stow, John. *A survey of London.* [First edition London, 1598, quarto.]
Vincent, Augustin, ed. *A discoverie of Errours in the first Edition of the Catalogue of Nobility, published by R. Brooke . . . 1619 . . . With a continuance of the succession.* London, 1622, folio.
William of Malmesbury. "De Gestis Regum Anglorum." In Henry Savile, *Rerum Anglicarum scriptores post Bedam praecipui, etc.* London, 1596, folio.

BIBLIOGRAPHY

Manuscript Sources

Oxford

Bodleian Library

MSS Add. A 115. "A Commonplace Book of Lionel Day, Fellow of Balliol College, Oxford."
———. A 380.
———. C 40. "Catalogue of Mr. Seldens Library, given to Oxford."
———. C 297. James Ussher. "Notes on Chronology."
MS Arch. Selden 88.
MS Ballard. 52. Sir Arthur Gorges. "Islands Voyage," ff. 64–122b.
MSS Bodleian. 313.
———. 666. Thomas Lydiat. "Chronological Papers," etc.
MS Dodsw. 141.
MS Douce. 363. Stephen Batman. Commonplace Book.
MSS Eng. misc. d.47. "Literary Commonplace Book of the Revd Lyonell Day, BD 1611–28."
MS Eng. poet. d.3. "Commonplace Book of Edward Pudsey."
MS Gough Staffordshire. 4. Sampson Erdeswick. *A View of Staffordshire* (1593).
MSS Rawl. Qe 31. Thomas James. "Catalogus in Librorum in Bibliothecae Bodl. 1602–3."
———. Quarto 263. William Lambarde. *A Perambulation of Kent: Contayning the Description, Hystorie, and Customes of that Shyre.* London, 1576, with manuscript additions.
MSS Savile. 48. "William Adam's journal of his voyage to Cochin China in the Gift of God, . . . 17 March–6 August 1617."
———. 95. List of some books owned by Henry Savile.
———. 106. List of books owned by Henry Savile.
MSS Smith. 41.
———. 71. Letters to Sir Robert B. Cotton.
———. 74. Latin Letters from William Camden.
———. 84. William Camden. "Notes on English Antiquities."
———. 86. Letters to William Camden.
———. 92. John Bainbridge. "Astronomical and Mathematical Papers."
MS Wood. D.10. Dr. John Reinold's books willed to friends (1607).
———. D.32. Brian Twine. "Notes collected about Oxford."

Oxford University Archives

"Box of Transcripts of all inventories mentioning books."
Vice-Chancellor's Court Inventories.

All Souls' College Library

"Catalogus librorum ad Bibliothecam Collegii Animarum omnium Fidelium defunct: de Oxon: Spectantium et donatorum eorundem." Benefactor's Register 1064–65.

Balliol College Library

"Catalogus Librorum et Benefactorum Bibliotheca Collegii de Baliolo Oxon."

Brasenose College Library

Benefactors 15.
Joyce Frankland's Inventory to go to College, etc. (1594).

Christ Church College

Donor's Register 1614–1841.

Corpus Christi College

B11/1/2. "Catalogus Librorum Collegii Corpus Christi" in book of Wills and Donations.
MS 254. Brian Twine. "Mathematical and Astronomical Notes."
MS D/3/1. Library Catalog 1589.
MS D/3/2. "Catalogus Librorum Collegii Corpus Christi."
MS D/3/3. Donors Book.

Jesus College

Benefactors' Books 1626–1712.
———. 1624–84.

Magdalen College

MS 777. "Catalogus Benefactorum Bibliotheca."

Merton College Library

P.3.35. Manuscript Catalogues of books ca. 1586–1600.
E.2.1.b. Extract from Will of John Chamber 1604.
Register 1.3.

New College Library

3582. Library Benefaction Book 1617–1909.
Library Catalog 1624.

Oriel College Library

Benefactors Book.
Library Catalog.

Queen's College Library

MS 556. Benefactor's Register from 1621; 1659 transcription.

St. Johns College Archives

"List of Books and Mathematical instruments in Library" (mid-seventeenth century) in back of Parchment Book of List of Fellows to 1617.
Muniments. X 17. Henry Price's Will, 1600.
———. X 19. Inventory of Henry Price, Fellow in 1601.
———. X 26. Mr. John Lees Will, October 1609.

St. Johns College Library

Library Benefactors Book.

Trinity College

Benefactors Book, mid-sixteenth century to mid-eighteenth century.

University College

Benefactors Book (abortive), prepared in 1674.
Roll HH.5.6. Inventory of Master's Lodging and Fellow's Chambers, May 12, 1587, 29 Eliz.

Cambridge
Cambridge University Archives

Registry MS 31.1, item 10.
Registry 21. MS Inventories 1560–1729.
Vice-Chancellor's Court Probate Inventories (2–9).
MS Add. 3308. Commonplace Book.
CUL Dd.3.84. Edward Waterhouse. *A Collection of the Description and Division of all the severall Shires and Townes in Ireland* [ca. 1600].
———. Dd.5.54. St. John's College Library Catalog ca. 1620.
———. Dd.11.40. William Patten. Commonplace Book.
———. Ee.2.32.
———. Oo.7.52. "Catalogus librorum ques habet Bibliotheca Publica Academ. Cantabrig."

Christ's College

Donation Book, 1623.

Corpus Christi College

MS 575. Library list by Matthew Parker, ca. 1593.
MS 576. Library list, ca. 1607.

Emmanuel College

Library Catalogs 1597, 1621, 1622, 1626, 1628, 1632.

Exeter College

Archives A.1 (2).

Gonville and Caius College

MS 645/352. Perhaps compiled by William Braithwaite, 1618.

King's College

Chained Donors Book
I. B. Catalog, transcribed by F. L. Clarke.

Pembroke College

Benefactors Book, 1617.

Peterhouse College

MS 400. "Nomina librorum qui erant in Bibliotheca Collii ante Doctorum Perne mortuum [1588]."
MS 405. Library list, ca. 1610.
Unnumbered MS, probably close copy of MS 405.

St. John's College

MS U. 3. Library List and Donors Book.

Sidney Sussex College

MS 91. Donors Book.

Trinity College Library

R.17.8. "Memoriale Collegio Stae. et Individuae Trinitatis in Academia Cantabrigiensi dicatum 1614."

British Library

MS Add. 4159.
———. 6038. Sir Julius Caesar's Commonplace Book.
———. 6788. Thomas Harriot. *Of the Manner to Observe the Variation of the Compasse.*
———. 10,126. William Burton. *The Description of Leicestershire, Containing Matters of Antiquity, Historye, Armorye, and Genealogy* (n.d.).
———. 14,925. Humphrey Lhuyd. *Mona, the Isle of the Druids* (1568).
———. 17,480. Thomas Dallam. Diary, 1599–1600.
———. 19,889. George Waymouth. *The Jewelle of Artes* (1604).
———. 38,139. Francis Bacon. Correspondence.
———. Eg. 2086. East India Company Papers 1611–18.
MS Burney. 368.
MS Cotton. Galba D.9.
———. Julius F.10.
MS Harley. 167.
———. 1990. Sampson Erdeswick. *A View of Staffordshire* (1593–1603).
MS Landsdowne. 10. Burghley Papers 1567–68.
———. 60. Giles Fletcher. *Of the Russe Commonwealth* [1583?].
———. 101.
———. 121.
MS Sloane. 2279. "A Manuscript Collected and Written at Several Times by Hen. Coley."
———. 2596. William Smith. *The Particular Description of England, with the Portratures of Certaine of the Cheiffest Citties and Townes* (1588).
MS Stowe. 163. Papers on Norfolk.

India Office Marine Records

IOR L/MAR/A I,II, etc. no. 411 Japan Series—Will Adam's log book, transcribed by A. J. Farrington.
———. no. 227 Japan Series.

Kent Record Office

MS Sackville. ON 6014. Letters from the East India Company.

Public Record Office

SP 14/3. #15.
———. /51. #11.
———. /71. #32, 42.
———. /96. #96.
———. /216. #110–13, 122.

PRIMARY SOURCES

Ascham, Roger. 1570. *The Scholemaster or plaine and perfite way of teachyng children the Latin tong.* London.

Bacon, Francis. [1890]. *Advancement of Learning.* Reprinted New York: P. F. Collier and Son.

Barlow, William. 1597. *The Navigator's Supply Conteining many things of Principall importance belonging to Navigation . . . but serving also for sundry other of Cosmography in generall.* London: G. Bishop.

———. 1616. *Magneticall Advertisements: or Divers Pertinent observations and approved experiments concerning the nature and properties of the Load-Stone.* London: E. Griffin.

Bent, J. Theodore, ed. 1893. *Early Voyages and Travels in the Levant. I. The Diary of Master Thomas Dallam, 1599–1600.* London: Hakluyt Society.

Blundeville, Thomas, 1594. *Blundeville his Exercises containing sixe Treatises.* London.

———. 1636. *Blundeville his Exercises containing sixe Treatises.* Ed. Robert Hartwell. 7th ed. London: R. Bishop.

Bond, Edward A., ed. 1856. *Russia at the Close of the Sixteenth Century.* London: Hakluyt Society.

Borough, William. 1581. *A Discourse on the Variation of the Cumpas or Magneticall Needle . . . to be annexed to The New Attractive of R.N.* London: J. Kingston. Facsimile Amsterdam: Theatrum Orbis Terrarum, 1974.

Bourne, William. 1574. *A Regiment of the Sea.* London: T. Dawson.

Bry, Theodore de. 1590. *America. Pars I. Admiranda narratio . . . de commodis et incolarum ritibus Virginiae . . . Anglico scripta sermone a Thoma Hariot.* Frankfurt.

Buck, Sir George. 1615. *The Third Universitie of England. Or a Treatise of the foundations of all the colledges, ancient schooles of priviledge, and of houses of learning, and liberall arts, within and about the most famous cittie of London.* London: T. Dawson. In John Stow, *The Annales, or Generall Chronicle of England.* London, 1615.

Caius, Johannes. 1576. *Of Englishe Dogges, the diversities, the names, the natures, and the properties.* London: Rychard Johnes. Facsimile Amsterdam: Theatrum Orbis Terrarum, 1969.

Camden, William. 1587. *Britannia, sive Florentissimorum Regnorum Angliae, Scotiae, Hiberniae et Insularum adiacentium ex intima antiquitate Chorographie descriptio.* 2nd ed. London: R. Newbery.

———. [1971.] *Britannia, 1695: A Facsimile of the 1695 Edition.* Introduction by Stuart Piggott. New York: Johnson Reprint Corp.

Carew, Richard. 1602. *The Survey of Cornwall.* London: S. S[tafford] for J. Jaggard.

Carpenter, Nathanael. 1625. *Geography delineated Forth in Two books, containing the Sphaericall and Topicall Parts Thereof.* Oxford: J. Lichfield and Wm. Turner.

Chaloner, Sir Thomas. 1584. *Virtue of Nitre, wherein is declared the sundry cures by the same effected.* London.

Clark, Andrew, and C. W. Boase. 1885–87. *Register of the University of Oxford.* Oxford: Clarendon Press.

Collection of Statutes for the Universities and Colleges of Cambridge. 1840. Trans. J. Heywood. London.

Cooper, Michael, S. J., ed. 1965. *They came to Japan. An Anthology of European Reports on Japan, 1543–1640.* London: Thames and Hudson.

Cornwallis, Sir C. 1641. *The Life and Death of our late more Incomparable and Heroique Prince, Henry Prince of Wales.* London.

Cortes, Martin. 1561. *The Arte of Navigation, Conteynyng a compendious description of the Sphere, with the makyng of certen Instrumentes and Rules for Navigations: and exemplified by manye Demonstrations.* Trans. Richard Eden. London.

Crashaw, William. 1610. *A Sermon Preached in London before the right honorable the Lord Lawarre, Lord Governour and Captaine Generall of Virginia . . . At the said Lord Generall his leave taking of England . . . Febr. 21 1609.* London: W. Welby.

Danvers, Frederick Charles, ed. 1896. *Letters Received by the East India Company from its Servants in the East. Transcribed from the "Original Correspondence" Series of the India Office Records. Vol. 1, 1602–1613.* London: Sampson, Low Marston and Co.

Davis, Captain John. 1595. *The Seaman's Secrets.* London: Thomas Dawson.

Dee, John. 1570. *The Mathematical Preface to the Elements of Geometrie of Euclid of Megara.* London: J. Daye. Facsimile New York: Science History Publications, 1975.

———. 1577. *General and Rare Memorials pertayning to the Perfect Arte of Navigation.* London: J. Daye.

———. 1842. *The Private Diary of Dr. John Dee.* Ed. J. O. Halliwell. London: John Bowyer Nichols and Sons.

Digges, Sir Dudley. 1612. *Of the circumference of the earth: or, a treatise of the northeast Passage.* London: W. White for J. Barnes.

Digges, Leonard. 1571. *A Geometrical Practise, named Pantometria ... lately finished by Thomas Digges his Sonne.* London: H. Bynneman.

———. 1576. *A Prognostication Everlastinge Corrected and Augmented by Thomas Digges.* London: Thomas Marsh. Facsimile Amsterdam: Theatrum Orbis Terrarum, 1975.

———. 1579. *An Arithmeticall Militare Treatise, named Stratioticus: Compendiously teaching the Science of Numbers.* London: H. Bynneman. Facsimile Amsterdam: Theatrum Orbis Terrarum, 1975.

Documents relating to the Universities and Colleges of Cambridge. 1852. London: G. E. Eyre and W. Spottiswoode, for H. M. Stationey Office.

Dodeons, Rembert. 1595. *A New Herball, or Histoire of Plants: Wherein is contained the whole discourse and perfect description of all sorts of Herbes and Plants ... now first translated out of French into English, by Henrie Lyte. Corrected and amended.* London: Edm. Bollifant.

Draper, John W. 1875. *History of the Conflict Between Religion and Science.* London.

Drayton, Michael. 1613. *Poly-Olbion. Of A Chorographicall Description of Tracts, Rivers, Mountaines, Forests & Other Parts of this renowned Isle of Greate Britaine.* London.

Elyot, Thomas. 1531. *The boke named the Governour.* London.

Erondelle, Pierre, trans. 1609. *Nova Francia: Or the Description of that Part of New France, which is one continent with Virginia [by Marc Lescarbot].* London: G. Bishop.

The First Printed Catalogue of the Bodleian Library 1605. A Facsimile. Catalogus Librorum Bibliothecae Publicae quam Vir Ornatissimus Thomas Bodleius Euques Auratus in Academia Oxoniensi Nuper Institut. 1986. Oxford: Oxford University Press.

Fletcher, Giles. 1966. *Of the Russe Commonwealth. Facsimile [1st] edition with variants.* Introduction by Richard Pipes. Cambridge, Mass.: Harvard University Press.

Gellibrand, Henry. 1633. *An Appendix Concerning Longitude.* In Capt. Thomas James, *The Strange and dangerous voyage of captaine Thomas James.* London: J. Legatt for J. Partridge.

———. 1635. *A Discourse Mathematicall on the variation of the magneticall needle; together with its admirable Diminution lately discovered.* London: W. Jones.

Gibson, S., ed. 1931. *Statuta Antiqua Universitatis Oxoniensis.* Oxford: Oxford University Press.

Gilbert, Humphrey. 1869. *Queen Elizabethe's Achademy.* London: Early English Text Society.

Gilbert, William. 1600. *De Magnete, Magneticisque Corporibus, et de Magno magnete tellure; physiologia nova, plurimis et argumentis, et experimentis demonstrata.* London: P. Short.

———. 1893. *On the Loadstone and Magnetic Bodies, and on the Great Magnet the Earth.* Trans. Paul Fleury Mottelay. London.

Greville, Sir Fulke. 1652. *The life of the renowned Sir Philip Sidney.* London: Henry Seile. Reprinted Delmar, N.Y.: Scholars' Facsmiles and Reprints, 1984.

Gunter, Edmund. 1620. *Canon Triangulorum, sive tabulae sinuum et tangentium Artificialium.* London: W. Jones.

Hakluyt, Richard. 1582. *Divers Voyages touching the discoverie of America, and the Ilands adiacent unto the same, made first of all by our Englishmen.* London: Thos. Woodcocke.

———. 1589. *The Principal Navigations, Voiages and Discoveries of the English Nation, made by Sea or over Land.* London: G. Bishop and R. Newberrie.

———. 1598–1600. *The Principal Navigations, Voiages, Traffiques and Discoveries of the English Nation.* 3 vols. London: G. Bishop, R. Newberie, and R. Barker.

Halliwell, James Orchard. 1841. *A Collection of Letters Illustrative of the Progress of Science in England from the Reign of Queen Elizabeth to that of Charles the Second.* London: History of Science Society. Reprinted London: Dawsons of Pall Mall, 1965.

Harriot, Thomas. 1588. *A Briefe and true Report of the New found land of Virginia: of the commodities there found and to be raysed, as well marchantable, as others for victuall, building and other necessarie uses.* London: [R. Robinson].

Hartwell, Abraham, trans. 1597. *A Report of the Kingdom of Congo, a Region of Africa. . . . Drawn out of the writings and discourses of Odoardo Lopez, a Portingall, by Philippo Pigafetta.* London.

Henisch, Georg. [1591]. *The Principles of Geometrie, Astronomie, and Geographie . . . Gathered out of the Tables of the Astronomicall institutions of Georgius Henischius. By Francis Cooke. Appointed publiquelye to be read in the Staplers Chappell at Leaden hall, by the Wor. Tho. Hood, Mathematicall Lecturer of the Cittie of London.* London.

Herbert, Edward. 1976. *The Life of Edward, First Lord Herbert of Cherbury, written by himself.* Ed. J. M. Shuttleworth. Oxford: Oxford University Press.

Heylyn, Peter. 1621. *Microcosmus: or, A little description of the great world.* Oxford: J. Lichfield. Facsimile Amsterdam: Theatrum Orbis Terrarum, 1975.

Hood, Thomas. n.d. *A Copie of the speache: made by the Mathematicall Lecturer unto the Worshipfull Companye present in Gracious Street the 4 of November 1588.* London.

Hues, Robert. 1594. *Tractatus de Globis et eorum Usu, accomodatus iis qui Londoni editi sunt, anno 1593.* London: T. Dawson.

Ker, N. R. 1971. *Records of All Souls College Library 1437–1600.* Oxford: Oxford University Press.

Knolles, Richard. 1603. *The Generall Historie of the Turks.* London.

Lamb, John, ed. 1838. *A Collection of Letters, Statutes, and other Documents from the Manuscript Library of Corpus Christi College Illustrative of the History of the University of Cambridge during the period of the Reformation from 1500 to 1572.* London: J. W. Parker.

Lambarde, William. 1576. *A Perambulation of Kent: Contayning the Description, Hystorie and Customes of that Shyre.* London.

Leigh, Valentine. 1577. *The Most Profitable and Commendable Science of Surveying.* London: [J. Kingston] for A. Maunsell.

Linschoten, Jan Huygen van. 1598. *A Discourse of voyages into ye Easte and Weste Indies, divided into Foure Bookes.* London.

Lithgow, William. 1614. *A Most Delectable, and true Discourse, of an admired and painefull peregrination from Scotland, to the most famous Kingdomes in Europe, Asia and Affricke.* London.

Lucar, Cyprian. 1590. *A Treatise Named Lucarsolace.* London: R. Field for J. Harrison.

Macray, W. D. 1897. *A Register of the Members of St. Mary Magdalen College Oxford.* London: Frowde.

Mandeville, Sir John. 1928. *The Travels of Sir John Mandeville and the Journal of Friar Odoric.* Introduction by Jules Bramont. London: Everyman's Library, J. M. Dent and Sons.

Mayor, J. E. B., ed. 1859. *Early Statutes of the College of St. John the Evangelist in the University of Cambridge.* Cambridge: Cambridge University Press.

Meierus, Albertus. 1589. *Certaine briefe, and speciall Instructions for Gentlemen, Merchants, Students, Souldiers, Marriners, etc. Employed in services abroade, or anie way occasioned to converse in the kingdomes, and governementes of forren Princes.* Trans. Philip Jones. London.

The Merchaunts New-Royall-Exchaunge. 1604. London: T. C[reede] for C. Burbey.

Monardes, Nicolas. 1577. *Ioyfull Newes out of the newe founde worlde.* Trans. John Frampton. London.

Münster, Sebastian. 1559. *Cosmographia Universalis Libri vi in quibus iuxta certioris fidei scriptorum traditionem describunter.* Basel.

Norden, John. 1840. *Speculi Britanniae Pars: An Historical and Chorographical Description of the County of Essex, by John Norden, 1594.* London: Printed for the Camden Society, 1840.

Norman, Robert. 1581. *The New Attractive, containying a short discourse of the Magnes or Lodestone, and amoungest other his vertues, of a newe discovered secret and subtill propertie, concernyng the Declinyng of the Needle, toucheth therewith under the plaine of the Horizon.* London: J. Kingston. Facsimile Amsterdam: Theatrum Orbis Terrarum, 1974.

Ortelius, Abraham. 1570. *Theatrum Orbis Terrarum.* Antwerp.

Pantaleon, Heinrich. 1550. *Chronographia Ecclesiae Christianae, qua Patrum et Doctorum*

Ordo, cum uariarum Haeresum origine, &, multiplici innouatione rituum in Ecclesia . . . Basel: N. Brylinger.

Parr, Richard. 1686. *The Life of the Most Reverend Father in God, James Ussher.* London.

Parry, William. 1601. *New and large discourse of the Travels of Sir Anthony Sherley.* London.

Pontanus, Johannes. 1611. *Rerum et Urbis Amstelodamensium Historia.* Amsterdam.

Postel, Guillaume. 1635. *Guillelmi Postell De universitate libri duo: In quibus Astronomiae, Doctrinaeve Coelestis Compendium. Reliqua quae hisce Libris continentur, Pagina tertia ostendit. Editio tertia.* Leiden: Joannes Maire.

Ptolemy, Claudius. 1540. *Geographia.* Ed. Sebastian Münster. Basel. Facsimile Amsterdam: Theatrum Orbis Terrarum, 1966.

Purchas, Samuel. 1613. *Purchas His Pilgrimage: or, Relations of the world and the religions observed in all ages and places discovered, from the creation unto this present . . . containing a theologicall and geographicall histoire of Asia, Africa and America.* London: William Stansby for Henrie Fetherstone.

———. 1625. *Hakluytus Posthumus: or, Purchas His Pilgrimes. Contayning a History of the World, in Sea voyages and lande Travells, by Englishmen and others.* 4 vols. London: for Henry Fetherston.

Purnell, C. J., ed. 1916. *The Log-Book of William Adams.* London.

Raleigh, Sir Walter. 1614. *The History of the World.* London. Reprint ed. C. A. Patrides, London: MacMillan Press, 1971.

Rigaud, Stephen Peter, and S. J. Rigaud, eds. 1841–62. *Correspondence of Scientific Men of the Seventeenth Century.* 2 vols. Oxford: Oxford University Press.

Sabellico, Marco Antonio Coccio. 1556. *M. Antonii Sabellici, Historiae Rerum Venetarum ab urbe condita, libri xxxiii. Eiusdem in singulos libros Epitomae. Additus in fine est index rerum memorabilium copiosus.* Basel: [Apud Nic. Episcopium Iuniorem].

Sandys, George. 1615. *A Relation of a Journey begun An. Dom. 1610.* London.

Serres, Jean de. 1607. *A General Inventorie of the History of France.* Trans. Edward Grimstone. London.

Sherley, Anthony. 1613. *His Relation of his Travels into Persia.* London.

Speed, John. 1611a. *The History of Great Britaine Under the Conquests of ye Romans, Saxons, Danes and Normans.* London: [W. Hall and J. Beale].

———. 1611b. *The Theatre of the Empire of Great Britaine. Presenting an exact Geography of the Kingdomes of England, Scotland, Ireland and the Iles adioyning.* London: Cum Privelegio.

Statutes of the Colleges of Oxford: with Royal Patents of Foundations, Injunctions of Visitors, etc. 1853. London: J. H. Parker.

Stow, John. 1598. *A Survey of London.* London.

Strabo. 1917. *The Geography of Strabo.* Introduction by J. R. S. Stennett. Trans. Horace L. Jones. London: William Heinemann.

Stubbings, Frank. 1983. *The Statutes of Sir Walter Mildmay for Emmanuel College.* Cambridge: Cambridge University Press.

Suidas. 1581. *Historica, caeteraque omnia . . . opera ac studio Hier. Wolfii . . . in Latinum sermonem conversa . . . nunc vero emendata, et aucta.* Basel: Ex officina Heruagiana, per Eusebium Episcopum.

Ussher, James. 1650. *Annales Veteris Testamenti a prima mundi origine deducti.* London.

Vives, Joannes. 1544. *An Introduction to Wysdome.* Trans. Richard Morysine. London.

Ward, G. R. M., trans. 1840. *Statutes of Magdalen College, Oxford.* Oxford: Oxford University Press.

White, Andrew D. 1896. *A History of the Warfare of Science with Theology in Christendom.* New York.

Wilson, Henry A. 1899. *Magdalen College.* London: F. E. Robinson.

Worsop, Edward. 1582. *A Discoverie of sundrie errours and faults daily committed by Landemeaters, ignorant of Arithmetike and Geometrie, to the Damage, and preiudice of many her Maiuesties subiects.* London: H. Middleton.

Wright, Edward, 1599a. *Certaine Errors in Navigation arising either of the ordinarie erroneous*

making or using of the sea chart, Compasse . . . Detected and Corrected by Edward Wright. Including The Voyage of the Right Ho. George Earle of Cumberland to The Azores. London: V. Sims. Facsimile Amsterdam: Theatrum Orbis Terrarum, 1974.
———. 1599b. *The Haven finding art by the Latitude and Variation.* From the Dutch of Simon Stevin. London: GBRN and RB. Facsimile Amsterdam: Theatrum Orbis Terrarum, 1968.
———. 1610. *Certaine Errors in Navigation, detected and corrected by Edw. Wright. With many additions that were not in the former edition as appeareth in the new pages.* London: F. Kingston.

SECONDARY SOURCES

Adams, Percy G. 1983. *Travel Literature and the Evolution of the Novel.* Lexington: University of Kentucky Press.
Adamson, Ian Richardson. 1976. "The Foundation and Early History of Gresham College, London, 1596–1704." Ph.D. dissertation, Cambridge University.
———. 1980. "The Administration of Gresham College and Its Fluctuating Fortunes as a Scientific Institution in the Seventeenth Century." *History of Education* 9:13–25.
Akerman, James, ed. Forthcoming. *Cartography, Statescraft, and Political Culture.* Proceedings of the 15th International Conference on the History of Cartography (1993). Chicago: Speculum Orbis Press.
Alpers, Svetlana. 1983. *The Art of Describing: Dutch Art in the Seventeenth Century.* Chicago: University of Chicago Press.
Alsop, James D., and Wesley Stevens. 1986. "William Lambarde and the Elizabethan Polity." *Studies in Medieval and Renaissance History* n. s., 8:231–65.
Ames, Joseph. 1810. *Typographical Antiquities: or The History of Printing in England Scotland and Ireland,* Vol. 1. London: William Miller.
Andrews, K. R. 1984. *Trade, Plunder and Settlement. Maritime Enterprise and the Genesis of the British Empire, 1480–1630.* Cambridge: Cambridge University Press.
Andrews, K .R., N. P. Canny, and P. E. H. Hair, eds. 1978. *The Westward Enterprise: English Activities in Ireland, the Atlantic, and America 1480–1650.* Liverpool: Liverpool University Press.
Apt, Adam. 1982. "The English Reception of Kepler's Astronomy: 1596–1650." D. Phil. Thesis, Oxford University.
Archer, Ian W. 1991. *The Pursuit of Stability: Social Relations in Elizabethan London.* Cambridge: Cambridge University Press.
Bagrow, Leo. 1964. *History of Cartography.* Rev. R. A. Skelton. London: C. A. Watts and Co.
Baker, J. N. L. 1928. "Nathaniel Carpenter and English Geography in the Seventeenth Century." *Geographical Journal* 71:261–71.
———. 1955. "The Geography of Bernhard Varenius." *Transactions of the Institute of British Geographers* 21:51–60.
———. 1963. *The History of Geography.* Oxford and New York: Basil Blackwell.
Bantock, G. H. 1980. *Studies in the History of Educational Theory,* Vol. 1: *Artifice and Nature, 1350–1765.* London: G. Allen and Unwin.
Barber, Charles. 1976. *Early Modern English.* London: Deutsch.
Barber, Peter. 1992a. "England I: Pageantry, Defense, and Government: Maps at Court to 1550." In *Monarchs, Ministers, and Maps: The Emergence of Cartography as a Tool of Government in Early Modern Europe,* ed. David Buisseret, 57–98. Chicago: University of Chicago Press.
———. 1992b. "England II: Monarchs, Ministers, and Maps 1550–1625." In *Monarchs, Ministers, and Maps: The Emergence of Cartography as a Tool of Government in Early Modern Europe,* ed. David Buisseret, 57–98. Chicago: University of Chicago Press.
———. 1995. "The Evesham World Map: A Late Medieval English View of God and the World." *Imago Mundi* 47:13–33.
Barton, John. 1986. "The Faculty of Law." In *The Collegiate University,* ed. James K.

McConica, 257–93. Vol. 3 of *The History of the University of Oxford,* ed. T. H. Aston. Oxford: Clarendon Press.

Bendall, Sarah. 1992. *Maps, Land and Society: A History, with a Carto-bibliography of Cambridgeshire Estate Maps, c. 1600–1836.* Cambridge: Cambridge University Press.

Bennett, James A. 1986. "The Mechanic's Philosophy and the Mechanical Philosophy." *History of Science* 24:1–28.

Biagioli, Mario. 1990. "Galileo's System of Patronage." *History of Science* 28:1–62.

———. 1993. *Galileo Courtier: The Practice of Science in the Culture of Absolutism.* Chicago: University of Chicago Press.

Birch, Thomas. 1760. *The Life of Henry, Prince of Wales, Eldest Son of King James I.* London.

Blair, Ann. 1990. "Humanist Methods in Natural Philosophy: The Commonplace Book." *Journal of the History of Ideas* 53:541–51.

Bowen, Margarita. 1981. *Empiricism and Geographical Thought: From Francis Bacon to Alexander von Humboldt.* Cambridge: Cambridge University Press.

Brandon, William. 1986. *New Worlds for Old. Reports from the New World and Their Effect on the Development of Social Thought in Europe, 1500–1800.* Athens, Ohio: Ohio University Press.

Buisseret, David, ed. 1992. *Monarchs, Ministers, and Maps: The Emergence of Cartography as a Tool of Government in Early Modern Europe.* Chicago: University of Chicago Press.

Bulmer-Thomas, Ivor. 1976. "Theodosius of Bithynia." In *Dictionary of Scientific Biography,* ed. Charles C. Gillespie, 13:320. New York: American Council of Learned Societies by Scribner.

Burke, Peter. 1969. *The Renaissance Sense of the Past.* London: Edward Arnold.

Buxton, John. 1964. *Sir Philip Sidney and the English Renaissance.* London: Macmillan.

Byrne, M. St. Clare, and Gladys Scott Thomas. 1931. "'My Lord's Books': The Library of Francis, Second Earl of Bedford, in 1584." *Review of English Studies* 7:385–405.

Campbell, Mary. 1992. "The Illustrated Travel Book and the Birth of Ethnography: Part I of De Bry's *America.*" In *The Work of Dissimilitude,* ed. David G. Allen and Robert White, 177–95. Newark.

Campbell, Tony. 1973. "The Drapers' Company and Its School of Seventeenth Century Chart-makers." In *My Head Is a Map,* ed. Helen Wallis and Sarah Tyacke, 81–106. London: Francis Edwards and Carta Press.

Caspari, Fritz. 1974. *Humanism and Social Order in Tudor England.* Chicago: University of Chicago Press.

Charlton, Kenneth. 1960. "Liberal Education and the Inns of Court in the Sixteenth Century." *British Journal of Educational Studies* 9:25–38.

———. 1965. *Education in Renaissance England.* London: Routledge and Kegan Paul.

Chilton, D. 1957–58. "Land Measurement in the Sixteenth Century." *Newcomen Society of the Study of History of Engineering and Technology Transactions* 31:111–29.

Christianson, Paul. 1978. *Reformers and Babylon: English Apocalyptic Visions from the Reformation to the Eve of the Civil War.* Toronto: University of Toronto Press.

Christopher Saxton's Sixteenth-Century Maps: The Counties of England and Wales. 1992. Introduction by William Ravenhill. Shrewsbury: Chatsworth Library.

Clulee, Nicholas. 1988. *John Dee's Natural Philosophy: Between Science and Religion.* London: Routledge.

Cohen, Floris H. 1994. *The Scientific Revolution: A Historiographical Inquiry.* Chicago: University of Chicago Press.

Coleman, Christopher, and David Starkey, eds. 1986. *Revolution Reassessed: Revision in the History of Tudor Government and Administration.* Oxford: Oxford University Press.

Collinson, Patrick. 1967. *The Elizabethan Puritan Movement.* Oxford: Oxford University Press.

———. 1983. *Godly People: Essays on English Protestantism and Puritanism.* London: Hambledon.

Colloque International d'Histoire Maritime. 1960. *Le Navire et l'Economie Maritime du Nord de l'Europe du Moyen Age au XVIIIe siècle,* ed. Michel Mollat. Paris: SEVPEN.

Colvin, H. M., ed. 1975. *The History of the King's Works.* Vol. IV. London: Her Majesty's Stationery Office.

Corbett, Margery, and Ronald Lightbown. 1979. *The Comely Frontispiece: The Emblematic Title-Page in England 1550–1660.* London: Routledge and Kegan Paul.

Cormack, Lesley B. 1991a. "'Good Fences Make Good Neighbors': Geography as Self-Definition in Early Modern England." *Isis* 82:639–61.

———. 1991b. "Twisting the Lion's Tail: Practice and Theory at the Court of Henry Prince of Wales." In *Patronage and Institutions: Science, Technology, and Medicine at the European Court,* ed. Bruce Moran, 67–84. Woodbridge: Boydell Press.

———. 1994a. "The Fashioning of an Empire: Geography and the State in Elizabethan England." In *Geography and Empire,* ed. Anne Godlewska and Neil Smith, 15–30. Oxford: Blackwells.

———. 1994b. "Flat Earth or Round Sphere: Misconceptions of the Shape of the Earth and the Fifteenth Century Transformation of the World." *Ecumene* 1 (4):365–85.

Costello, William T. 1958. *The Scholastic Curriculum at Early Seventeenth Century Cambridge.* Cambridge, Mass.: Harvard University Press.

Crone, Gerald Roe. 1953. *Maps and Their Makers: An Introduction to the History of Cartography.* London: Hutchinson's University Library.

Cunningham, Andrew, and Perry Williams. 1993. "De-centring the 'Big Picture': *The Origins of Modern Science* and the Modern Origins of Science." *British Journal for the History of Science* 26:407–32.

Curtis, Mark. 1959. *Oxford and Cambridge in Transition 1558–1642.* Oxford: Clarendon Press.

Daston, Lorraine, and Peter Galison. 1992. "The Image of Objectivity." *Representations* 40:81–128.

Davis, Ralph. 1973. *English Overseas Trade 1500–1700.* London: MacMillan.

Deacon, Richard. 1968. *John Dee: Scientist, Geographer, Astrologer and Secret Agent to Elizabeth I.* London: Muller.

DeMolen, Richard L. 1984. "The Library of William Camden." *Proceedings of the American Philosophical Society* 128:327–409.

Dibner, Bern. 1947. *Doctor William Gilbert.* New York: Burndy Library.

Dickens, A. G. 1964. *The English Reformation.* New York: Schocken Books.

———. 1977. *The Courts of Europe: Politics, Patronage and Royalty 1400–1800.* London: Thames and Hudson.

Dictionary of National Biography: From the Earliest Times to 1900. 1921. 22 vols. Ed. Sir Leslie Stephen and Sir Sidney Lee. Oxford: Oxford University Press; 1964 imprint.

Dictionary of Scientific Biography. 1970–80. 16 vols. Ed. Charles C. Gillespie. New York: American Council of Learned Societies by Scribner.

Dilke, Oswald A. W. 1985. *Greek and Roman Maps.* London: Thames and Hudson.

Dorsten, Jan Adrianus van. 1962. *Poets, Patrons, and Professors: Sir Philip Sidney, Daniel Rogers and the Leiden Humanists.* Leiden, London: Oxford University Press.

Dowling, Maria. 1986. *Humanism in the Age of Henry the Eighth.* London: Croom Helm.

Drake, Stillman. 1970. "Early Science and the Printed Book: The Spread of Science Beyond the Universities." *Renaissance and Reformation* 6:43–52.

Duffy, Eamon. 1992. *The Stripping of the Altars: Traditional Religion in England 1400–1580.* New Haven: Yale University Press.

Eamon, William. 1994. *Science and the Secrets of Nature: Books of Secrets in Medieval and Early Modern Culture.* Princeton: Princeton University Press.

Ede, Andrew. 1993. "When Is a Tool Not a Tool? Understanding the Role of Laboratory Equipment in the Early Colloidal Chemistry Laboratory." *Ambix* 40:11–24.

Eden, Peter, ed. 1975. *Dictionary of Land Surveyors and Local Cartographers of Great Britain and Ireland, 1550–1850.* Folkestone: William Dawson and Sons.

———. 1983. "Three Elizabethan Estate Surveyors: Peter Kempe, Thomas Clerke and Thomas Langdon." In *English Map-Making 1500–1650,* ed. Sarah Tyacke, 68–84. London: British Library.

Edgerton, Samuel, Jr. 1987. "From Mental Matrix to *Mappamundi* to Christian Empire: The Heritage of Ptolemaic Cartography in the Renaissance." In *Art and Cartography: Six Historical Essays,* ed. David Woodward, 10–50. Chicago: University of Chicago Press.

Eisenstein, Elizabeth. 1979. *The Printing Press as an Agent of Change.* Cambridge: Cambridge University Press.

Elton, Geoffrey. 1953. *The Tudor Revolution in Government: Administrative Changes in the Reign of Henry VIII.* Cambridge: Cambridge University Press.

———. 1959. *England under the Tudors.* London: Methuen.

———. 1985. *Policy and Police.* Cambridge: Cambridge University Press.

———. 1986. "Revisionism Reassessed: The Tudor Revolution a Generation Later." *Encounter* 67 (2):37–42.

Evans, Ifor, and Heather Lawrence. 1979. *Christopher Saxton: Elizabethan Map-Maker.* Wakefield: Wakefield Historical Publications and the Holland Press.

Evans, Joan. 1956. *A History of the Society of Antiquaries.* Oxford: Oxford University Press.

Evans, R. J. W. 1973. *Rudolf II and His World: A Study in Intellectual History 1576–1612.* Oxford: Clarendon Press.

Fehrenbach, R. J., and Elisabeth Leedham-Green, eds. 1996. *The Private Libraries in Renaissance England,* Vols. 1–3. *Medieval and Renaissance Texts and Studies* 87, 105, 117. Marlborough: Adam Matthew Publications.

Feingold, Mordechai. 1984a. *The Mathematicians' Apprenticeship: Science, Universities, and Society in England, 1560–1640.* Cambridge: Cambridge University Press.

———. 1984b. "The Occult Tradition in the English Universities of the Renaissance: A Reassessment." In *Occult and Scientific Mentalities in the Renaissance,* ed. Brian Vickers, 73–94. Cambridge: Cambridge University Press.

———. 1989. "The Universities and the Scientific Revolution: The Case of England." In *New Trends in the History of Science,* ed. R. P. W. Visser et al., 29–48. Amsterdam: Rodolpi.

Ferguson, Arthur. 1965. *The Articulate Citizen and the English Renaissance.* Durham: Duke University Press.

Fernandez-Armesto, Felipe. 1991. *Columbus.* Oxford: Oxford University Press.

Fisher, R. M. 1975. "William Crashawe's Library of the Temple 1605–1615." *The Library* ser. 5, 30:116–24.

Fletcher, J. M. 1986. "The Faculty of Arts." In *The Collegiate University,* ed. James K. McConica, 157–99. Vol. 3 of *The History of the University of Oxford,* ed. T. H. Aston. Oxford: Clarendon Press.

Foster, Joseph. [1892]. *Alumni Oxonienses. The Members of the University of Oxford, 1500–1714: Their Parentage, Birthplace and Year of Birth, with a Record of Their Degrees.* Oxford: Parker and Company.

Foster, Michael. 1981. "Thomas Allen, Gloucester Hall and the Survival of Catholicism in Post-Reformation Oxford." *Oxoniensis* 46:99–128.

Fox, Alistair. 1986. *Reassessing the Henrician Age: Humanism, Politics, and Reform, 1500–1550.* New York: B. Blackwell.

Frank, Robert G., Jr. 1973. "Science, Medicine and the Universities of Early Modern England." *History of Science* 11:194–216, 239–69.

French, Peter J. 1972. *John Dee: The World of an Elizabethan Magus.* London: Routledge and Kegan Paul.

Fussner, F. Smith. 1962. *The Historical Revolution, English Historical Thought and Writing, 1580–1640.* London: Routledge.

Gascoigne, John. 1985. "The Universities and the Scientific Revolution: The Case of Newton and Restoration Cambridge." *History of Science* 23:391–434.

Gaskell, Philip. 1980. *Trinity College Library: The First 150 Years.* Cambridge: Cambridge University Press.

Gatti, Hilary. 1989. *The Renaissance Drama of Knowledge: Giordano Bruno in England.* London: Routledge.

Gilbert, Edmund W. 1972. *British Pioneers in Geography.* Newton Abbot: David and Charles.
Gillies, John. 1994. *Shakespeare and the Geography of Difference.* Cambridge: Cambridge University Press.
Gingerich, Owen, and Robert S. Westman. 1988. *The Wittich Connection: Conflict and Priority in Late Sixteenth-Century Cosmology.* Philadelphia: American Philosophical Society.
Glacken, Clarence J. 1967. *Traces on the Rhodian Shore: Nature and Culture in Western Thought from Ancient Times to the End of the Eighteenth Century.* Berkeley and Los Angeles: University of California Press.
Glick, Thomas F. 1983. "In Search of Geography." *Isis* 74:92–97.
Gough, Richard. 1770. "An Historical Account of the Origin and Establishment of the Society of Antiquaries." *Archaeologia,* 2–39.
Grafton, Anthony. 1975. "Joseph Scaliger and Historical Chronology: The Rise and Fall of a Discipline." *History and Theory* 14 (2):156–85.
———. 1985. "From *De Die Natali* to *De Emendatione Temporum:* The Origins and Setting of Scaliger's Chronology." *Journal of the Warburg and Courtauld Institutes* 48:100–144.
———. 1993. *Joseph Scaliger: A Study in the History of Classical Scholarship. II: Historical Chronology.* Oxford: Clarendon Press.
Grafton, Anthony, and Lisa Jardine. 1986. *From Humanism to the Humanities: Education and the Liberal Arts in Fifteenth- and Sixteenth-Century Europe.* Cambridge, Mass.: Harvard University Press.
Grafton, Anthony, et al. 1992. *New Worlds, Ancient Texts: The Power of Tradition and the Shock of Discovery.* Cambridge, Mass: Harvard University Press.
Gransden, Antonia. 1982. *Historical Writing in England.* Vol. 2: *C. 1307 to the Early Sixteenth Century.* London: Routledge and Kegan Paul.
Greaves, Richard L. 1969. "Puritanism and Science: The Anatomy of a Controversy." *Journal of the History of Ideas* 30:345–63.
Greenblatt, Stephen. 1980. *Renaissance Self-Fashioning from More to Shakespeare.* Chicago: University of Chicago Press.
———. 1991. *Marvelous Possessions: The Wonder of the New World.* Chicago: University of Chicago Press.
Gumilev, L. N. 1987. *Searches for an Imaginary Kingdom: The Legend of the Kingdom of Prester John.* Cambridge: Cambridge University Press.
Guy, John A. 1980. "The Tudor Commonwealth: Revising Thomas Cromwell." *Historical Journal* 23:681–87.
Haigh, Christopher. 1993. *English Reformations: Religion, Politics and Society under the Tudors.* Oxford: Clarendon Press.
Hale, John. 1993. *The Civilization of Europe in the Renaissance.* London: HarperCollins.
Hall, A. Rupert. 1959. "The Scholar and the Craftsman in the Scientific Revolution." In *Critical Problems in the History of Science,* ed. Marshall Clagett, 1–22. Madison: University of Wisconsin Press.
———. 1963. "Merton Revisited or Science and Society in the Seventeenth Century." *History of Science* 2:1–16.
Hall, Marie Boas. 1962. *The Scientific Renaissance 1450–1630.* London: Collins.
Hallowes, D. M. 1962. "Henry Briggs, Mathematician." *Transactions of the Halifax Antiquarian Society,* 79–92.
Hammer, Paul E. J. 1991. "'The Bright Shininge Sparke': The Political Career of Robert Devereux, 2nd Earl of Essex, c. 1585–1597." Ph.D. dissertation, Cambridge University.
Harley, J. B. 1983. "Meaning and Ambiguity in Tudor Cartography." In *English Map-Making 1500–1650,* ed. Sarah Tyacke, 22–45. London: British Library.
———. 1988. "Maps, Knowledge, and Power." In *The Iconography of Landscape: Essays on the Symbolic Representation, Design and Use of Past Environments,* ed. Denis Cosgrove and Stephen Daniels. Cambridge: Cambridge University Press.
———. 1989. "Deconstructing the Map." *Cartographica* 26 (2):1–19.

———. 1990. "Cartography, Ethics and Social Theory." *Cartographica* 27:1–23.

Harley, J. B., and David Woodward, eds. 1987. *The History of Cartography.* Vol. I: *Cartography in Prehistoric, Ancient and Medieval Europe and the Mediterranean.* Chicago: University of Chicago Press.

Harrison, George B. [1937]. *The Life and Death of Robert Devereux, Earl of Essex.* New York: H. Holt.

Harvey, P. D. A. 1993. *Maps in Tudor England.* Chicago: University of Chicago Press.

Hay, Denys. 1977. *Annalists and Historians: Western Historiography from the Eighth to the Eighteenth Centuries.* London: Methuen.

Helgerson, Richard. 1986. "The Land Speaks: Cartography, Chorography and Subversion in Renaissance England." *Representations* 16:51–85.

———. 1992. *Forms of Nationhood: The Elizabethan Writing of England.* Chicago: University of Chicago Press.

Heninger, S. K. 1977. *The Cosmographical Glass: Renaissance Diagrams of the Universe.* San Marino: Huntington Library Press.

Hesse, Mary B. 1960. "Gilbert and the Historians." *British Journal of the Philosophy of Science* 11:1–10, 130–42.

Hibbert, Christopher. 1969. *The Grand Tour.* London: Weidenfeld and Nicolson.

Hill, Christopher. 1965a. *Intellectual Origins of the English Revolution.* Oxford: Clarendon Press.

———. 1965b. "Puritanism, Capitalism, and the Scientific Revolution." *Past and Present* 29:88–97.

———. 1966. "Science, Religion and the Society in the Sixteenth and Seventeenth Centuries." *Past and Present* 32:110–12.

Hill, Lamar M. 1988. *Bench and Bureaucracy: The Public Career of Sir Julius Caesar, 1580–1636.* Cambridge: James Clarke and Co.

Horton-Smith, L. G. H. 1952. *Dr. Walter Bailey, or Bayley c. 1529–1592: Physician to Queen Elizabeth.* London.

Howson, A. G. 1982. *A History of Mathematics Education in England.* Cambridge: Cambridge University Press.

Hudson, Winthrop S. 1980. *The Cambridge Connection and the Elizabethan Settlement of 1559.* Durham: Duke University Press.

Huffman, William H. 1989. *Robert Fludd and the End of the Renaissance.* London: Routledge.

Hunt, W. 1983. *The Puritan Moment: The Coming of Revolution in an English Country.* Cambridge, Mass.: Harvard University Press.

Hunter, Michael. 1981. *Science and Society in Restoration England.* Cambridge: Cambridge University Press.

Jacquot, Jean. 1974. "Harriot, Hill, Warner and the New Philosophy." In *Thomas Harriot: Renaissance Scientist,* ed. John W. Shirley, 107–28. Oxford: Clarendon Press.

James, M. R. 1921. "Lists of Manuscripts Formerly Owned by Dr. John Dee." *Bibliographical Society, Supplement to Transactions* i.

Jardine, Lisa. 1974. "The Place of Dialectic Teaching in Sixteenth-Century Cambridge." *Studies in the Renaissance* 21:31–62.

Jardine, Lisa, and Anthony Grafton. 1990. "'Studies for Action': How Gabriel Harvey Read His Livy." *Past and Present* 129:30–78.

Jayne, Sears R. 1956. *Library Catalogues of the English Renaissance.* Berkeley: University of California Press.

Jeffrey, Reginald W. 1909. "History of the College 1547–1603." *Brasenose College Quartercentenary Monographs* 2, part 10. Oxford: Clarendon Press.

Johannesson, Kurt. 1991. *The Renaissance of the Goths in Sixteenth-Century Sweden: Johannes and Olaus Magnus as Politicians and Historians.* Los Angelos: University of California Press.

Johnson, Francis R. 1937. *Astronomical Thought in Renaissance England: A Study of the English Scientific Writings from 1500 to 1645.* New York: Octagon Books; reprint 1968.

———. 1940. "Gresham College: Precursor of the Royal Society." *Journal of the History of Ideas* 1: 413–38.

Jones, Richard Foster. 1953. *The Triumph of the English Language: A Survey of Opinions Concerning the Vernacular.* Stanford: Stanford University Press.

———. 1961. *Ancients and Moderns: A Study of the Rise of the Scientific Movement in Seventeenth Century England.* St. Louis: Washington University Press.

Kain, Robert J. P., and Elizabeth Baigent. 1992. *The Cadastral Map in the Service of the State.* Chicago: University of Chicago Press.

Kearney, Hugh. 1964. "Puritanism, Capitalism, and the Scientific Revolution." *Past and Present* 28:81–101.

———. 1965. "Puritanism and Science: Problems of Definition." *Past and Present* 31:104–10.

———. 1970. *Scholars and Gentlemen: Universities and Society in Pre-Industrial Britian 1500–1700.* London: Faber and Faber.

Kelly, Suzanne. 1965. *The De Mundo of William Gilbert.* Amsterdam: M. Hertzberger.

Ker, N. R. 1986. "The Provision of Books." In *The Collegiate University,* ed. James K. McConica, 441–519. Vol. 3 of *The History of the University of Oxford,* ed. T. H. Aston. Oxford: Clarendon Press.

Knox, R. Buick. 1967. *James Ussher: Archbishop of Armagh.* Cardiff: University of Wales Press.

Kocher, Paul H. 1969. *Science and Religion in Elizabethan England.* New York: Octagon Books.

Kristeller, Paul Oskar, and F. Edward Cranz. 1960. *Catalogas Translationum et Commentariorum: Medieval and Renaissance Latin Translations and Commentaries.* Washington, D.C.: Catholic University Press.

Kupperman, Karen Ordahl. 1982. "The Puzzle of the American Climate in the Early Colonial Period." *American Historical Review* 87:1262–89.

Lacey, Robert. 1971. *Robert, Earl of Essex: An Elizabethan Icarus.* London: Weidenfeld and Nicolson.

Lach, Donald F. 1965. *Asia in the Making of Europe.* Vol. 1: *The Century of Discovery.* Chicago: University of Chicago Press.

Laird, W. R. 1991. "Archimedes among the Humanists." *Isis* 82:629–38.

Lattis, James M. 1994. *Between Copernicus and Galileo: Christoph Clavius and the Collapse of Ptolemaic Cosmology.* Chicago: University of Chicago Press.

Laslett, Peter. 1965. *The World We Have Lost.* London: Methuen.

Lawson, John, and Harold Silver. 1973. *A Social History of Education in England.* London: Methuen.

Leach, A. F. 1899. *A History of Winchester College.* London: Duckworth.

Leader, John Temple. 1895. *Life of Sir Robert Dudley, Earl of Warwick and Duke of Northumberland.* Florence: G. Barbera.

Lechner, Joan Marie. 1962. *Renaissance Concepts of the Commonplaces.* New York: Pageant Press.

Leedham-Green, Elisabeth S. 1986. *Books in Cambridge Inventories: Booklists from Vice-Chancellor's Court Probate Inventories in the Tudor and Stuart Periods.* 2 vols. Cambridge: Cambridge University Press.

Lestringant, Frank. 1994. *Mapping the Renaissance World.* Trans. David Fausett. Cambridge: Polity Press.

Levy, F. J. 1959. "William Camden as a Historian." Ph.D. dissertation, Harvard University.

———. 1964. "The Making of Camden's *Britannia.*" *Bibliographie d'Humanisme et Renaissance* 26:76–92.

Lewis, Gillian. 1986. "The Faculty of Medicine." In *The Collegiate University,* ed. James K. McConica, 213–56. Vol. 3 of *The History of the University of Oxford,* ed. T. H. Aston. Oxford: Clarendon Press.

Liddell, J. R. 1937–38. "The Library of Corpus Christi College, Oxford, in the Sixteenth Century." *The Library* ser. 4. 18:385–416.

Lindberg, David, and Robert Westman, eds. 1990. *Reappraisals of the Scientific Revolution.* Cambridge: Cambridge University Press.

Livingstone, David N. 1984. "The History of Science and the History of Geography: Interactions and Implications." *History of Science* 22:271–302.

———. 1992. *The Geographical Tradition.* Oxford: Blackwells.

Long, Pamela. 1991. "Invention, Authorship, 'Intellectual Property,' and the Origin of Patents: Notes toward a Conceptual History." *Technology and Culture* 32:846–84.

———. 1994. "Military Secrecy in Antiquity and Early Medieval Europe: A Critical Reassessment." *History and Technology* 11:259–90.

Looney, Jefferson. 1981. "Undergraduate Education in Early Stuart Cambridge." *History of Education* 10:9–19.

Lynam, Edward, ed. 1946. *Richard Hakluyt and His Successors: A Volume Issued to Commemorate the Centenary of the Hakluyt Society.* London: Hakluyt Society.

———. 1950. "English Maps and Map-Makers of the Sixteenth Century." *Geographical Journal* 116:7–28.

Lytle, Guy Fitch, and Stephen Orgel, eds. 1981. *Patronage in the Renaissance.* Princeton: Princeton University Press.

Mallet, Charles Edward. 1924–27. *The History of the University of Oxford.* 3 vols. London: Methuen.

Marshall, P. J., and Glyndwr Williams. 1982. *The Great Map of Mankind: British Perceptions of the World in the Age of Enlightenment.* London, Melbourne and Toronto: J. M. Dent and Sons.

Martin, R. Julian. 1992. *Francis Bacon, the State, and the Reform of Natural Philosophy.* Cambridge: Cambridge University Press.

Mason, S. F. 1974. "Science and Religion in Seventeenth Century England." In *The Intellectual Revolution of the Seventeenth Century,* ed. Charles Webster, 197–217. London: Routledge and Kegan Paul. Reprinted from *Past and Present* 3 (1953):28–44.

McConica, James K. 1974. "Scholars and Commoners in Renaissance Oxford." In *The University in Society,* ed. Lawrence Stone, 1:151–81. Princeton: Princeton University Press.

———. 1977. "The Social Relations of Tudor Oxford." *Transactions of the Royal Historical Society* 5th series, 27:115–34.

———. 1979. "Humanism and Aristotle in Tudor Oxford." *English Historical Review* 94:291–317.

———. 1986a. *The Collegiate University.* Vol. 3 of *The History of the University of Oxford,* ed. T. H. Aston. Oxford: Clarendon Press.

———. 1986b. "The Rise of the Undergraduate College." In *The Collegiate University,* ed. James K. McConica, 1–68. Vol. 3 of *The History of the University of Oxford,* ed. T. H. Aston. Oxford: Clarendon Press.

———. 1986c. "Elizabethan Oxford: The Collegiate Society." In *The Collegiate University,* ed. James K. McConica, 645–732. Vol. 3 of *The History of the University of Oxford,* ed. T. H. Aston. Oxford: Clarendon Press.

McGee, J. Sears. 1976. *The Godly Man in Stuart England.* New Haven: Yale University Press.

McGurk, J. 1988. "William Camden: Civil Historian or Gloriana's Propagandist." *History Today* 38:47–53.

McLean, Antonia. 1972. *Humanism and the Rise of Science in Tudor England.* London: Heinemann Edal Books.

McMullin, Ernan. 1987. "Bruno and Copernicus." *Isis* 78:55–74.

Mendyk, Stanley J. G. 1983. "Painting the Landscape: Regional Study in Britain during the Seventeenth Century." Ph.D. dissertation, McMaster University.

———. 1986. "Early British Chorography." *Sixteenth Century Journal* 17:459–81.

———. 1989. *"Speculum Britanniae": Regional Study, Antiquarianism, and Science in Britain to 1700.* Toronto: University of Toronto Press.

Merchant, Carolyn. 1980. *The Death of Nature: Women, Ecology and the Scientific Revolution.* San Fransisco: Harper and Row.

Merton, Robert King. 1938. "Science, Technology and Society in Seventeenth-Century England." *Osiris* 4; reprinted New York: H. Fertig, 1970.
Millar, Oliver. 1963. *The Tudor-Stuart and Early Georgian Pictures in the Collection of Her Majesty the Queen.* 2 vols. London: Phardon Press.
Mirrlees, Hope. 1962. *A Fly in Amber: Being an Extravagant Biography of the Romantic Antiquary Sir Robert Cotton.* London: Faber and Faber.
Moran, Bruce T. 1981. "German Prince-Practitioners: Aspects in the Development of Courtly Science, Technology, and Procedures in the Renaissance." *Technology and Culture* 22:253–74.
———. 1991. *Patronage and Institutions: Science, Technology, and Medicine at the European Court.* Woodbridge: Boydell Press.
Morgan, John. 1979. "Puritanism and Science: A Reinterpretation." *Historical Journal* 22:535–60.
Morgan, Paul. 1980. *Oxford Libraries outside the Bodleian: A Guide.* Oxford: Bodleian Library.
Morison, Samuel Eliot. 1935. *The Founding of Harvard College.* Cambridge, Mass.: Harvard University Press.
Morrish, P. S. 1982. "Griffin Higgs: A Study in Seventeenth-Century Book Collecting and Librarianship." Unpublished manuscript, Merton College Library, Oxford.
Moseley, C. W. R. D. 1975. "The Availability of *Mandeville's Travels* in England, 1356–1750." *The Library* ser. 5, 30:125–33.
Moss, Ann. 1991. "Printed Commonplace Books in the Renaissance." In *Acta Conventus Neo-Latini Torontonensis,* ed. Alexander Dalzell, Charles Fantazzi, and Richard Schoeck, 509–18. Binghamton: Medieval and Renaissance Texts and Studies.
Mullinger, J. B. 1873–1911. *The University of Cambridge.* 3 vols. Cambridge: Cambridge University Press.
Neale, John Ernest. [1949]. *The Elizabethan House of Commons.* London: Cape.
Needham, Joseph. 1954–86. *Science and Civilisation in China.* Cambridge: Cambridge University Press.
Oakshott, Walter. 1968. "Sir Walter Raleigh's Library." *The Library* ser. 5, 23:285–327.
Opfell, Olga S. 1982. *The King James Bible Translators.* Jefferson and London: McFarland.
Orgel, Stephen, and Roy Strong. 1973. *Inigo Jones: The Theater of the Stuart Court.* Los Angeles: University of California Press.
Ornstein, Martha. 1928. *The Role of Scientific Societies in the Seventeenth Century.* Chicago: University of Chicago Press.
Pagden, Anthony. 1982. *The Fall of Natural Man.* Cambridge: Cambridge University Press.
———. 1993. *European Encounters with the New World.* New Haven: Yale University Press.
Paine, Gustavus S. 1977. *The Men behind the King James Version.* Grand Rapids, Mich.: Baker Book House.
Palumbo, Rina. 1986. "The Path to Fame, the Proofe of Zeale . . . English Political Patronage and the Religious Factor in Colonial Virginia and Maryland, 1606–1632." M. A. dissertation, Queen's University (Canada).
Parker, John. 1965. *Books to Build an Empire: A Bibliographical History of English Overseas Interests to 1620.* Amsterdam: N. Israel.
———. 1972. *Discovery: Developing Views of the Earth from Ancient Times to the Voyages of Captain Cook.* New York: Charles Scribner's Sons.
Parks, George Bruner. 1930. *Richard Hakluyt and the English Voyages.* New York: American Geographical Society.
———. 1974. "Tudor Travel Literature: A Brief History." In *The Hakluyt Handbook,* ed. D. B. Quinn, 97–132. London: Hakluyt Society.
Parry, G. J. R. 1984. "Puritanism, Science, and Capitalism: William Harrison and the Rejection of Hermes Trismegistus." *History of Science* 22:245–70.
———. 1987. *A Protestant Vision: William Harrison and the Reformation of Elizabethan England.* Cambridge: Cambridge University Press.

Peck, Linda Levy. 1982. *Northampton: Patronage and Policy at the Court of James I.* London: Allen and Unwin.

———. 1990. *Court Patronage and Corruption in Early Stuart England.* London: Routledge.

Pennington, Donald, and Keith Thomas, eds. 1978. *Puritans and Revolutionaries: Essays in Seventeenth Century History Presented to Christopher Hill.* Oxford: Clarendon Press.

Perver, Margery. 1967. *The Royal Society: Concept and Creation.* Cambridge, Mass.: MIT Press.

Piggott, Stuart. 1976. *Ruins in a Landscape: Essays in Antiquarianism.* Edinburgh: Edinburgh University Press.

Pocock, J. G. A. 1957. *The Ancient Constitution and the Feudal Law.* Cambridge: Cambridge University Press.

Porter, Harry C. 1972. *Reformation and Reaction in Tudor Cambridge.* London: Archan Books.

———. 1979. *The Inconstant Savage: England and the North American Indian 1500–1660.* London: Duckworth.

Porter, Roy. 1980. "The Terraqueous Globe." In *The Ferment of Knowledge: Studies in the Historiography of Eighteenth Century Science,* ed. G. S. Rosseau and Roy Porter, 285–324. Cambridge: Cambridge University Press.

Power, M. J. 1985. "John Stow and His London." *Journal of Historical Geography* 11:1–20.

Powicke, Maurice. 1948. "William Camden." *Essays and Studies* n. s., 1:74.

Prest, Wilfrid R. 1972. *The Inns of Court under Elizabeth and the Early Stuarts 1590–1640.* Totowa, N.J.: Longman.

Preston, Joseph H. 1977. "Was There an Historical Revolution?" *Journal of the History of Ideas* 33:353–64.

Pumfrey, Stephen. 1987. "William Gilbert's Magnetic Philosophy 1580–1684: The Creation and Dissolution of a Discipline." Ph.D. dissertation, University of London.

———. 1989. "'O Tempora, O Magnes': A Sociological Analysis of the Discovery of Secular Magnetic Variation in 1634." *British Journal for the History of Science* 22:181–214.

Quinn, David Beers. 1955. *The Roanoke Voyages, 1584–1589: Documents to Illustrate the English Voyages to North America.* London: Hakluyt Society.

———. 1974a. *The Hakluyt Handbook.* 2 vols. London: Hakluyt Society.

———. 1974b. "Thomas Harriot and the New World." In *Thomas Harriot: Renaissance Scientist,* ed. John W. Shirley, 36–53. Oxford: Clarendon Press.

Rabb, Theodore K. 1962. "Puritanism and the Rise of Experimental Science in England." *Journal of World History* 7:46–67.

———. 1965. "Religion and the Rise of Modern Science." *Past and Present* 31:111–26.

———. 1966. "Science, Religion and Society in the Sixteenth and Seventeenth Centuries." *Past and Present* 33:148. Reprinted in *The Intellectual Revolution of the Seventeenth Century,* ed. Charles Webster, 284–85. London: Routledge and Kegan Paul, 1974.

Rappaport, Steve. 1989. *Worlds within Worlds: Structures of Life in Sixteenth-Century London.* Cambridge: Cambridge University Press.

Rashdall, Hastings, and Robert S. Rait. 1901. *New College: University of Oxford College Histories.* London: F. E. Robinson and Co.

Richeson, Allie Wilson. 1966. *English Land Measuring to 1800: Instruments and Practices.* Cambridge, Mass.: Society for the History of Technology and MIT Press.

Ritchie, Neil. 1976. "Sir Robert Dudley: Expatriate in Tuscan Service." *History Today* 26:385–92.

Roberts, Julian, and Andrew G. Watson, eds. 1990. *John Dee's Library Catalogue.* London: Oxford University Press.

Rockett, William. 1990. "Historical Topography and British History in Camden's *Britannia.*" *Renaissance and Reformation* n. s., 14:71–80.

———. 1995. "The Structural Plan of Camden's *Britannia.*" *Sixteenth Century Journal* 26:829–42.

Roller, Duane H. du Bose. 1959. *The De Magnete of William Gilbert.* Amsterdam: M. Hertzberger.

Rose, Paul L. 1977. "Erasmians and Mathematicians at Cambridge in the Early Sixteenth Century." *Sixteenth Century Journal* 8 (supp.):47–59.

Rosenberg, Eleanor. 1955. *Leicester: Patron of Letters.* New York: Columbia University Press.

Rücker, Elisabeth. 1980. "Nürnberger Frühhumanisten und ihre Beschäftigung mit Geographie. Zur Frage einer Mitarbeit von Hieronymus Münzer und Conrad Celtis am Text der Schedelschen Weltchronik." In *Humanismus und Naturwissenschaften. Beitrage zur Humanismus-forschung,* ed. Rudolf Schmitz and Fritz Krafft, 6:181–92.Boppard: Boldt.

Russell, Elizabeth. 1977. "The Influx of Commoners into the University of Oxford before 1581, an Optical Illusion." *English Historical Review* 92:721–45.

Russell, John L. 1973. "The Copernican System in Great Britain." In *Colloquia Copernicana II, Etudes sur l'audience de la théorie héliocentrique* (*Studia Copernicana* 5), ed. Jerzy Dobrzycki, 189–239. Wroclaw.

Scarisbrick, J. J. 1984. *The Reformation and the English People.* Oxford: Oxford University Press.

Schmitt, Charles B. 1973. "Towards a Reassessment of Renaissance Aristotelianism." *History of Science* 11:159–93.

———. 1975. "Philosophy and Science in Sixteenth-Century Universities: Some Preliminary Comments." In *The Cultural Context of Medieval Learning,* ed. J. E. Murdoch and E. G. Sylla. Dordrecht-Holland: D. Reidel Co.

———. 1983. *John Case and Aristotelianism in Renaissance England.* Kingston and Montreal: McGill-Queen's University Press.

Schoeck, R. J. 1954. "The Elizabethan Society of Antiquaries and Men of Law." *Notes and Queries* n. s., 1:417–21.

Schuyler, R. L. 1946. "The Antiquaries and Sir Henry Spelmen." *Proceedings of the American Philosophical Society* 90:91–103.

Severin, Timothy. 1976. *The Oriental Adventure: Explorers of the East.* Boston: Little, Brown.

Shapin, Steven. 1988. "The House of Experiment in Seventeenth-Century England." *Isis* 79:373–404.

———. 1994. *A Social History of Truth: Civility and Science in Seventeenth-Century England.* Chicago: University of Chicago Press.

Shapin, Steven, and Simon Schaffer. 1985. *The Leviathan and the Air Pump: Hobbes, Boyle, and the Experimental Life.* Princeton: Princeton University Press.

Shapiro, Barbara J. 1974. "Latitudinarianism and Science in Seventeenth-Century England." In *The Intellectual Revolution of the Seventeenth Century,* ed. Charles Webster, 286–316. London: Routledge and Kegan Paul. Reprinted from *Past and Present* 40 (1968).

———. 1979. "History and Natural History in Sixteenth- and Seventeenth-Century England: An Essay on the Relationship between Humanism and Science." In *English Scientific Virtuosi in the Sixteenth and Seventeenth Centuries,* ed. Barbara Shapiro and Robert G. Frank, Jr., 1–55. Los Angeles: William Andrews Clark Memorial Library.

———. 1991. "Early Modern Intellectual Life: Humanism, Religion and Science in Seventeenth-Century England." *History of Science* 29:45–71.

Sharpe, Kevin. 1978. *Sir Robert Cotton, 1586–1631: History and Politics in Early Modern England.* Oxford: Oxford University Press.

Sherman, William H. 1995. *John Dee: The Politics of Reading and Writing in the English Renaissance.* Amherst: University of Massachusetts Press.

Shirley, John W. 1974. *Thomas Harriot: Renaissance Scientist.* Oxford: Clarendon Press.

———. 1983. *Thomas Harriot: A Biography.* Oxford: Clarendon Press.

———. 1985. "Science and Navigation in Renaissance England." In *Science and the Arts in the Renaissance,* ed. John W. Shirley and F. David Hoeniger, 74–93. Washington, D.C.: Folger Shakespeare Library.

Shirley, John W., and F. David Hoeniger, eds. 1985. *Science and the Arts in the Renaissance.* Washington, D.C.: Folger Shakespeare Library.

Shirley, Rodney W. 1983. *The Mapping of the World: Early Printed World Maps 1472–1700.* London: Holland Press.
Simon, Joan. 1966. *Education and Society in Tudor England.* Cambridge: Cambridge University Press.
Skelton, R. A. 1971. "The Maps of a Tudor Statesman." In *Description of the Maps and Architectural Drawings in the Collection . . . at Hatfield House,* ed. R. A. Skelton and J. Summerson. Oxford: Roxburghe Club.
Smith, Pamela H. 1994. *The Business of Alchemy: Science and Culture in the Holy Roman Empire.* Princeton: Princeton University Press.
Smith, Thomas. 1952. "On the Life and Works of Henry Briggs." In *Logarithmetica Brittanica,* ed. A. J. Thompson, appendix I. Cambridge: Cambridge University Press.
Smith, Thomas R. 1978. "Manuscript and Printed Sea Charts in Seventeenth-Century London: The Case of the Thames School." In *The Compleat Plattmaker,* ed. Norman J. Thrower, 45–100. Berkeley: University of California Press.
Smuts, R. Malcolm. 1981. "The Political Failure of Stuart Cultural Patronage." In *Patronage in the Renaissance,* ed. Guy Fitch Lytle and Stephen Orgel, 165–87. Princeton: Princeton University Press.
———. 1987. *Court Culture and the Origins of a Royalist Tradition in Early Stuart England.* Philadelphia: University of Pennsylvania Press.
Snyder, John P. 1993. *Flattening the Earth.* Chicago: University of Chicago Press.
Stannard, David E. 1992. *American Holocaust: Columbus and the Conquest of the New World.* Oxford: Oxford University Press.
Steele, C. R. 1974. "From Hakluyt to Purchas." In *The Hakluyt Handbook,* ed. D. B. Quinn, 74–96. London: Hakluyt Society.
Sterne, Virginia F. 1979. *Gabriel Harvey: His Life, Marginalia, and Library.* Oxford: Clarendon Press.
Stimson, Dorothy. 1935. "Puritanism and the New Philosophy in Seventeenth-Century England." *Bulletin of the Institute of the History of Medicine* 3:321–24.
———. 1948. *Scientists and Amateurs: A History of the Royal Society.* New York: Henry Schuman.
Stoddart, David Ross, ed. 1981. *Geography, Ideology and Social Concern.* Oxford: Basil Blackwell.
Stone, Lawrence. 1964. "The Educational Revolution in England, 1560–1640." *Past and Present* 28:41–80.
———. 1965. *The Crisis of the Aristocracy: 1558–1641.* Oxford: Clarendon Press.
———. 1974a. *The University in Society.* Vol. 1: *Oxford and Cambridge from the Fourteenth to the Early Nineteenth Century.* Princeton: Princeton University Press.
———. 1974b. "The Size and Composition of the Oxford Student Body 1580–1909." In *The University in Society,* ed. Lawrence Stone, 1:3–110. Princeton: Princeton University Press.
Stoyre, John. 1952. *English Travellers Abroad 1604–1667.* Oxford: Cape.
Strachan, Michael. 1962. *The Life and Adventures of Thomas Coryate.* London: Oxford University Press.
Strachey, Giles Lytton. [1928]. *Elizabeth and Essex: A Tragic History.* New York: Harcourt and Brace.
Strong, Roy. 1986. *Henry, Prince of Wales and England's Lost Renaissance.* London: Thames and Hudson.
Studies in the History of English Commerce in the Tudor Period. I: *The Organization and Early History of the Muscovy Company.* 1912. Philadelphia: University of Pennsylvania Press.
Taylor, E. G. R. 1930. *Tudor Geography 1485–1583.* London: Methuen.
———. 1934. *Late Tudor and Early Stuart Geography 1583–1650.* London: Methuen.
———. 1935. *The Original Writings and Correspondence of the Two Richard Hakluyts.* 2 vols. London: Hakluyt Society.
———. 1947. "Richard Hakluyt." *Geographical Journal* 109:165–74.

———. 1954. *Mathematical Practitioners of Tudor and Stuart England.* Cambridge: Cambridge University Press.

———. 1957. *The Haven-Finding Art: A History of Navigation from Odysseus to Captain Cook.* New York: Abelard-Schuman.

Taylor, John. 1966. *The* Universal Chronicle *of Ranulf Higden.* Oxford: Clarendon Press.

Thirsk, Joan. 1978. *Economic Policy and Projects: The Development of a Consumer Society in Early Modern England.* Oxford: Clarendon Press.

Thrower, Norman J. W., ed. 1978. *The Compleat Plattmaker: Essays on Chart, Map and Globe Making in England in the Seventeenth and Eighteenth Centuries.* Berkeley: University of California Press.

Tite, Colin G. C. 1994. *The Manuscript Library of Sir Robert Cotton: The Panizzi Lectures 1993.* London: British Library.

Todd, Margo. 1986. "Providence, Chance and the New Science in Early Stuart Cambridge." *Historical Journal* 29:697–711.

Trevor-Roper, Hugh. 1971. *Queen Elizabeth's First Historian: William Camden and the Beginnings of English Civil History.* London: Cape.

———. 1975. "John Stow." *London and Middlesex Antiquarian Society Journal* 26:337–42.

———. 1976. *Princes and Artists: Patronage and Ideology at Four Habsburg Courts, 1517–1633.* London: Thames and Hudson.

Turner, J. G. 1978. "The Matter of Britain: Topographical Poetry in English 1600–1660." *Notes and Queries* (December):514–24.

Tyacke, Nicholas. 1978. "Science and Religion at Oxford before the Civil War." In *Puritans and Revolutionaries: Essays in Seventeenth-Century History Presented to Christopher Hill,* ed. Donald Pennington and Keith Thomas, 73–93. Oxford: Clarendon Press.

Tyacke, Sarah, ed. 1983. *English Map-Making 1500–1650.* London: British Library.

Tyacke, Sarah, and John Huddy. 1980. *Christopher Saxton and Tudor Map-Making.* London: British Library, series 2.

Van Norden, Linda. 1949–50. "Sir Henry Spelman on the Chronology of the Elizabethan Society of Antiquaries." *Huntingdon Library Quarterly* 13:131–60.

Venn, John. 1922. *Alumni Cantabrigienses: A Biographical List of All Known Students, Graduates and Holders of Office at the University of Cambridge, from the Earliest Times to 1900.* Part I: *From the Earliest Times to 1751.* 4 vols. Cambridge: Cambridge University Press.

Wakeling, G. H. 1909. "History of the College 1603–1660." *Brasenose College Quartercentenary Monographs* 2, part 11. Oxford: Clarendon Press.

Wallace, William, O. P. 1978. "The Philosophical Setting of Medieval Science." In *Science in the Middle Ages,* ed. David C. Lindberg, 91–119. Chicago: University of Chicago Press.

———. 1984. *Galileo and His Sources: The Heritage of the Collegio Romano in Galileo's Science.* Princeton: Princeton University Press.

Wallis, Helen M. 1951. "The First English Globe: A Recent Discovery." *Geographical Journal* 117:275–90.

———. 1978. "'Geographie Is Better than Divinitie': Maps, Globes and Geography in the Days of Samuel Pepys." In *The Compleat Plattmaker,* ed. Norman J. Thrower, 1–43. Berkeley: University of California Press.

Wallis, P. J. 1956. "The Library of William Crashawe." *Transactions of the Cambridge Bibliographic Society* 2:213–28.

Waters, David Watkin. 1958. *The Art of Navigation in England in Elizabethan and Early Stuart Times.* London: Hollis and Carter.

Watson, Foster. 1909. *The Beginnings of the Teaching of Modern Subjects in England.* London: Sir Isaac Pitman and Sons; reprint London: S. R. Publishers, 1971.

Webster, Charles, ed. 1974. *The Intellectual Revolution of the Seventeenth Century.* London: Routledge and Kegan Paul.

———. 1975. *The Great Instauration: Science, Medicine and Reform 1626–1660.* London: Gerald Duckworth.

Weiner, Andrew D. 1980. "Expelling the Beast: Bruno's Adventures in England." *Modern Philology* 78:1–13.

Westfall, Richard S. 1977. *The Construction of Modern Science: Mechanisms and Mechanics.* Cambridge: Cambridge University Press.

Westman, Robert S. 1975. "The Melanchthon Circle, Rheticus, and the Wittenberg Interpretation of the Copernican Theory." *Isis* 66:165–93.

———. 1980a. "The Astronomer's Role in the Sixteenth Century: A Preliminary Study." *History of Science* 18:105–47.

———. 1980b. "Humanism and Scientific Roles in the Sixteenth Century." In *Humanismus und Naturwissenschaften. Beitrage zur Humanismusforschung,* ed. Rudolf Schmitz and Fritz Krafft, 6:83–99. Boppard: Boldt.

White, Jeffrey A. 1984. "Towards a Critical Edition of Biondo Flavio's 'Italia Illustrata'; A Survey and Evaluation of the Manuscripts." In *Umanesimo a Roma nel Quattrocento,* Instituto di Studi Romani Roma, Barnard College, 210–48. Conference on "Umanesimo a Roma Nel Quattrocento," New York, December 1981. Printed at Citta di Castello.

Wiener, Philip P., and Aaron Noland, eds. 1957. *Roots of Scientific Thought.* New York: Basic Books.

Wilcox, Donald. 1987. *The Measure of Times Past: Pre-Newtonian Chronologies and the Rhetoric of Relative Time.* Chicago: University of Chicago Press.

Willan, Thomas Stuart. 1953. *The Muscovy Merchants of 1555.* Manchester: Manchester University Press.

Williamson, George Charles. 1920. *George, Third Earl of Cumberland, 1558–1605, His Life and His Voyages.* Cambridge: Cambridge University Press.

Williamson, J. W. 1978. *The Myth of the Conqueror: Prince Henry Stuart, a Study in Seventeenth Century Personation.* New York: AMS Press.

Wilson, E. C. 1946. *Prince Henry and English Literature.* Ithaca: Cornell University Press.

Wood, Denis. 1993. *The Power of Maps.* London: Routledge.

Woodward, David. 1987. "Medieval *Mappaemundi.*" In *The History of Cartography.* Vol. I: *Cartography in Prehistoric, Ancient and Medieval Europe and the Mediterranean,* ed. J. B. Harley and David Woodward, 290–367. Chicago: University of Chicago Press.

Woolf, D. R. 1990. *The Idea of History in Early Stuart England.* Toronto: University of Toronto Press.

Wormald, Francis, and C. E. Wright, eds. 1958. *The English Library before 1700.* London: Athlone Press.

Wright, John Kirtland. 1966. *Human Nature in Geography.* Cambridge, Mass.: Harvard University Press.

Wright, Louis B. 1958. *Middle-Class Culture in Elizabethan England.* Ithaca: Cornell University Press.

Yates, Frances A. 1934. *John Florio: The Life of an Italian in Shakespeare's England.* Cambridge: Cambridge University Press.

———. 1938–39. "Giordano Bruno's Conflict with Oxford." *Journal of the Warburg Institute* 2:227–42.

———. 1964. *Giordano Bruno and the Hermetic Tradition.* London: Routledge and Kegan Paul.

———. 1969. *Theatre of the World.* London: Routledge and Kegan Paul.

———. 1972. *The Rosicrucian Enlightenment.* London: Routledge and Kegan Paul.

Zilsel, Edgar. 1942. "The Sociological Roots of Science." *American Journal of Sociology* 47:544–62.

INDEX

Page numbers for illustrations are in *italics.*